## Chronic Myeloproliferative Disorders

# Chronic Myeloproliferative Disorders

## Cytogenetic and Molecular Genetic Abnormalities

Editor *Barbara J. Bain,* London

15 figures, 4 in color, and 18 tables, 2003

KARGER

Basel · Freiburg · Paris · London · New York · Bangalore · Bangkok · Singapore · Tokyo · Sydney

**Barbara J. Bain,** FRACP, FRCPATH

Department of Haematology
St. Mary's Hospital
Praed Street
London W2 1NY (UK)

**Updated reprint of reviews originally published in *Acta Haematologica***

**Vol. 107, No. 2, 2002 and Vol. 108, No. 2, No. 3 and No. 4, 2002.**

Drug Dosage. The authors and the publisher have exerted every effort to ensure that drug selection and dosage set forth in this text are in accord with current recommendations and practice at the time of publication. However, in view of ongoing research, changes in government regulations, and the constant flow of information relating to drug therapy and drug reactions, the reader is urged to check the package insert for each drug for any change in indications and dosage and for added warnings and precautions. This is particularly important when the recommended agent is a new and/or infrequently employed drug.


P.O. Box, CH–4009 Basel (Switzerland)
www.karger.com
Printed in Switzerland on acid-free paper by
Reinhardt Druck, Basel
ISBN 3–8055–7307–3

# Contents

VII **Preface**

1 **An Overview of Translocation-Related Oncogenesis in the Chronic Myeloid Leukaemias**
Bain, B.J. (London)

9 **Molecular Basis of Chronic Myeloid Leukaemia**
Barnes, D.J.; Melo, J.V. (London)

32 **Cytogenetic and Molecular Monitoring of Residual Disease in Chronic Myeloid Leukaemia**
Kaeda, J. (London); Chase, A. (Salisbury); Goldman, J.M. (London)

44 **Cytogenetic and Molecular Genetic Evolution of Philadelphia-Chromosome-Positive Chronic Myeloid Leukaemia**
Johansson, B.; Fioretos, T.; Mitelman, F. (Lund)

62 **The 8p11 Myeloproliferative Syndrome: A Distinct Clinical Entity Caused by Constitutive Activation of *FGFR1***
Macdonald, D. (London); Reiter, A. (Mannheim); Cross, N.C.P. (Salisbury/Southampton)

69 **Myeloproliferative Disorders with Translocations of Chromosome 5q31–35: Role of the Platelet-Derived Growth Factor Receptor Beta**
Steer, E.J.; Cross, N.C.P. (Salisbury/Southampton)

80 **Cytogenetic and Molecular Genetic Aspects of Childhood Myeloproliferative/ Myelodysplastic Disorders**
Hall, G.W. (Oxford)

90 **Polycythaemia vera: Will New Markers Help Us Answer Old Questions?**
Pahl, H.L. (Freiburg)

102 **Cytogenetic and Molecular Genetic Aspects of Idiopathic Myelofibrosis**
Reilly, J.T. (Sheffield)

109 **Cytogenetic and Molecular Genetic Aspects of Essential Thrombocythemia**
Steensma, D.P.; Tefferi, A. (Rochester, Minn.)

120 **Cytogenetic and Molecular Genetic Abnormalities in Systemic Mastocytosis**
Gupta, R.; Bain, B.J.; Knight, C.L. (London)

127 **Author Index**

128 **Subject Index**

# Preface

The initial descriptions of the myeloproliferative disorders were of clinicopathological entities with characteristic haematological features. A new era in our understanding of these diseases was heralded by Nowell and Hungerford's discovery, in 1960, of a 'minute chromosome' consistently associated with the condition they called 'chronic granulocytic leukaemia'. A little over 10 years later, Rowley, utilizing chromosome banding for the more accurate characterization of chromosomal rearrangements, reported that the 'minute chromosome', the Philadelphia or Ph chromosome, in what she termed 'chronic myelogenous leukaemia', was a derivative of the normal chromosome 22 and resulted from a reciprocal translocation between chromosomes 9 and 22. Major advances continued about once a decade with Bartram and colleagues in 1983 reporting that the translocation in what they termed 'chronic myelocytic leukaemia' involved the *ABL* gene on chromosome 22; *ABL* was a homologue of a viral oncogene, *v-abl*, implicated in murine leukaemia, and was thus a proto-oncogene. The partner gene was identified a year later. By the time another decade had passed the dysregulated tyrosine kinase function of the protein derived from the *BCR-ABL* fusion gene had been discovered and animal models of chronic myeloid leukaemia had been developed. The subsequent decade saw the development of a specific, clinically efficacious drug capable of inhibiting the function of the abnormal gene product.

Investigation of this one disease with many names has had a profound impact on our understanding of neoplasia and has opened up the exciting prospect of specific anticancer drugs based on an understanding of the molecular basis of individual neoplasms. The scientific advances in the much rarer chronic myeloid leukaemias associated with the *ZNF198-FGFR1* and *ETV6-PDGFRB* fusion genes have paralleled those in chronic granulocytic leukaemia. Scientific progress in other, more common, myeloproliferative disorders has lagged behind that in the chronic myeloid leukaemias, but advances are now being made and these are discussed in detail in individual chapters.

In this book we have summarized all that is now known of the cytogenetic and molecular genetic basis of the myeloproliferative disorders and have set the most recent advances in their historical context. It may reasonably be predicted that the next decade will unravel more of the mysteries of these diseases for the ultimate benefit of patients.

*Barbara J. Bain*

Bain BJ (ed): Chronic Myeloproliferative Disorders.
Basel, Karger, 2003, pp 1–8

# An Overview of Translocation-Related Oncogenesis in the Chronic Myeloid Leukaemias

Barbara J. Bain

Department of Haematology, Faculty of Medicine, Imperial College of Science, Technology and Medicine, St. Mary's Hospital, London, UK

**Key Words**
Chronic myeloid leukaemia · Myeloproliferative disorders · Chromosomal translocation · Cytogenetics · Molecular genetics · Oncogenes

**Abstract**
The demonstration of the *BCR-ABL* fusion gene in patients with chronic granulocytic leukaemia and t(9;22)(q34;q11) represents the first recognition, in a human neoplasm, of a translocation leading to formation of an oncogenic fusion gene. Since this initial observation, this leukaemogenic mechanism has been increasingly recognized in chronic myeloid leukaemias (CMLs). The fusion gene has often incorporated part of a gene encoding a receptor or cytoplasmic tyrosine kinase, particularly *ABL*, *PDGFRB* and *FGFR1*. This contrasts with the frequent involvement of genes encoding transcription factors or other nuclear proteins in acute myeloid leukaemia. Nevertheless, genes encoding tyrosine kinases have also been implicated in some cases of acute leukaemia. With the exception of the *BCR-ABL* fusion gene in chronic granulocytic leukaemia, all these fusion genes are uncommon or rare among cases of CML. The molecular mechanisms underlying the great majority of cases of Philadelphia-negative CML remain to be discovered.

This article is based on a review first published in *Acta Haematologica* 2002;107:57–63.

The term 'chronic myeloid leukaemia' (CML) covers a heterogeneous group of haematological neoplasms of which the best known and best characterized is that associated with the Philadelphia (Ph) chromosome. In this review, I shall refer to this condition as chronic granulocytic leukaemia (CGL), in order to distinguish it from other types of CML in which the mechanism of leukaemogenesis differs. The discovery of the Ph chromosome in 1960 was the first recognition of the association of a recurrent cytogenetic abnormality with a human neoplasm [1]. This 'minute chromosome' was named the Philadelphia[1] (Ph[1]) chromosome for the city of its discovery. Subsequently, the terminology applied to chromosomes altered and the preferred designation is now Philadelphia (Ph) [2]. Subsequently, the Ph chromosome was demonstrated to be a deleted chromosome 22 [3] and later, with the development of banding techniques that made recognition of individual chromosomes more reliable, it was found to result from a reciprocal translocation between chromosomes 9 and 22, t(9;22)(q34;q11) [4]. Investigation of CGL also provided the first evidence, for a human leukaemia, that the molecular mechanism of leukaemogenesis could be the formation of a fusion gene, in this case a *BCR-ABL* fusion gene on the Ph chromosome (22q–). The *ABL* gene on chromosome 9 was found to be homologous with a known viral oncogene, v-*abl*, operating in some murine leukaemias. Subsequently, the *BCR-ABL* fusion gene on the Ph chromosome was shown to

**Table 1.** Molecular variants of *BCR-ABL* and associated clinicopathological features

| Protein | Breakpoint | Disease association |
|---|---|---|
| p210$^{BCR\text{-}ABL}$ | M-bcr | the great majority of cases of typical CGL<br>about a third of cases of Ph-positive ALL |
| p190$^{BCR\text{-}ABL}$ | m-bcr | a minority of cases of CGL, often with monocytosis or dysplastic features<br>about two thirds of cases of Ph-positive ALL<br>rare cases of AML |
| p230$^{BCR\text{-}ABL}$ | μ-bcr | neutrophilic variant of CGL, or CGL with marked thrombocytosis |

have the capacity to transform cells in culture and to be oncogenic in transfection experiments in mice [5].

The normal *ABL* gene encodes a tyrosine kinase that is active in signal transduction. The chimeric *BCR-ABL* gene encodes an aberrant tyrosine kinase of higher molecular weight than that encoded by *ABL* (p210 in the majority of cases, in comparison with p145). There are also examples of t(9;22)(q34;q11) in which the breakpoints differ at a molecular level, encoding either a smaller or a larger protein. The p190 BCR-ABL protein is associated particularly with Ph-positive acute lymphoblastic leukaemia (ALL) but, when it is associated with CGL, the haematological features often differ from those of cases with a p210 protein; monocytosis is common and there may be dysplastic features. The p230 protein is associated with prominent neutrophilia with relatively few circulating neutrophil precursors; this condition is best designated 'the neutrophilic variant of chronic granulocytic leukaemia' or 'the neutrophilic variant of chronic myeloid leukaemia', since the term 'neutrophilic leukaemia' has been in use for many years to designate a Ph-negative condition. In each of these subtypes of CGL, an aberrant ABL protein with enhanced tyrosine kinase activity is encoded. The variable association with Ph-positive ALL and the haematological differences between cases with a different molecular abnormality indicate that there is a subtle difference in the effect of these tyrosine kinases on signal transduction. Constitutive activation of tyrosine kinase appears to result from oligomerization of the BCR-ABL protein, attributable to the coiled coil motif of BCR [6]. The oligomerized BCR-ABL protein causes constitutive activation of proteins of many signalling pathways including those involving p21 RAS (through GRB/SOS), phosphatidylinositol-3-kinase (PI3K), JUN, MYC and STAT5. BCR-ABL also induces expression of the anti-apoptotic mitochondrial protein, $BCL_X$. Constitutive activation of the transcription regulator, STAT5, appears to be important in the transforming ability of BCR-ABL [7]. STAT5 is a member of the STAT (signal transducer and activator of transcription) family of cytoplasmic transcription regulators which form hetero- or homo-dimers and subsequently migrate to the nucleus where they interact with the promoter region of their target genes. In CGL, STAT5 is tyrosine phosphorylated, causing it to dimerize, translocate to the nucleus and bind to target DNA motifs. The molecular variants of *BCR-ABL* and their clinicopathological correlations are summarized in table 1.

Subsequent to the discovery of *BCR-ABL*, other oncogenic events in chronic myeloid leukaemias (CMLs) have been found to involve fusion genes encoding tyrosine kinases. These include *ETV6-PDGFRB, HIP1-PDGFRB, ETV6-JAK2, ETV6-ABL, ETV6-EVI1* and four fusion genes resulting from rearrangement of the *FGFR1* gene.

The second association recognized between a specific translocation and a subtype of CML was the association between t(5;12)(q31–33;p13) and CML with prominent eosinophilia. This translocation results in fusion of part of the *ETV6* gene at 12p13, also known as the *TEL* gene, with part of the *PDGFRB* gene (also known as *PDGFRβ*, encoding the receptor for platelet-derived growth factor β), from 5q33 [8]. The *ETV6* gene, a member of the *ETS* family, was first identified in this specific translocation [8], although it was subsequently found to have a much wider role in oncogenesis. The t(5;12) translocation leads to constitutive activation of PDGFRB. The fusion gene phosphorylates and activates STAT5, phospholipase Cγ and PI3K, all of which contribute to transformation [9]. It appears possible that there are also other genes in the 5q31–q33 region that can be involved in rearrangements with similar oncogenic effects. In one case, described as atypical CML with eosinophilia, rearrangement of *ETV6* was demonstrated but rearrangement of *PDGFRB* was not [10]. Several alternative leukaemogenic mechanisms also involve *PDGFRB*. A case of chronic myelomonocytic leukaemia (CMML) associated with t(5;7)(q33;q11.2) was found to have fusion of *PDGFRB* gene with *HIP1* (the gene encoding Huntingtin Interacting Protein 1) to form an *HIP1-PDGFRB* fusion gene encoding a protein with constitutive tyrosine kinase activity [11]. Two patients have been described with t(5;10)(q33;q21.2–22) and fusion of *H4/D10S170*, a ubiquitously expressed gene, to *PDGFRB* to form a *H4/*

*D10S170-PDGFRB* fusion gene [12–16]. One of these patients had haematological features suggestive of CML in accelerated phase with an increase in cells of neutrophil and eosinophil lineage but not in monocytes or basophils [15]. The other was described as having atypical CML [13]. A further fusion gene involving *PDGFRB, rabaptin5-PDGFRB,* was described in a patient with CMML associated with t(5;17)(q33;p13) [17]. Finally, another fusion gene involving *PDGFRB* has been described only as a second event in a patient with acute myeloid leukaemia (AML) associated with t(7;11)(p15;p15). At relapse there was also t(5;14)(q33;q32) associated with formation of a *CEV14-PDGFRB* fusion gene [18]. Interestingly, eosinophilia developed at relapse. The fusion genes involving *PDGFRB* found in CMLs are summarized in table 2.

**Table 2.** Fusion genes involving *PDGFRB* implicated in CML

| Translocation | Fusion gene | Reference |
|---|---|---|
| t(5;7)(q33;q11.2)[a] | *HIP1-PDGFRB* | 11 |
| t(5;10)(q33;q21)[b] | *H4/D10S170-PDGFRB* | 12–16 |
| t(5;12)(q33;p13) | *ETV6-PDGFRB* | 8 |
| t(5;17)(q33;p13)[a] | *Rabaptin5-PDGFRB* | 17 |

[a] Single cases.
[b] Two cases.

The *ETV6/TEL* gene has been involved in three other translocations associated with CML or a chronic myeloproliferative disorder, with formation of *ETV6-JAK2, ETV6-ABL* and *ETV6-MDS/EVI1* fusion genes. An *ETV6-JAK2* fusion gene has been recognized in single cases of B-lineage (common) ALL and T-lineage ALL, both associated with t(9;12)(p24;p13), and in a single case of atypical CML with bone marrow eosinophilia in transformation, associated with a variant translocation – t(9;15;12)(p24;q15;p13) [19]. JAK2, encoded by a gene at 9p24 is one of four JAK or (Janus) kinases that are activated in response to cytokines and phosphorylate a number of tyrosine residues in the JAK kinase itself and in the cytoplasmic domain of the associated cytokine receptor [20]. JAK2 is the predominant JAK kinase activated in response to interleukin (IL)-3, IL-5, granulocyte-macrophage colony-stimulating factor (GM-CSF), erythropoietin and thrombopoietin and is also activated, together with other members of the JAK family, by IL-6, granulocyte colony-stimulating factor (G-CSF) and γ-interferon [20]. Tyrosine phosphorylation of these residues leads in turn to activation of members of the STAT family of cytoplasmic transcription regulators. Formation of the *ETV6-JAK2* fusion gene leads to synthesis of an ETV6-JAK fusion protein that undergoes homotypic oligomerization with consequent constitutive JAK2 activation [21, 22] and activation of STAT1 and STAT5 [20]. Transfection of the fusion gene leads to growth factor independence of cell lines [23] and a fatal myelolymphoproliferative disorder [23] or T-cell leukaemia [20] in mice. An *ETV6-ABL* fusion gene has been reported in a four cases of CML [24–27] as well as in single cases of B-lineage ALL [28], T-lineage ALL [27] and AML [28]. This fusion gene can only result from a complex rearrangement with at least three breakpoints since the two genes are oriented in opposite directions. In one case of CML associated with t(12;14)(p12;q11–13), the fusion gene resulted from the cryptic insertion of *ETV6* into *ABL* [25]. In another patient with Ph-negative CML, *ETV6-ABL* fusion was demonstrated but no cytogenetic data were given [26]. In a further patient with atypical CML with eosinophilia associated with t(9;12)(q34;p13) there was either an inversion or an insertion in addition to the translocation [27]. The case of *ETV6-ABL* fusion in B-lineage ALL was associated with t(9;12)(q34;p13) [28]. The single case of T-lineage ALL had no detectable cytogenetic abnormality of chromosome 9 or chromosome 12, but appeared to have had an insertion of *ETV6* at 9q34 followed by an inversion of chromosome 9 [27]. Interestingly, this patient had neutrophilia, eosinophilia and basophilia, but no studies were performed to elucidate which lineages formed part of the leukaemic clone. The case of AML with *ETV6-ABL* fusion was association with a complex karyotype which included t(9;12;15)(q34;p13;q22) [29]. It seems likely that the leukaemogenic mechanism associated with *ETV6-ABL* fusion involves abnormal tyrosine kinase activity, probably consequent on ETV6-mediated oligomerization [22, 29]. Transfection of *ETV6-ABL* into a cell line has been shown to result in activation of STAT1, STAT3 and STAT5 and transformation to growth factor independence [22]. Another fusion gene involving *ETV6* was reported in a patient with a myeloproliferative disorder associated with t(3;12)(q26;p13) who was found to have an *ETV6-MDS/EVI1* fusion gene [30]. *ETV6* was also rearranged in a patient with t(9;12)(q22;p12) associated with refractory anaemia with excess of blasts in transformation showing very rapid evolution to AML [31]; there were myeloproliferative features at presentation including neutrophilia and eosinophilia (eosinophil count was $1.6 \times 10^9$/l). This was subse-

**Table 3.** Fusion genes involving *ETV6 (TEL)* implicated in CML

| Translocation | Fusion gene | Reference |
|---|---|---|
| t(3;12)(q26;p13), t(9;15;12)(p21;q15;p13)[a, b] | *ETV6-MDS/EVI1* | 30 |
| t(5;12)(q33;p13) | *ETV6-PDGFRB* | 8 |
| t(9;15;12)(p24;q15;p13)[a] | *ETV6-JAK2* | 19 |
| t(9;12)(q22;p12)[a] | *ETV6-SYK* | 32 |
| t(12;14)(p12;q11-13) or cryptic or uncharacterized chromosomal rearrangement | *ETV6-ABL* | 24–27 |

[a] Single cases.
[b] Second patient reported had refractory anaemia with excess of blasts in transformation (French-American-British classification of myelodysplastic syndrome) rather than CML.

quently shown to have resulted in an *ETV6-SYK* fusion gene leading to constitutive activation of *SYK*, a gene that encodes a non-receptor tyrosine kinase [32]. Fusion genes involving *ETV6* that are implicated in CML are summarized in table 3.

*JAK2* has been involved not only in the *ETV6-JAK2* fusion described above but also in a *BCR-JAK2* fusion associated with t(9;22)(p24;q11) in a patient with CML with eosinophilia but no basophilia or monocytosis [33].

STAT5 activation has been observed in association with transfection of various fusion genes observed in human leukaemias – *BCR-ABL, ETV6-PDGFRB* and *ETV6-JAK2* – and also in association with transfection of other constructed genes that are capable of transforming cell lines – *ETV6-JAK1, ETV6-TYK2* and *ETV6-JAK1* – suggesting that STAT5 activation may be a crucial step in tyrosine-kinase-mediated leukaemogenesis [22].

A unique syndrome, designated the 8p11 syndrome [34], the 8p12 stem cell myeloproliferative disorder or the stem cell leukaemia/lymphoma syndrome, has been associated with t(8;13)(p11;q12) and two variant translocations. This syndrome, which results from a lymphoid-myeloid stem cell mutation, is characterized by a chronic myeloproliferative disorder with eosinophilia (either chronic eosinophilic leukaemia, CMML with eosinophilia or atypical CML with eosinophilia) with a high probability of acute transformation. The transformation may be myeloid, T-lymphoblastic or, least often, B-lymphoblastic. There are clear analogies with Ph-chromosome-related malignancies. The relevant gene at 8p11–12 is *FGFR1*, encoding fibroblast growth factor receptor 1, one of a group of four receptor tyrosine kinases. In t(8;13)(p11;q12) there is formation of a fusion gene, *ZNF198-FGFR1* [35]. At a molecular level, t(8;13) is not a simple translocation; there is also inversion of the 8p11–12 region that includes the *ZNF198* gene [36]. *ZNF198* (also known as *FIM* and *RAMP*) encodes a protein with a repeated domain resembling a zinc finger domain [36]. The fusion gene also includes the tyrosine kinase domain of *FGFR1*. The *ZNF198-FGFR1* fusion protein is localized in the cytoplasm, thus showing neither the cell surface localization of FGFR1 nor the nuclear/nucleolar localization of ZNF198 protein [37]. The fusion protein has been demonstrated to be constitutively phosphorylated and the fusion gene transforms cells in vitro, transformation being associated with high levels of tyrosine phosphorylation of STAT1 and STAT5 [38, 39]. A proline-rich oligomerization domain in *ZNF198* has been shown to be of critical importance in causing oligomerization and FGFR1 tyrosine kinase activation [40]. In the variant translocation, t(6;8)(q27;p12), there is formation of a *FOP-FGFR1* fusion gene [41]. The partner gene at 6q27, designated *FOP* (*FGFR1* Oncogene Partner), encodes a cytoplasmic protein. The FOP protein has leucine-rich regions which, in the fusion protein, are likely to lead to activation of tyrosine kinase. The FOP-FGFR1 fusion protein is located in the cytoplasm and is constitutively phosphorylated [39]. In the t(8;9)(p11–12;q32–34) variant translocation there is formation of a fusion gene *CEP110-FGFR1* [39, 42]. The *CEP110* (CEntrosome Protein 110) gene has leucine zipper motifs and encodes a protein associated with the centrosome [39]. The fusion protein is an aberrant tyrosine kinase, which is constitutively activated and localized in the cell cytoplasm [39], thus showing neither the membrane localization of FGFR1 nor the centrosomal localization of CEP110. The same myeloproliferative/T-lymphoid myeloid stem cell syndrome with eosinophilia has also been observed in single patients with *FGFR1* rearrangement and, respectively, t(8;12)(p11;q15) and ins(12;8)(p11;p11p21) [43]. *FGFR1* was also found to be rearranged in a case of CML associated with systemic mastocytosis with t(8;17)(p11;q25) [44].

There is a possibility of other myeloproliferative features in the 8p11 stem cell syndrome since two patients with t(6;8)(q27;p11) have been reported with polycythaemia rubra vera, one presenting with polycythaemia at the age of 12 years (6 years in advance of other features of the syndrome) and one developing polycythaemia during the course of the disease [45].

It is likely that myeloblastic and lymphoblastic transformation in the 8p11 syndrome result from molecular genetic if not cytogenetic evolution. Evidence of this was provided by a patient who presented with common or pre-B ALL with associated myeloproliferative features and with a complex karyotype including t(8;13)(p10;q10); after remission of the ALL and subsequent relapse as a myeloproliferative disorder with eosinophilia, only t(8;13) was present [46].

Three patients have been described with CML associated with t(8;22)(p11;q11) and a *BCR-FGFR1* fusion gene [47, 48]. The haematological features were similar, although perhaps not identical, to those of *BCR-ABL*-positive CGL. In each instance there was neutrophilia, eosinophilia, basophilia and circulating immature granulocytes, but the platelet count was reduced in one patient and not increased in either of the others and monocytosis was prominent in two of the patients. It is of note that the dominant eosinophilia that characterizes most patients with the 8p11 syndrome was not present. However, one of the three patients progressed to lymphoid blast crisis at 6 months, with populations of both B-lymphoblasts and T-lymphoblasts being present [49], suggesting that these cases also represent a variant of the 8p11 syndrome.

Chromosomal rearrangements associated with *FGFR1* are summarized in table 4.

**Table 4.** Chromosomal rearrangements associated with the 8p11 stem cell syndrome and *FGFR1* rearrangement

| Chromosomal rearrangement | Molecular event, if known | Reference |
|---|---|---|
| t(6;8)(q27;p12) | *FOP-FGFR1* | 41 |
| t(8;9)(p12;q33) | *CEP110-FGFR1* | 39 |
| t(8;12)(p11;q15)[a] | *FGFR1* disrupted | 43 |
| t(8;13)(p12;q12) | *ZNF198-FGFR1* | 35 |
| t(8;17)(p12;q25)[a] | *FGFR1* disrupted | 43 |
| t(8;22)( p11;q11)[b] | *BCR-FGFR1* | 47, 48 |
| ins(12;8)(p11;p11p21)[a] | *FGFR1* disrupted | 43 |

[a] Single patients.
[b] Three patients.

*BCR* contributes not only to *BCR-ABL*, *BCR-JAK2* and *BCR-FGFR1* but also to a recently described *BCR-PDGFRA* fusion gene in atypical CML associated with t(4;22) (see page 75).

The CMLs that result from dysregulation of genes encoding tyrosine kinases appear to result from mutations in either a pluripotent lymphoid-myeloid stem cell (Ph-positive conditions and the 8p11 syndrome) or in a multipotent myeloid stem cell (other conditions). With the exception of the Ph-positive conditions, a notable increase of the basophil count is uncommon. However, the eosinophil count is very often increased, many cases being designated chronic eosinophilic leukaemia, CMML with eosinophilia or atypical CML with eosinophilia. A prominent eosinophilic component is usual in the 8p11 syndrome and in CML associated with t(5;12)(q33;p13) and its variants. The molecular explanation of the eosinophilic differentiation is not known.

Another recurring chromosomal abnormality in Ph-negative CML is t(7;11)(p15;p15). Most patients with this translocation have been Chinese or Japanese and have had M2 or M4 AML but at least six patients have been described with either atypical CML or CMML [50–56]. This translocation usually results in formation of a *NUP98-HOXA9* fusion gene but sometimes *NUP98* is fused with an alternative *HOX* gene. In the first reports of patient with CML and t(7;11)(p15;p15) the molecular mechanism was not defined [50–52] but more recently several different molecular mechanisms have been identified. A Chinese patient reported by Wong et al. [53] with CMML had the typical *NUP98-HOXA9* fusion gene. However, a Japanese patient reported by Hatano et al. [54] with CMML evolving within a year to AML, had NUP98-HOXA9 fusion transcripts although no *NUP98-HOXA9* fusion gene was demonstrable. A possible explanation of this is provided by a Japanese patient with atypical CML transforming rapidly to M2 who was found to have a *NUP98-HOXA11* fusion gene but transcripts of both *NUP98-HOXA9* and *NUP98-HOXA11*, possibly because of alternative splicing [55, 56]. This group of patients appears to represent a novel leukaemogenic mechanism in CML, unrelated to genes encoding tyrosine kinases but involving fusion of a nucleoporin gene with one member of the family of homeobox transcription factor genes. It is possible that other translocations with an 11p15 breakpoint, such as inv(11)(p15q23), represent a related leukaemogenic mechanism. This subtype of CML is generally associated with a poor prognosis because of rapid evolution to AML.

Miscellaneous mechanisms associated with CMLs include activation of *EVI1*, a sequence-specific transcriptional repressor, demonstrated in a patient with chronic myelomonocytic leukaemia with marked eosinophilia associated with the recurring translocation, t(2;3)(p23;q26) [57]. It will be noted that *EVI1* dysregulation is also implicated in t(3;12)(q26;p13) [30] (see above).

## Conclusions

It is clear from the translocations and associated molecular abnormalities described in various types of CML that disruption and dysregulation of genes encoding tyrosine kinases is an important mechanism of leukaemogenesis in these chronic leukaemias. This is in contrast to observations in AML, in which genes encoding transcription factors or other nuclear proteins are much more commonly implicated. There are, however, some cases of acute leukaemia or lymphoblastic lymphoma associated with these same chronic-leukaemia-related translocations. These include:

(1) B-lineage ALL or, less often, AML associated with t(9;22)(q34;q11); (2) AML, T-lineage lymphoblastic lymphoma and, less often, B-lineage ALL associated with the 8p11 syndrome, and (3) single cases of T-lineage ALL, B-lineage ALL and AML associated with t(9;12)(p24;p13) and its variants.

It appears likely that in these instances acute transformation is consequent on the occurrence of further molecular events. This appears to be so for acute transformation of CGL associated with t(9;22)(q34;q11) and a p210 protein and for acute transformation of CMML/chronic eosinophilic leukaemias associated with t(8;13)(p11;q12). As for CGL, some patients with the 8p11 syndrome present already in transformation, very probably because further genetic events have already occurred.

There are other related molecular abnormalities that are usually associated with acute leukaemia but can occasionally be associated with a chronic leukaemia. This is so for t(9;22)(q34;q11) associated with a p190 protein. It could be speculated that further genetic events are very common and occur early, usually leading to presentation as acute leukaemia but that in occasional cases further molecular events have not occurred and the patient presents in chronic phase.

In addition to the translocations associated with dysregulation of tyrosine kinase genes that lead to various forms of CML, there are occasional cases in which translocations involving these genes have been associated with acute leukaemia. Whether this implies that further genetic events have occurred or whether dysregulation of tyrosine kinase genes alone is sometimes associated with the phenotype of acute leukaemia is unknown. Sometimes, it is clear that the translocation involving the tyrosine kinase gene was a second event. This is so, for example, for the *ETV6-ARG* fusion gene associated with t(1;12)(q25;p13) which has been described as a second event in both M4Eo AML associated with inv(16) [58] and in M3 AML associated with t(15;17) [59].

However, the consistent association of large cell anaplastic lymphoma with t(2;5)(p23;q31) and *NPM-ALK* fusion suggests that dysregulation of some tyrosine kinase genes, including *ALK*, is sufficient to produce the phenotype of an acute haematological malignancy. This might also be so for AML and ALL.

Despite the advances that have been made in defining translocation-related oncogenesis in CMLs, the molecular mechanism of leukaemogenesis in the majority of cases of Philadelphia-negative CML remains to be defined.

## References


1 Nowell PC, Hungerford DA: A minute chromosome in human granulocytic leukaemia. Science 1960;132:1497.

2 Sandberg AA, Hecht BK, Hecht F: Nomenclature: The Philadelphia chromosome or Ph without superscript. Cancer Genet Cytogenet 1985;14:1.

3 Caspersson T, Gahrton G, Lindsten J, Zech L: Identification of the Philadelphia chromosome as a number 22 by quinacrine mustard fluorescence analysis. Exp Cell Res 1970;63:238–240.

4 Rowley J: A new consistent chromosomal abnormality in chronic myelogenous leukemia identified by quinacrine fluorescence and Giemsa staining. Nature 1973;243:290–293.

5 Daley GQ, van Etten RA, Baltimore D: Induction of chronic myelogenous leukemia in mice by the P210bcr/abl gene of the Philadelphia chromosome. Science 1990;247:824–830.

6 McWhirter JR, Galasso DL, Whang JY: A coiled-coil oligomerization domain of Bcr is essential for the transforming function of Bcr-Abl oncoproteins. Mol Cell Biol 1993;13:7587–7595.

7 Sillaber C, Gesbert F, Frank DA, Sattler M, Griffin JD: STAT5 activation contributes to growth and viability in Bcr/Abl-transformed cells. Blood 2000;95:2118–2125.

8 Golub TR, Barker GF, Lovett M, Gilliland DG: Fusion of PDGF receptor β to a novel ets-like gene, tel, in chronic myelomonocytic leukemia with t(5;12) chromosomal translocation. Cell 1994;77:307–316.

9 Sternberg DW, Tomasson MH, Carroll M, Curley DP, Barker G, Caprio M, Wilbanks A, Kazlauskas A, Gilliland G: The TEL/PDGFβR fusion in chronic myelomonocytic leukemia signals through STAT5-dependent and STAT5-independent pathways. Blood 2001;98:3390–3397.

10 Wlodarska I, La Starza R, Baens M, Dierlamm J, Uyttebroeck A, Selleslag D, Francini A, Mecucci C, Hagemeijer A, van den Berghe H, Marynen P: Blood 1998;91:1399–1406.

11 Ross TS, Bernard OA, Berger R, Gilliland DG: Fusion of Huntingtin Interacting Protein 1 to platelet-derived growth factor β receptor (PDGFβR) in chronic myelomonocytic leukemia with t(5;7)(q33;q11.2). Blood 1998;91:4419–4426.

12 Kulkarni S, Parker S, Pocock CF, Kaeda J, Cwynarski K, Goldman JM, Cross NCP: Fusion of H4/D10S170 to the platelet-derived growth factor receptor β in a patient with a BCR-ABL negative myeloproliferative disorder and a t(5;10)(q33;q21). Blood 1999;94(suppl 1):51a.

13 Anastasiadou E, Schwaller J, Stenberg DW, Cain D, Grieco M, Siena S, Mecucci C, Gilliland DG: H4(10S170). A gene frequently rearranged in primary thyroid carcinoma is fused to the platelet-derived β receptor (PDGFβR) in atypical chronic myeloid leukemia with t(5;10)(q33;q22). Blood 1999;94(suppl 1):51a.

14 Siena S, Sammarelli G, Grimoldi MG, Schiavo R, Carlo-Stella C: New reciprocal translocation t(5;10)(q33;q22) associated with atypical chronic myeloid leukemia. Haematologica 1999;84:369–372.

15 Kulkarni S, Heath C, Parker S, Chase A, Iqbal S, Pocock CF, Kaeda J, Cwynarski K, Goldman J, Cross NCP: Fusion of H4/D10S170 to the platelet-derived growth factor receptor B in BCR-ABL negative myeloproliferative disorders with a t(5;10)(q33;q21). Cancer Res 2000;60:3592–3598.

16 Schwaller J, Anastasiadou E, Cain D, Kutok J, Wojiski S, Williams IR, LaStarza R, Crescenzi B, Sternberg DW, Andreasson P, Schiavo R, Siena S, Mecucci C, Gilland DG: H4(D10S170), a gene frequently rearranged in papillary thyroid carcinoma, is fused to the platelet-derived growth factor receptor β gene in atypical chronic myeloid leukemia with t(5;10)(q33;q22). Blood 2001;97:3910–3918.

17 Magnusson MK, Brown KE, Krueger LA, Arthur DC, Barrett J, Dunbar KE: Rabaptin-5, a novel fusion partner to platelet-derived growth factor beta receptor in chronic myelomonocytic leukemia. Blood 2000;96:692a.

18 Abe A, Tanimoto M, Towatari M, Matsuoka A, Kitaori K, Kato H, Toyozumi H, Takeo T, Adachi K, Emi N, Kawashima K, Saito H: Acute myeloblastic leukemia (M2) with translocation followed by marked eosinophilia and additional abnormalities of chromosome 5. Cancer Genet Cytogenet 1995;37:37–41.

19 Peeters P, Raynaud SD, Cools J, Wlodarska I, Grosgeorge J, Philip P, Monpoux F, Van Rompaey L, Baens M, Van den Berghe H, Marynen P: Fusion of TEL, the ETS variant gene 6 (ETV6) to the receptor-associated kinase JAK2 as a result of t(9;12) in a lymphoid and t(9;15;12) in a myeloid leukemia. Blood 1997;90:2535–2540.

20 Carron C, Cormier F, Janin A, Lacronique V, Giovannini M, Daniel M-T, Bernard O, Ghysdael J: TEL-JAK2 transgenic mice develop T-cell leukemia. Blood 2000;95:3891–3899.

21 Lacronique V, Boureux A, Monni R, Dumon S, Mauchauffé M, Mayeux P, Gouilleux F, Berger R, Gisselbrecht S, Ghysdael J, Bernard OA: Transforming properties of chimeric TEL-JAK proteins in Ba/F3 cells. Blood 1997;95:2076–2083.

22 Lacronique V, Boureux A, Valle VD, Poirel H, Quang CT, Mauchauffé M, Berthou C, Lessard M, Berger R, Ghysdael J, Bernard OA: A TEL-JAK2 fusion protein with constitutive kinase activity in peripheral blood cells from acute leukemia patients. Science 1997;278:1309–1312.

23 Schwaller J, Frantse J, Aster J, Williams IR, Tomasson MH, Ross TS, Peeters P, van Rompaey L, van Etten RA, Ilaria R, Marynen P, Gilliland DG: Transformation of hematopoietic cell lines to growth-factor independence and induction of a fatal myelo- and lymphoproliferative disorder in mice by retrovirally transduced TEL/JAK2 fusion genes. EMBO J 1998;17:5321–5333.

24 Brunel V, Sainty D, Carbuccia N, Mozzicolazi M, Fernandez F, Simonetti J, Gabert J, Dubreuil P, Lafage-Pochitaloff M, Birg F: A TEL/ABL fusion gene on chromosome 12p13 in a case of Ph-, BCR-atypical CML. Leukemia 1996;10:2003.

25 Andreasson P, Johansson B, Carlsson M, Jarlsfelt I, Fioretos T, Mitelman F, Höglund M: BCR/ABL-negative chronic myeloid leukaemia with ETV6/ABL fusion. Genes Chromosomes Cancer 1997;20:1–6.

26 Lin H, Guo J, Andreff M, Arlinghaus RB: Dual TEL-ABL fusion in a Ph-negative chronic myeloid leukemia patient. Blood 2000:96;255b.

27 Van Limbergen H, Beverloo HB, van Drunen E, Janssens A, Hählen K, Poppe B, Van Roy N, Marynen P, De Paepe A, Slater R, Speleman F: Molecular cytogenetic and clinical findings in ETV6/ABL1-positive leukemia. Genes Chromosomes Cancer 2001:31;274–282.

28 Papadopoulos P, Ridge SA, Boucher CA, Stocking C, Wiedemann LM: The novel activation of ABL by fusion to an ets-related gene, TEL. Cancer Res 1995;55:34–38.

29 Golub TR, Goga A, Barker GF, Afar DE, McLaughlin J, Bohlander SK, Rowley JD, Witte ON, Gilliland DG: Oligomerization of the ABL tyrosine kinase by the Ets protein TEL in human leukemia. Mol Cell Biol 1996;16:4107–4116.

30 Peeters P, Wlodarska J, Baens M, Criel A, Selleslag D, Hagemeijer A, van den Berghe, Marynen P: Fusion of ETV6 to MDS/EVI1 as a result of t(3;12)(q26;p13) in myeloproliferative disorders. Cancer Res 1997;57:564–569.

31 Kuno Y, Abe A, Emi N, Iida M, Yamamori T, Tanimoto M, Saito H: An atypical myelodysplastic syndrome with t(9;12)(q22;p12) and TEL rearrangement. Br J Haematol 1999;106:570–571.

32 Kuno Y, Abe A, Emi N, Iida M, Yokozawa T, Towatari M, Tanimoto M, Saito H: Constitutive kinase activation of the TEL-SYK fusion gene in myelodysplastic syndrome with t(9;12)(q22;p12). Blood 2001;97:1050–1055.

33 Griesinger F, Podleschny M, Steffens R, Bohlander S, Woermann B, Haase D: A novel BCR-JAK2 fusion gene is the result of a translocation (9;22)(p24;q11) in a case of CML. Blood 2000;96:352a.

34 Macdonald D, Aguiar RCT, Mason PJ, Goldman JM, Cross NCP: A new myeloproliferative disorder associated with chromosomal translocations involving 8p11. Leukemia 1995;9:1628–1630.

35 Xiao S, Nalbolu SR, Aster JC, Ma J, Abruzzo L, Jaffe ES, Stone R, Weissman SM, Hudson T, Fletcher JA: FGFR1 is fused with a novel zinc-finger gene, ZNF198, in the t(8;13) leukaemia/lymphoma syndrome. Nat Genet 1998;18:84–87.

36 Reiter A, Sohal J, Kulkarni S, Chase A, Macdonald DHC, Aguiar RCT, Gonçalves C, Hernandez JM, Jennings BA, Goldman JM, Cross NCP: Consistent fusion of ZNF198 to the fibroblast growth factor receptor-1 in the t(8;13)(p11;q12) myeloproliferative syndrome. Blood 1998;92:1735–1742.

37 Ollendorff V, Guasch G, Isnardon D, Galindo R, Birnbaum D, Pébusque M-J: Characterization of FIM-FGFR1, the fusion product of the myeloproliferative disorder associated-t(8;13) translocation. J Biol Chem 1999;274:922–930.

38 Smedley D, Demiroglu A, Abdul-Rauf M, Heath C, Cooper C, Shipley J, Cross NCP: ZNF198-FGFR1 transforms Ba/F3 cells to growth factor independence and results in high level tyrosine phosphorylation of STATs 1 and 5. Neoplasia 1999;1:349–355.

39 Guasch G, Mack GJ, Popovici C, Dastague N, Birnbaum D, Rattner JB, Pébusque M-J: FGFR1 is fused to the centrosome-associated protein CEP110 in the 8p12 stem cell myeloproliferative disorder with t(8;9)(p12;q33). Blood 2000;95:1788–1796.

40 Xiao S, McCarthy JG, Aster JC, Fletcher JA: ZNF198-FGFR1 transforming activity depends on a novel proline-rich ZNF198 oligomerization domain. Blood 2000;96:699–704.

41 Popovici C, Zhang B. Grégoire M-J, Jonveaux P, Lafage-Pochitaloff M, Birnbaum D, Pébusque M-J: The t(6;8)(q27;p11) translocation in a stem cell myeloproliferative disorder fuses a novel gene, FOP, to fibroblast growth factor receptor 1. Blood 1999;93:1381–1389.

42 Pébusque MJ, Popovic C, Guasch G, Adélaide J, Chaffanet M, Zhang B, Ollendorff V, Birnaum D: Fibroblast growth factor receptor 1 is fused to different unrelated genes in a stem cell myeloproliferative disorder linked to 8p11–12 chromosomal region. Blood 1998;92(suppl 1): 594a.

43 Sohal J, Chase A, Mould S, Corcoran M, Oscier D, Wilkins B, Iqbal S, Parker S, Welborn J, Harris RI, Martinelli G, Montefusco V, Sinclair P, van den Berg H, Goldman JM, Cross NCP: Identification of four new variant translocations involving FGFR1 in myeloproliferative disorders with a propensity to evolve into AML. Blood 2000;96:511a.

44 Sohal J, Chase A, Iqbal S, Parker S, Goldman JM, Cross NCP: Two colour FISH to detect FGFR1 rearrangements in myeloproliferative disorders: Identification of the t(8;17)(p11;q25) as a new cytogenetic variant. Blood 1999;94(suppl 1):180b.

45 Chaffanet M, Popovici C, Leroux D, Jacrot M, Adélaîde J, Dastugue N, Grégoire M-J, Hagemeijer A, Lafage-Pochitaloff M, Birnbaum D, Pébusque M-J: t(6;8), t(8;9) and t(8;13) translocations associated with stem cell myeloproliferative disorders have close or identical breakpoints in chromosome region 8p11–12. Oncogene 1998;16:145–149.

46 Roy S, Juneja SK, Campbell LJ, Szer J: A rare myeloproliferative disorder with t(8;13) presenting as B-lineage ALL. Blood 2000;96: 268b.

47 Fioretos T, Panagopoulos J, Larsen C, Swedin A, Billström R, Isaksson M, Strömbeck B, Olofsson T, Mitelman F, Johansson B: Fusion of the BCR and the fibroblast growth factor receptor-1 (FGFR1) genes as a result of t(8;22)(p11;q11) in a myeloproliferative disorder: The first fusion gene with BCR but not ABL. Genes Chromosomes Cancer 2001;32: 302–310.

48 Demiroglu A, Steer EJ, Heath C, Taylor K, Bentley M, Allen SL, Koduru P, Brody JP, Hawson G, Rodwell R, Doody M-L, Camicero F, Reiter A, Goldman JM, Melo JV, Cross NCP: The t(8;22) in chronic myeloid leukemia fuses BCR to FGFR1: Transforming activity and specific inhibition of FGFR1 fusion proteins. Blood 2001;98:3778–3783.

49 Brody J, Basham K, Stanek A, Liu J, Koduru P, Cross NCP, Allen SL: Rapid transformation to biphenotypic (B and T cell) lymphoid blast crisis in a patient with CML associated with the fusion of BCR to FGFR1. Blood 2001;98: 255b.

50 Kaneko Y, Maseki N, Takasaki N, Sakurai M, Kawai K, Sakurai M: Possible association of a new translocation, t(7;11)(p15;p15), with Ph[1] chromosome-negative chronic myelogenous leukemia. Int J Cancer 1985;36:657–659.

51 Takeda T, Ikebuchi K, Zaike Y, Mori M, Ohyashiki K, Ikeuchi T: Ph-negative chronic myelocytic leukemia with a complex translocation involving chromosomes 7 and 11. Cancer Genet Cytogenet 1986;21:123–127.

52 Sato Y, Abe S, Mise K, Sasaki M, Kamada N, Kouda K, Musashi M, Saburi Y, Horikoshi A, Minami Y: Reciprocal translocation involving the short arms of chromosomes 7 and 11, t(7p–;11p+), associated with myeloid leukemia with maturation. Blood 1987;70:1654–1658.

53 Wong KF, So CC, Kwong YL: Chronic myelomonocytic leukemia with t(7;11)(p15;p15) and NUP98/HOXA9 fusion. Cancer Genet Cytogenet 1999;115:70–72.

54 Hatano Y, Miura I, Nakamura T, Yamazaki Y, Takahashi N, Miura AB: Molecular heterogeneity of the NUP98/HOXA9 fusion transcript in myelodysplastic syndromes associated with t(7;11)(p15;p15). Br J Haematol 1999;107: 600–604.

55 Fujino T, Suzuki A, Ito Y, Ohyashiki K, Hatano Y, Miura I, Nakamura T: Single-translocation and double-chimeric transcripts: Detection of NUP98-HOXA9 in myeloid leukemias with HOXA11 or HOXA13 breaks of the chromosomal translocation t(7;11)(p15;p15). Blood 2002;99:1428–1433.

56 Suzuki A, Ito Y, Sashida G, Honda S, Katagiri T, Fujino T, Nakamura T, Ohyashiki K: t(5;11)(p15;p15) chronic myeloid leukaemia developing into blastic transformation showing a novel NUP98/HOXA11 fusion. Br J Haematol 2002;116:170–172.

57 Poppe B, Yigit N, Marynen P, Hagemeijer A, Van Limbergen H, de Smet B, de Paepe A, Speleman F: Molecular characterization of a t(2;3)(p23;q26); a recurrent translocation involving the EVI1 gene. Blood 2001;98:965a.

58 Cazzaniga G, Tosi S, Aloisi A, Giudici G, Daniotti M, Pioltelli P, Kearney L, Biondi A: The tyrosine kinase Abl-related gene ARG is fused to ETV6 in an AML-M4Eo patient with t(1;12)(q25)(p13): Molecular cloning of both reciprocal transcripts. Blood 1999;94:4370–4373.

59 Iijima Y. Ito T, Oikawa T, Eguchi M, Eguchi-Ishimae M, Kamada N, Kishi K, Asano S, Sakaki Y, Sato Y: A new ETV6/TEL partner gene, ARG (ABL-related gene or ABL2), identified in an AML-M3 cell line with a t(1;12)(q25;p13) translocation. Blood 2000;95: 2126–2131.

Dr. Barbara J. Bain
Department of Haematology
St Mary's Hospital
Praed Street, London W2 1NY (UK)
E-Mail b.bain@ic.ac.uk

Bain BJ (ed): Chronic Myeloproliferative Disorders.
Basel, Karger, 2003, pp 9–31 / Acta Haematol 2002;108:180–202

# Molecular Basis of Chronic Myeloid Leukaemia

David J. Barnes Junia V. Melo

Department of Haematology, Faculty of Medicine, Imperial College of Science, Technology and Medicine, Hammersmith Hospital, London, UK

**Key Words**
Chronic myeloid leukaemia · Cytogenetics · Molecular genetics · Oncogenes

**Abstract**
Chronic myeloid leukaemia (CML) is caused by the product of the *BCR-ABL* oncogene, located on the Philadelphia (Ph) chromosome. *BCR-ABL* is generated as a result of a reciprocal t(9;22) chromosomal translocation. The mechanisms responsible for this illegitimate recombination event remain elusive but are presumed to require a close spatial association of the translocation partners (chromosomes 9 and 22). *BCR-ABL* fusion transcripts can be detected by a sensitive reverse transcription-polymerase chain reaction (RT-PCR) in the leucocytes of some healthy individuals suggesting that chromosomal translocations may occur frequently in the general population. The presence of *BCR-ABL* fusion transcripts does not imply that the individual will inevitably develop CML since other conditions must be favourable for expansion of the abnormal clone. Breakpoints in the *ABL* gene occur within a 5′ segment. *BCR-ABL* fusion transcripts lack *ABL* exon a1 and consist of *BCR* exons fused directly to *ABL* exon a2. The breakpoints in the *BCR* gene on chromosome 22 are found within three defined regions. Depending on the position of the *BCR* breakpoint, fusion genes are generated that encode 190-, 210- or 230-kD forms of the Bcr-Abl tyrosine kinase. Since the *ABL* component of the fusion gene is largely invariant, it follows that variability in disease phenotype may be due to protein sequences encoded by the translocation partner, *BCR*. Different disease phenotypes are associated with each of the three Bcr-Abl oncoproteins, $p190^{Bcr\text{-}Abl}$, $p210^{Bcr\text{-}Abl}$ and $p230^{Bcr\text{-}Abl}$. Mechanisms associated with malignant transformation include altered cellular adhesion, activation of mitogenic signalling pathways, inhibition of apoptosis and proteasomal degradation of physiologically important cellular proteins. CML is subject to an inexorable progression from an 'indolent' chronic phase to a terminal blast crisis. Disease progression is presumed to be associated with the phenomenon of genomic instability.

## Introduction and Historical Perspective

Chronic myeloid leukaemia (CML), also known as chronic granulocytic and chronic myelogenous leukaemia, may well be the most extensively studied human neoplasm. The discovery of the minute Philadelphia (Ph) chromosome in 1960 [1] represented the first link between a consistent chromosomal abnormality and a specific human cancer. The origin of the Ph chromosome remained unknown until 1973 when it was established that it was the product of a reciprocal translocation between chromosomes 9 and 22 [t(9;22)(q34;q11)] [2] (fig. 1). The next major discovery occurred in 1983 when the translocation was shown to involve the *ABL* proto-oncogene normally found on chromosome 9 [3]. The partner gene in the translocation was identified a year later

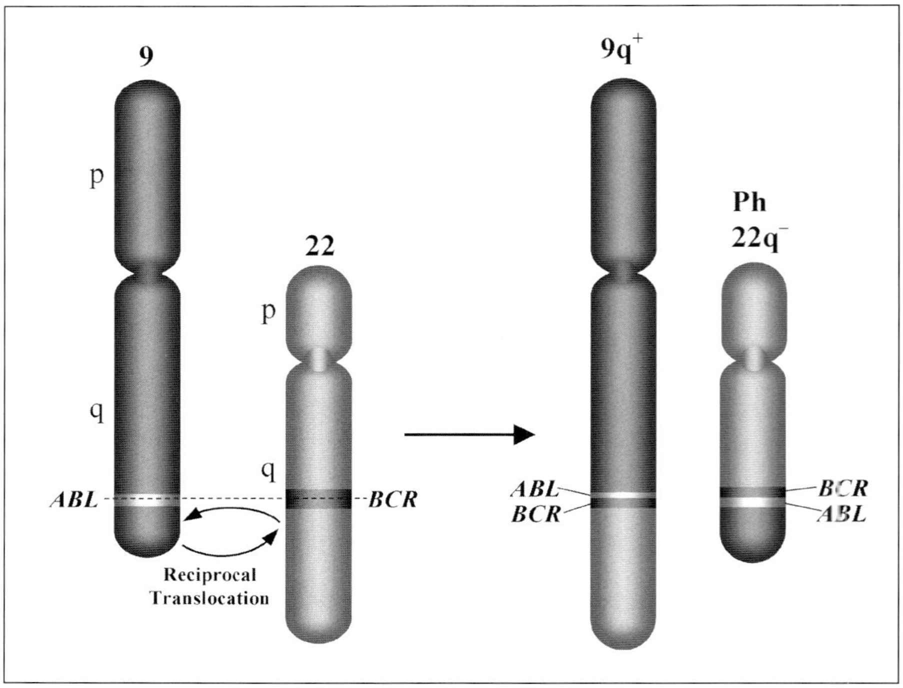


**Fig. 1.** The t(9;22) reciprocal translocation. The Ph chromosome is a deleted chromosome 22 (22q⁻) that is produced as a result of a reciprocal translocation between the long (q) arms of chromosomes 9 and 22. Breakpoints occur within the *ABL* gene situated at 9q34 and the *BCR* gene situated at 22q11. The Ph chromosome bears the *BCR-ABL* oncogene formed from the juxtaposition of 5′ *BCR* sequences and 3′ *ABL* sequences. The phenotype of the corresponding translocation partner gene, *ABL-BCR*, on chromosome 9q+ is unknown.

and was mapped to chromosome 22 [4]. In 1984, the function of this gene was unknown and consequently the name 'breakpoint cluster region' (*BCR*) was proposed to describe its most characteristic feature, the formation of variable, clustered chromosomal breakpoints. By 1990, the transforming potential of the *BCR-ABL* gene product (Bcr-Abl) was shown to derive from deregulation of the tyrosine kinase activity of Abl [5]. The same year saw the development of the first animal models of CML in which the introduction of *BCR-ABL*-positive haemopoietic cells into mice caused a myeloproliferative disorder that recapitulated features of the human disease [6]. The demonstration of the key role of Bcr-Abl in the aetiology of CML led members of the pharmaceutical industry to regard this tyrosine kinase as a suitable target for drug development. A systematic programme of synthesis and screening was begun in the 1990s to develop specific inhibitors of oncogenic tyrosine kinases [7]. The most promising compound to emerge from this work, imatinib mesylate, entered clinical trials in 1998 and was approved for the treatment of CML patients in the United States in May 2001 [8].

In this review we will summarise the current state of knowledge regarding the cytogenetic and molecular genetic aspects of CML. The physiological roles of the Bcr and Abl proteins will be discussed with reference to how these functions may be subverted in the pathological Bcr-Abl oncoprotein. Particular emphasis will be given to consideration of how the structure of the Bcr-Abl fusion protein can affect the phenotype of the disease.

## The t(9;22) Reciprocal Chromosomal Translocation

The mechanisms responsible for the t(9;22) reciprocal chromosomal translocation remain elusive. Chromosomal translocations are initiated by DNA double-strand (ds) breaks occurring on pairs of non-homologous chromosomes [9]. The erroneous recombination that follows represents a failed attempt by genomic repair enzymes to re-ligate the severed chromatin [10]. The question as to why the t(9;22) with the specific 9q34 and 22q11 involvement, rather than other reciprocal translocations involving different partners, is seen at such a high frequency has been variously regarded from opposing viewpoints. At one extreme, it is possible that the breaking and re-ligating process is entirely random and occurs with relatively equal frequency between any two chromosomes in the cell. In this case, the 'success' of the t(9;22) (q34;q11) is due to it being the only (or nearly only) recombination process that generates a fusion gene with oncogenic properties that will endow the cell with a selective growth advantage (i.e., leukaemia). Any other chromosomal rearrangement that does not fulfil this essential criterion for survival in an expanded clone of cells will never become visible and will disappear during natural cell wastage. Conversely, it has been suggested that the consistent involvement of specific and apparently non-random regions of the genome in reciprocal translocations results from chromatin at particular loci being especially susceptible to DNA damage [11].

Ionising radiation is one of the best-known agents capable of inducing DNA ds breaks. It has long been known from epidemiological studies that ionizing radiation is a risk factor for CML [12–14]. Furthermore, it has been shown that high-dose ionising radiation can lead to the in vitro formation of a *BCR-ABL* gene [15], although the frequency of this phenomenon may not be significantly higher than the background observed in control, non-irradiated cells [16]. It should also be noted in this context that both 'functional' (in frame) and truncated BCR-ABL fusions occur during high-dose ionising radiation, possibly supporting the contention that at the simple genomic level the location of ds breaks and re-joining is random.

It seems likely that the topographical arrangement of chromosomes 9 and 22 within the cell nucleus favours their reciprocal translocation. Experiments in which human lymphocytes were irradiated with neutrons revealed that the incidence of exchange aberrations between chromosomes 9 and 22 was 11 times higher than the predicted frequency [17]. The investigators attributed this finding to the co-localisation of the territories of chromosomes 9 and 22 within a central region of the cell nucleus. In fact, mapping of the positions of the *ABL* and *BCR* genes within the nuclei of human haemopoietic cells by fluorescent in situ hybridisation (FISH) and confocal microscopy [18] demonstrated that juxtaposition of *ABL* and *BCR* occurred at the transition from the S to the $G_2$ phase of the cell cycle. The genes remained associated throughout $G_2$ and prophase before separating at metaphase. A subsequent report on FISH analysis of blasts obtained from a patient with acute lymphoblastic leukaemia (ALL) has confirmed the proximity of the *ABL* and *BCR* genes during late S phase [19]. The authors made the interesting observation that the fluorescent *BCR* and *ABL* signals appear to merge at the time of replication of the second *BCR* allele and that this may represent the point at which there is the greatest risk of *BCR-ABL* fusion. Thus, a picture is emerging of the juxtaposition of the *ABL* and *BCR* genes at a critical period in the cell cycle. During this interval, the two respective chromosomal regions are most vulnerable to illegitimate recombination.

The involvement of sequences potentially relevant to chromosomal translocations such as Alu repeats, Chi-like octamers, immunoglobulin gene heptamers and others surrounding the breakpoint positions on both the Ph (22q–) and the derivative 9q+ chromosome in a number of cloned *BCR-ABL* rearrangements has been inconclusive [20]. More recently, a large segment of duplicated genomic sequences ('duplicon') was found to map close to the *BCR* and the *ABL* genes on chromosomes 22 and 9, respectively. Such a duplicon was postulated to facilitate the homologous recombination between the two chromosomes to originate a functional *BCR-ABL* fusion gene [21].

Another interesting aspect of the mechanics of the t(9;22) translocation is the recent finding that substantial deletions surrounding the fusion point on the 9q+ derivative occur in 15–30% of patients with CML [22]. Significantly, the presence of these deletions was shown to carry an especially bad prognosis [23]. The underlying molecular basis of such an adverse effect has not yet been identified, but it has been shown that non-expression of the most obvious candidate, the reciprocal *ABL-BCR* gene spanning the actual 9q34–22q11 junction, is not the cause of the ominous outcome [24].

## Structure of the *BCR-ABL* Fusion Genes and Their Transcripts

### *Breakpoints on ABL*

Breakpoints within the *ABL* gene can occur anywhere within a 5′ segment that extends for over 300 kb (fig. 2). Typically, breakpoints form within intronic sequences that may be located upstream of exon 1b, downstream of exon 1a or, more frequently, between the two [25]. Thus, *BCR-ABL* fusion genes may contain exons 1b and 1a, exon 1a alone or neither of the alternative first exons. *BCR-ABL* messenger RNA (mRNA) lacks exon 1, regardless of the structure of the fusion gene, with the transcript consisting of *BCR* exons fused directly to *ABL* exon a2.

### *Breakpoints on BCR*

The breakpoints within the *BCR* gene on chromosome 22 are found within three defined regions (fig. 2). In 95% of patients with CML [26] and approximately one third of patients with ALL, the *BCR* gene is truncated within a 5.8-kb region known as the major breakpoint cluster region (M-*bcr*) [4]. This region contains five exons that were originally named b1–b5, although they are now known to reside within the central region of the *BCR* gene

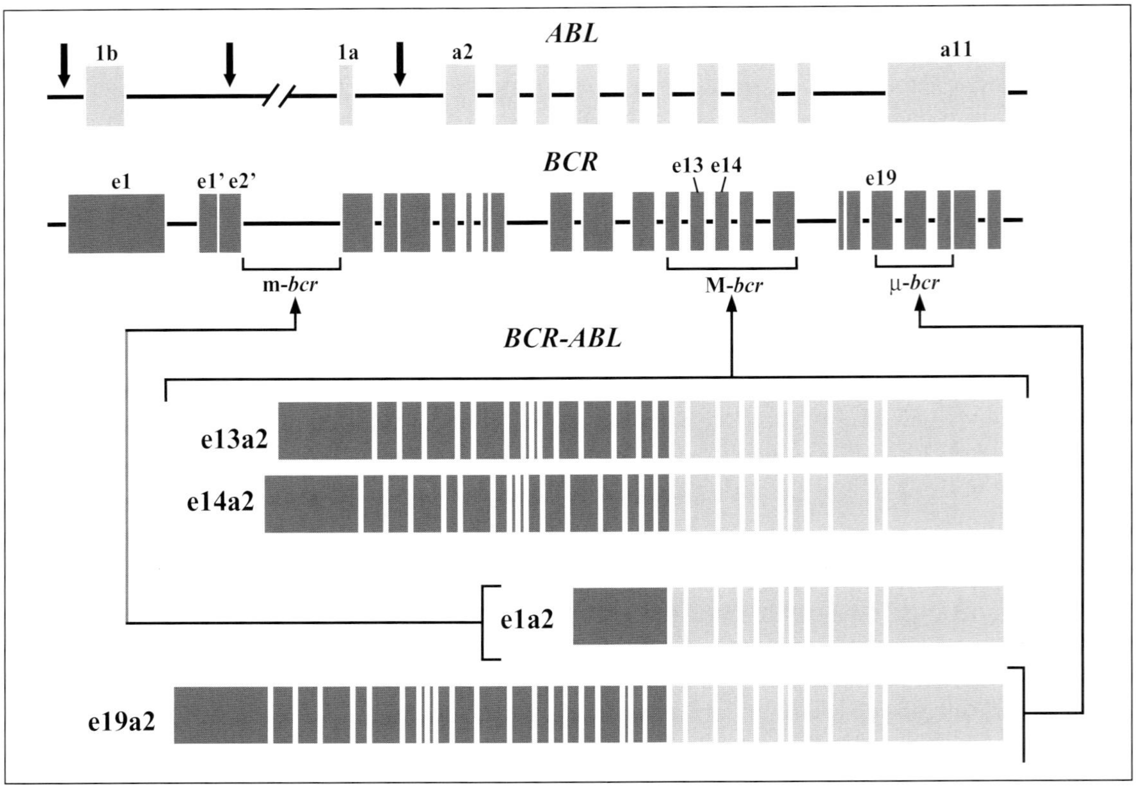


**Fig. 2.** Breakpoints in the *ABL* gene (indicated by arrows) and the three BCRs regions in the *BCR* gene (m-*bcr*, M-*bcr* and μ-*bcr*) (top). The fusion transcripts derived from each of the respective BCRs are also shown (bottom).

and to correspond to exons 12–16 (e12–e16) [20]. The majority of breakpoints form within introns immediately downstream of either exon 13 (b2) or exon 14 (b3). Since processing of *BCR-ABL* mRNA results in the joining of *BCR* exons to *ABL* exon a2, hybrid transcripts are produced that have either an e13a2 (b2a2) or e14a2 (b3a2) junction. In approximately 20% of patients, both e13a2 and e14a2 transcripts can be detected, due to alternative splicing of exon e14 [27]. In both cases the mRNA consists of an 8.5-kb sequence that encodes a 210-kD fusion protein, $p210^{Bcr\text{-}Abl}$.

In two thirds of patients with Ph+ ALL and in rare cases of CML and AML, the breakpoint in *BCR* occurs in a region upstream of M-*bcr* known as the minor breakpoint cluster region (m-*bcr*). This consists of the 54.4-kb intron between the two alternative second exons of the *BCR* gene e2′ and e2. *BCR-ABL* fusion genes having breakpoints within m-*bcr* contain both *BCR* alternative first exons (e1 and e1′) together with the alternative second exon e2′. Chimaeric RNA derived from these fusion genes is evidently subject to splicing since the mature transcripts lack exons e1′ and e2′. The hybrid mRNA consists of sequences approximately 7 kb in length in which exon e1 from *BCR* is joined to exon a2 of *ABL*. The translated product is a 190-kD fusion protein, $p190^{Bcr\text{-}Abl}$ (also referred to as $p185^{Bcr\text{-}Abl}$) [28].

When highly sensitive, nested reverse transcription-polymerase chain reactions (RT-PCR) are used, it is possible to detect transcripts with an e1a2 junction in cells from patients with a M-*bcr* breakpoint where the expressed fusion protein is $p210^{Bcr\text{-}Abl}$. The appearance of *BCR-ABL* mRNA encoding the p190 form of Bcr-Abl in addition to mRNA for p210 led to early speculation that this was somehow connected with disease progression to the acute phase or blast crisis [29]. Subsequent studies have revealed, however, that these p190-type *BCR-ABL* mRNA molecules can be detected in the cells of the majority (>90%) of CML patients with a single and classical M-*bcr* breakpoint, irrespective of the phase of disease [30, 31]. Their low level of expression in comparison with the predominant p210[Bcr-Abl] message suggests that the e1a2 transcripts are most likely the result of alternative splicing of e13a2 or e14a2 transcripts [31, 32]. *ABL* exon 2 has been shown to be essential for the 'exon-skipping' that occurs with *BCR-ABL* transcripts and a model has been proposed to account for this phenomenon [33].

The third defined breakpoint cluster region within the *BCR* gene was named 'micro' breakpoint cluster region (μ-*bcr*) [30]. In this case, the breaks occur within a 3′ segment of the *BCR* gene between exons e19 and e20 (known as c3 and c4 in the original nomenclature). Transcription of the hybrid gene yields an e19a2 *BCR-ABL* fusion transcript that encodes a 230-kD protein, p230[Bcr-Abl].

*'Unusual' Breakpoints*

Several atypical *BCR-ABL* transcripts having junction configurations other than e13a2, e14a2, e1a2 or e19a2 have been reported. Although they may first appear to represent an apparently disparate collection of cases, these rare transcripts may be assigned to four groups according to common features of the *BCR* and *ABL* breakpoints within the parental fusion gene. The first group consists of transcripts derived from fusion genes in which the *BCR* breakpoints originate within introns that lie outside the recognised breakpoint cluster regions (M-*bcr*, m-*bcr* or μ-*bcr*). Fusion of *BCR* sequences to *ABL* occurs at conventional *ABL* breakpoints in introns upstream of exon 2. Transcripts with an e6a2 [34] and an e2a2 junction [35] have been reported. In the latter case, the transcript was detected in a CML patient who had an additional Ph chromosome, monosomy 7 and a disease with an especially aggressive clinical course.

A second group consists of transcripts derived from a series of novel fusion genes in which the breaks occurred within *exons*. Two patients were identified as having e8a2 and e13a2 transcripts obtained from fusion genes with truncated *BCR* exons [36]. In two other patients, both the *BCR* and the *ABL* breakpoints originated within exons and yielded transcripts with an e15a2 junction [37] and an e2a1a junction [38]. The latter report was the first to describe a *BCR-ABL* mRNA containing any part of an Abl first exon. This fusion transcript was associated with a CML with an aggressive clinical course.

The third group of unusual transcripts are derived from fusion genes in which the *BCR* breakpoints occur within recognised cluster regions but in which the *ABL* breakpoints are located downstream of exon a2, giving rise to a hybrid mRNA with an a3 junction. Fusion transcripts with e1a3 [39–41], e13a3 [39, 40] and e14a3 [42] junctions have been described. The absence of *ABL* exon 2 sequences in these transcripts demonstrates that the 58 amino acids encoded by this exon are not essential for leukaemogenesis. It has been suggested that unusual transcripts may be more common than previously supposed [40] and that a widespread use of improved multiplex RT-PCR techniques [43] is the reason why novel transcripts are only now coming to light.

The fourth group consists of aberrant transcripts that contain 'bizarre' insertions or intervening sequences between *BCR* and *ABL*. Thus, transcripts that contain part of *ABL* intron 1b upstream of *ABL* exon 2 have been reported [44, 45]. In the former case, 55 bp of *ABL* intron 1b were detected and were joined in-frame to *BCR* exon e8. Curiously, the intronic sequence was inverted. In the latter case, 44 bp of *ABL* intron 1b were present and the *BCR* breakpoint occurred within exon e3. The clinical course of this patient's disease was aggressive, an observation that is consistent with the notion that shorter forms of *BCR-ABL* transcripts are associated with a more severe, acute disease phenotype. In another case, an abnormally large *BCR-ABL* mRNA transcript was detected [46]. Sequence analysis revealed that the intervening 589-bp insert between *BCR* exon e13 and *ABL* exon a2 consisted of a 5′ portion of the intron immediately downstream of *BCR* exon e13 and a 3′ segment of *ABL* intron 1b.

*Detection of BCR-ABL Fusion Transcripts in Healthy Individuals*

Evidence of either a t(9;22) chromosomal translocation or the detection of the *BCR-ABL* fusion gene in the leucocytes of a patient presenting with a chronic myeloproliferative disease is usually sufficient to confirm a diagnosis of CML. In healthy individuals, however, the mere presence of the *BCR-ABL* oncogene in infrequent leucocytes does not imply that these individuals will inevitably develop CML. This principle was established by two landmark studies in which highly sensitive, two-step RT-PCR was used to amplify *BCR-ABL* mRNA from the leucocytes of normal, healthy individuals. In the first report, *BCR-ABL* fusion transcripts could be detected in leucocytes from 30% of a sample of normal adults [47]. These findings were confirmed and enlarged upon by the second study in which hybrid *BCR-ABL* mRNA with either an e13a3 or an e14a3 junction was detected in 27% of the healthy individuals tested [48]. When primers designed to amplify transcripts with the e1a2 junction were used, the majority (69%) of normal individuals could be shown to have *BCR-ABL* mRNA in their leucocytes.

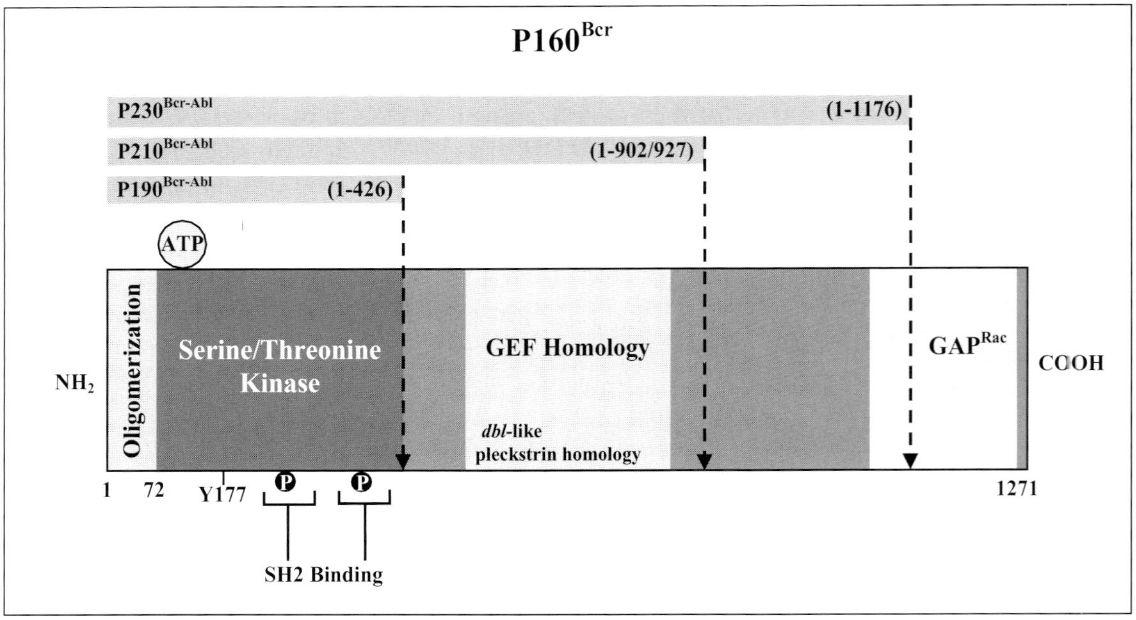


**Fig. 3.** Structure of the P160^Bcr protein. An oligomerisation domain consisting of the first 72 amino acids of the protein is located at the N-terminus. Proximal to the oligomerisation domain is a serine/threonine kinase domain. Also shown are the serine- and threonine-rich regions (amino acids 192–242 and 293–413) that, upon phosphorylation, can function as binding sites for the SH2 domains of other proteins. Tyrosine 177 undergoes autophosphorylation and serves as the binding site for the adaptor protein, Grb2. The centre of the protein contains a region with homology to guanine nucleotide exchange factors (GEFs) as well as '*dbl*-like' and pleckstrin homology domains. At the C-terminus is a domain that can function as a GTPase-activating protein (GAP) for the GTP-binding protein Rac. The dashed arrows indicate the position of the breakpoints in the Bcr-Abl fusion proteins and the shaded bars (top) indicate the extent of the Bcr amino acid sequences that are present in each of these proteins. Numbers refer to the position of amino acids in the primary sequence.

The biological and clinical implications of this work are profound. Both studies provide evidence that the generation of a specific oncogene as a result of a chromosomal translocation is not, in itself, sufficient for malignant transformation. On the contrary, these data suggest that recombination of specific pairs of non-homologous chromosomes is a commonplace occurrence within haemopoietic cells, perhaps reflecting an innate, basal level of genomic instability. In fact, many of the *BCR-ABL* transcripts detected in the healthy population have aberrant junctions and would be non-functional as they encode truncated proteins [48]. However, 'correct', leukaemia-type *BCR-ABL* genes are also found in these individuals, and how they can be generated, transcribed and possibly translated and yet prove to be innocuous is somewhat of an enigma. One possible explanation for the non-appearance of CML is that the *BCR-ABL* fusion genes are generated within relatively mature progenitors that are lost upon differentiation or cell death. According to this hypothesis, expansion of the malignant clone is only possible if the fusion genes are generated in primitive progenitors with a high capacity for self-renewal. Alternatively, the *BCR-ABL*-positive cells may be recognised and destroyed by the individual's immune system before clonal expansion can occur. It should be remembered that the RT-PCR used in these studies achieves a very high level of sensitivity and that the prevalence of *BCR-ABL* bearing leucocytes in the average, healthy individual is extremely low, probably accounting for no more than 1–10 per $10^8$ white blood cells [48].

## Structure and Known Functions of the Bcr Protein

The *BCR* gene is ubiquitously expressed. The predominant gene product is a 160-kD cytoplasmic protein with several functional domains (recently reviewed by Laurent et al. [49]) (fig. 3). The region of the Bcr protein consisting of the first 426 amino acids from the N-terminus has a special significance because it is encoded by the first exon of the *BCR* gene and consequently is the only Bcr peptide sequence that is retained in all Bcr-Abl fusion protein isoforms (p190, p210 and p230). A serine-threonine kinase domain was identified within this region [50]. Aside from autophosphorylation of serine and threonine residues within Bcr itself, the only known substrate for this kinase is Bap-1 (Bcr associated protein-1), a member of the 14-3-3 family of proteins [51]. Tyrosine 177, when phosphorylated, is a key residue since it forms a binding site for Grb-2, an adapter molecule that links Bcr to the Ras pathway [52].

Also encoded by *BCR* exon 1 are two regions consisting of amino acids 192–242 and 298–413 that are especially rich in serine and threonine residues. Phosphorylation of serine and threonine within these regions generates activated domains that bind to the Src homology 2 (SH2) domains of other proteins. The SH2-binding domains of Bcr were first identified by their ability to bind the SH2 domain of Abl with high affinity [53]. Binding of $p190^{Bcr\text{-}Abl}$ and $p210^{Bcr\text{-}Abl}$ to the Abl SH2 domain could also be demonstrated, suggesting that the Bcr SH2-binding domains are functionally active in Bcr-Abl fusion proteins. These sites are unique in that they are the only known SH2-binding domains in which binding may be mediated by phosphoserine and phosphothreonine as well as phosphotyrosine. Indeed, binding of the Bcr SH2-binding domains to Abl SH2 does not require tyrosine phosphorylation [53]. Mutation and deletion studies have shown that the proximal SH2-binding domain between amino acids 192 and 242 is essential for transformation of rat fibroblasts by Bcr-Abl [53].

The N-terminus of Bcr contains a coiled-coil oligomerisation domain. This region has been subject to intense scrutiny ever since it was discovered that amino acids 1–63 were essential for the oncogenicity of Bcr-Abl [54]. Its analysis revealed a heptad repeat – a repeated seven-amino-acid pattern – with hydrophobic residues at the first and fourth positions between amino acids 28 and 68 [54]. The resemblance of this pattern to others found in coiled-coil amphipathic alpha helices led to the suggestion that this amino acid stretch forms a coiled-coil domain [54]. Moreover, evidence that the N-terminal domain of Bcr has a role in oligomerisation comes from mutation and co-immunoprecipitation studies in which metabolically labelled Bcr-Abl proteins have been shown to form oligomers in vivo [54]. A crystal structure for the oligomerisation domain of Bcr has recently been published [55]. The complete domain encompasses amino acids 1–72 of the N-terminus of Bcr and consists of a long alpha helix equivalent to the putative coiled-coil domain (amino acids 28–67), a second shorter alpha helix (amino acids 5–15) and a flexible loop (amino acids 16–27) that links them. These three segments adopt an 'N'-shaped conformation with the flexible loop allowing the longer helix of one monomer to form an antiparallel coiled-coil with the longer helix of a second monomer. Interestingly, the oligomerisation domain crystallised as eight protein chains forming two identical tetramers. It has previously been reported that the predominant oligomer of the Bcr N-terminal domain in vitro is a homotetramer [54]. This finding led the authors to propose a model of activation of the Abl tyrosine kinase domain in Bcr-Abl based on tetramerisation of the fusion protein. The discovery of the *TEL-ABL* fusion gene, expressed in rare cases of ALL and AML lends weight to this hypothesis [56, 57]. Tel, the product of the *TEL (ETV6)* gene, contains a helix-loop-helix domain that mediates self-oligomerisation. In Tel-Abl this motif is essential for transformation. Tel-Abl-mediated, like Bcr-Abl-mediated, transformation requires activation of the Abl kinase domain. The primary importance of the Bcr oligomerisation domain in malignant transformation is underscored by a recent report [58] in which bone marrow cells that had been retrovirally transduced with mutant *BCR-ABL* constructs were introduced into mice. The development of a fully fledged myeloproliferative disorder was only noted in mice that had received cells expressing a mutant with an intact coiled-coil domain and Tyr177.

The central region of the Bcr protein, encoded by exons 3–10, shares homology with several guanine nucleotide exchange factors including Cdc24, Dbl [59], Vav and CDC25Mm [60]. Guanine nucleotide exchange factors activate guanine nucleotide-binding proteins by stimulating the exchange of guanosine triphosphate (GTP) for guanosine diphosphate (GDP).

The C-terminus of Bcr contains a domain that functions as a GTPase-activating protein (GAP) for Rac($p21^{rac}$). Rac is a member of the superfamily of small guanine nucleotide-binding proteins, of which the Ras ($p21^{ras}$) oncoprotein is the best known. Guanine nucleotide-binding proteins have an intrinsic GTPase activity that catalyses the hydrolysis of the bound GTP to GDP. Upon complete hydrolysis of GTP, the guanine nucleotide-binding protein becomes inactivated. GAPs hasten the inactivation of guanine nucleotide-binding proteins by accelerating the rate of GTP hydrolysis. The Bcr GAP domain interacts with Rac in vitro to stimulate GTP hydrolysis [61]. Bcr would thus appear to serve a dual function in guanine nucleotide-binding protein signalling,

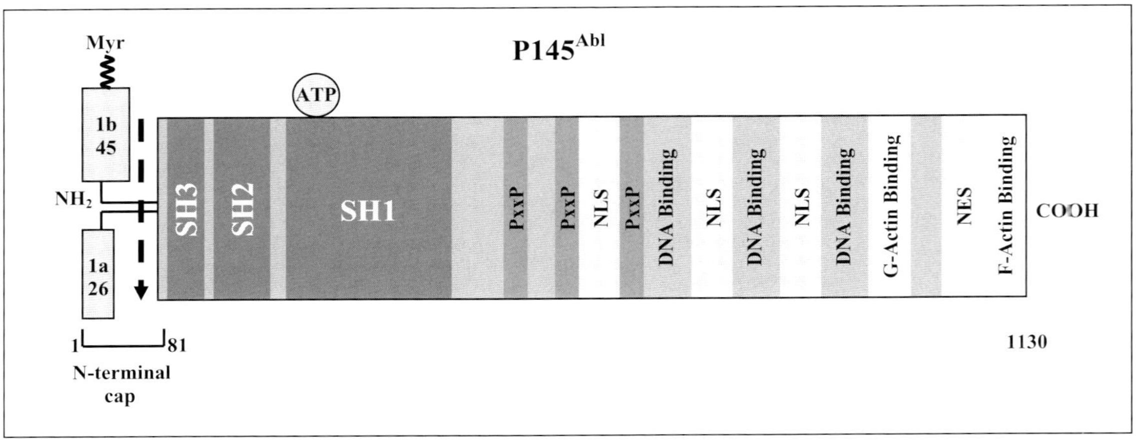


**Fig. 4.** Structure of the P145^Abl protein. The Abl protein exists as two isoforms, 1a and 1b, generated as a result of splicing of the alternative first exons of the *ABL* gene. The 1b isoform is 25 amino acids larger and is myristoylated (Myr) at its N-terminus. Only sequences downstream of the dashed arrow are present in the Bcr-Abl fusion protein. Proximal to the N-terminus are the three Src homology domains (SH1, SH2 and SH3). The centre of the protein contains proline-rich regions (PxxP) that function as binding sites for the SH3 domains of other proteins. Also present are three nuclear localisation signals (NLS) and a single nuclear export signal (NES). The C-terminus contains domains that are required for binding to DNA and G and F actin. Numbers refer to the position of amino acids in the primary sequence.

the central domain having an excitatory role and the C-terminus having an inhibitory role. The results of a study in which *BCR* knockout mice were generated [62] suggests that the functions of Bcr are not essential for normal cell physiology. The *BCR*-null mutants were viable and the only defect that could be detected was an increased neutrophilic oxidative burst. This finding probably reflects an inherent redundancy in cell-signalling pathways.

## Structure and Known Functions of the Abl Protein

The *ABL* gene is the human homologue of v-*abl*, the oncogene of the Abelson murine leukaemia virus (A-MuLV) [63]. Abl, the product of the *ABL* gene, is a 145-kD non-receptor tyrosine kinase that is ubiquitously expressed (fig. 4). The two alternative first exons of *ABL*, 1a and 1b, are subject to splicing and mRNA transcripts of 5 and 6.5 kb are generated. Translation of the splice variants yields two isoforms of the Abl peptide that differ only with respect to their N-terminal sequences [20]. The Abl type 1b protein contains a myristoylation site and can therefore be anchored to the plasma membrane.

Three domains located towards the N-terminus of Abl have a high degree of homology to domains found within Src, the prototype non-receptor, tyrosine kinase. Src homology 1 (SH1) is the kinase domain of Abl. Src homology domains 2 (SH2) and 3 (SH3) are non-catalytic and function as binding sites for other proteins. As described earlier, the SH2 domain of Abl normally binds to sites containing phosphotyrosine, but it can also bind to the unique phosphoserine-containing sites within Bcr [53]. The SH3 domain is essential for regulation of the kinase activity of the SH1 domain. The central region of the Abl molecule contains proline-rich sequences that provide binding sites for the SH3 domains of other proteins, such as Crk [64] and Crkl [65].

Abl differs markedly from Src in having a long (~90 kD) C-terminal region that contains multiple functional domains. Within this region are motifs that allow Abl to associate with actin filaments [66] and to bind to DNA sequences [67]. The C-terminus also contains sequences that act as signals to allow Abl to shuttle between the nucleus and the cytoplasm. Three nuclear localisation signals [68] are found within this region, counterbalanced by a single nuclear export signal that is sensitive to leptomycin B [69]. Abl is predominantly expressed in the nucleus [70]. Deletion of nuclear localisation signals [70] causes Abl to undergo translocation to the cytoplasm. Conversely, treatment of cells with leptomycin B causes the nuclear accumulation of Abl [69].

The functions of the Abl protein are many and complex (for a comprehensive review see Van Etten [71]). In the nucleus, Abl acts as a negative regulator of growth by inducing cell cycle arrest in the $G_1$ phase [67, 72]. Nuclear Abl also has a role in the cellular response to genotoxic stress [73] whereas cytoplasmic Abl is involved in the transduction of signals from integrins [74].

## Deregulation of the Abl Tyrosine Kinase Activity

Wild-type Abl does not transform fibroblasts or haemopoietic cells even when overexpressed by means of retroviral transduction [70]. This finding suggests that the kinase activity of Abl is tightly regulated in vivo, a property common to all members of the Src family of non-receptor, protein tyrosine kinases. Mutational and biochemical studies have shown, however, that the mechanism responsible for the regulation of the kinase activity of Abl is not entirely comparable to that of the Src kinases, although equivalent protein domains are involved. The regulatory mechanism that inhibits the kinase activity of the SH1 domain of Src has been shown to require phosphorylation of tyrosine 527 in its C-terminus [75]. The SH2 domain of Src binds to the phosphotyrosine residue forming an intramolecular interaction that causes the protein to assume an inactive conformation [76]. In this inactive state, the SH3 domain binds to the linker region between the SH2 and SH1 (kinase) domains via an interaction with proline 253. Thus, the SH2 and SH3 domains in Src function in *cis* to govern the behaviour of the SH1 domain.

In contrast to members of the Src family, inactive Abl is not phosphorylated [77] and its C-terminus lacks a tyrosine residue homologous to tyrosine 527 of Src. There is some controversy regarding the role, if any, of the C-terminus of Abl in regulating its kinase activity. Deletion mutations introduced into the last exon of the *ABL* gene – the exon that encodes the entire C-terminus – are sufficient to activate the tyrosine kinase activity of Abl in rodent fibroblasts [78]. The deregulated kinase activity produced by C-terminal mutation is qualitatively different from that produced by N-terminal mutation. Mutation of the N-terminus of Abl generates an activated kinase with a prodigious capacity for catalysing the phosphorylation of tyrosine residues on itself as well as on other substrate molecules. C-terminal mutants of Abl exhibit high kinase activity for other cellular substrates but only a low level of autophosphorylation [78]. Expression of last exon mutants in rodent fibroblasts, but not primary bone marrow cells, resulted in a transformed phenotype [78]. Contrary to these findings are data from a recent study [79] in which deletion of the last exon region of Abl failed to activate its tyrosine kinase activity in human fibroblasts.

Curiously, the SH2 domain of Abl has been shown to have a positive role in the transformation of fibroblasts. Mutants of Abl in which the SH2 domain has been either deleted or mutated at residues that are essential for binding to phosphotyrosine are unable to transform fibroblasts [80]. This is not the case in haemopoietic cells where the SH2 domain of Abl has been shown to be dispensable for transformation [54, 81]. The apparent inconsistency in the behaviour of the SH2 domain of Abl has been attributed to different modes of transformation in fibroblasts and haemopoietic cells possibly involving different cell-signalling pathways [81].

Certain features of the regulatory mechanism of the Src family of kinases are shared by Abl. In both Src and Abl there appears to be an absolute requirement for the SH3 domain in limiting the activity of the SH1 domain. Deletion or introduction of point mutations within the SH3 domain of Abl activates its transforming ability in vivo [80]. Furthermore, the inhibitory influence exerted by the SH3 on the SH1 domain is highly dependent upon its position within the molecule. Insertion of as few as three amino acids between the SH3 and SH2 domains is sufficient to eliminate the repression of the SH1 region and to unleash the transforming potential of Abl [80].

Mutational studies have proven to be extremely useful in dissecting the structural domains of Abl in order to determine their role in maintaining tight control over its kinase activity. There have been reports, however, in which the regulation of wild-type Abl has apparently been abolished. The tyrosine kinase activity of Abl can be restored by overexpression at levels greatly in excess (>500-fold) of its endogenous expression [77]. This finding has been interpreted as evidence that a cellular inhibitor is required to suppress the kinase activity of the SH1 domain. According to this hypothesis, deregulation of overexpressed Abl is a consequence of its titration with a limited pool of cellular inhibitory molecules. In addition, the Abl protein exhibits kinase activity when purified [77]. It has been postulated that Abl escapes negative regulation in vitro because of a failure of the inhibitor to copurify with it. Furthermore, SH3 domain-deleted mutants and intact Abl, when purified, have been shown to have nearly identical kinase activities [80]. The functional equivalence of the SH3 domain and its deletion can be explained if the SH3 domain is the binding site for an inhibitor. If inhibition mediated by the SH3 domain requires an inhibitory peptide as a co-factor then the tyrosine kinase activity of Abl would be expected to be restored upon its removal from the cellular environment containing the inhibitor. Several proteins that bind to the SH3 domain of Abl have been identified, including two related peptides, Abl interactor proteins 1 and 2 (Abi-1 and Abi-2) [82, 83]. Although, their precise role has yet to be defined, they may function to antagonise the kinase domain since their proteasomal degradation is enhanced by Abl [84]. Another protein that binds to the SH3 domain of Abl is the antioxidant enzyme, Pag/MSP23 [85]. Exposure of cells to oxidative stress causes this protein to dissociate from Abl, leading to activation of its kinase activity.

A recent report [79] casts doubt on the hypothesis that a cellular inhibitor is essential for regulation of the tyrosine kinase activity of Abl. Wild-type Abl and mutants with activating mutations were purified and their in vitro kinase activities were determined. In all cases, the mutants had significantly higher kinase activity than the wild-type, indicating that Abl retained its regulated state in vitro. The reason for the discrepancy between these data and those obtained in the earlier reports of purified Abl proteins is unclear. The ionic composition of the buffers used for isolation of the proteins may represent a source of variation. Differences in tyrosine kinase activity could be detected in the presence of $MgCl_2$ but not when $MnCl_2$ was used [79]. A series of Abl mutants were generated in an effort to determine the minimum requirement for regulation of its tyrosine kinase activity. As mentioned previously, the last exon of Abl, encoding the C-terminus, was found to be dispensable for activation of its kinase domain. Surprisingly, a mutant lacking the first 81 amino acids of the N-terminus exhibited deregulated kinase activity. GST fusion proteins containing this region exhibited significant binding affinity for the SH1 domain of Abl. This observation led the authors to propose that the N-terminus of Abl functions as an inhibitory 'cap' that clamps intramolecular domains into a conformation that suppresses kinase activity. According to this model, regulation of tyrosine kinase activity is a self-contained property of the Abl molecule and binding of cellular inhibitors to the SH3 domain is not required for inhibition of the catalytic SH1 domain.

## Structure of the Bcr-Abl Fusion Protein and Disease Phenotype

Bcr-Abl proteins are encoded by fusion genes in which *ABL* exon 1 is replaced by variable portions of the *BCR* gene. Since the *ABL* component of the fusion gene is largely invariant, it follows that variability in disease phenotype may be due to protein sequences encoded by the translocation partner, *BCR*. Each of the three Bcr-Abl oncoproteins, $p190^{Bcr\text{-}Abl}$, $p210^{Bcr\text{-}Abl}$ and $p230^{Bcr\text{-}Abl}$ will be considered in turn with particular reference to how their respective structures relate to the disease phenotypes with which they are associated.

The smallest Bcr-Abl oncoprotein is $p190^{Bcr\text{-}Abl}$, derived from fusion genes that have *BCR* breakpoints within the minor breakpoint cluster. In most cases, the expression of $p190^{Bcr\text{-}Abl}$ in progenitor cells is manifest as ALL but a minority of patients have a form of CML with a prominent monocytosis. This particular clinical feature, which is often accompanied by a low ratio of neutrophils to monocytes, is more often associated with chronic myelomonocytic leukaemia (CMML) than with CML. Other features, however, are less reminiscent of CMML, such as a high proportion of circulating immature granulocytes, varying degrees of basophilia and low neutrophil alkaline phosphatase (NAP) scores. A considered appraisal of these cases has led to the suggestion that they represent a disease phenotype that is intermediate between CML and CMML [27, 28].

Since $p190^{Bcr\text{-}Abl}$ is translated from mRNA having an e1a2 junction, the contribution made by Bcr to the overall oncogenicity of the fusion protein is limited to the peptide sequence encoded by *BCR* exon e1. Consequently, the only Bcr domains that are present in $p190^{Bcr\text{-}Abl}$ are the oligomerisation domain, the serine-threonine kinase domain, the SH2-binding domains and the Grb2 binding site at Tyr177. The prevalence of ALL as the dominant disease phenotype associated with $p190^{Bcr\text{-}Abl}$ implies that a selection pressure operates to favour the disturbance or perturbation of particular metabolic pathways in lymphoid progenitors. Bcr domains encoded by exons downstream of exon e1 that are not present in the fusion protein, such as those responsible for modulating the function of guanine nucleotide-binding proteins, would appear to be dispensable for the transformation of lymphoid precursors by $p190^{Bcr\text{-}Abl}$. Neoplastic expansion of CML progenitors is limited, possibly because disruption of a different set of cell-signalling pathways is less severe, or indeed may not be occurring at all, in the absence of these downstream domains. Thus, restriction of the oncogenic potential of $p190^{Bcr\text{-}Abl}$ to the lymphoid lineage suggests that transformation is mediated by interference with pathways that are specifically activated during lymphoid and monocytic differentiation. Interestingly, $p190^{Bcr\text{-}Abl}$ has a higher autophosphorylation activity than $p210^{Bcr\text{-}Abl}$, although the biological significance of this is unknown [5].

'Classical' CML is associated with the 210-kD form of the Bcr-Abl oncoprotein, $p210^{Bcr\text{-}Abl}$. In its chronic phase, this disease is marked by neoplastic expansion of the granulocytic and megakaryocytic lineages. The excessive production of immature myelocytes and metamyelocytes indicates that the presence of $p210^{Bcr\text{-}Abl}$ in CML progenitors is sufficient to arrest the majority of cells at intermediate stages of differentiation. Although the exact mechanism is unknown, $p210^{Bcr\text{-}Abl}$ is presumed to interfere with the cell-signalling pathways involved in granulocytic differentiation and maturation. This disruption must be relatively moderate, however, as it is insufficient to prevent a small proportion of cells from reaching the terminal stages to become mature neutrophils, eosinophils and basophils [27].

Although, by convention, $p210^{Bcr\text{-}Abl}$ is considered as a single entity there are of course two forms of this oncoprotein encoded by fusion transcripts having either an e13a2 (b2a2) junction or an e14a2 (b3a2) junction. *BCR* exon e14 consists of 75 nucleotides. Consequently, translation of e14a2 mRNA yields a protein containing 25 more amino acids than the protein translated from e13a2 mRNA. Fusion transcripts with an e13a2 junction occur when the breakpoints within M-*bcr* are 5′, whereas the e14a2 junction is formed when the breakpoints are 3′. Studies on the effect of the position of the M-*bcr* breakpoint on disease phenotype suggested that the onset of the acute phase of CML occurred later for patients with a 5′ breakpoint than for those with a 3′ breakpoint [86–88]. Debate was stimulated by subsequent studies that failed to confirm this finding [89, 90]. The controversy has been largely resolved by prospective studies that used larger representative samples of patients [26, 91]. These demonstrated that the position of the M-*bcr* breakpoint was not predictive of disease outcome. Hence, the nature of the transcript that encodes $p210^{Bcr\text{-}Abl}$ does not have any prognostic value.

Although it is doubtful whether any aspect of disease progression in CML may be attributed to the position of the breakpoint within M-*bcr,* there is evidence that at least one haematological phenotype, the platelet count at diagnosis, is related to the presence of either e13a2 transcripts or e14a2 transcripts. Significantly higher numbers of platelets have been reported in patients with e14a2 fusion transcripts [92, 93] and 3′ M-*bcr* breakpoints [94] than in those with e13a2 fusion transcripts and 5′ breakpoints. In contrast, two studies found no correlation between the number of platelets and the type of *BCR-ABL* transcript [26, 95]. In the latter report, however, the data were also stratified according to the white blood cell count at diagnosis. When this was taken into consideration, significantly higher numbers of platelets were evident in patients having both e14a2 transcripts and white blood cell counts of $<100 \times 10^9/l$ [26]. Taken together these findings suggest that the larger form of $p210^{Bcr\text{-}Abl}$ encoded by the e14a2 fusion transcript has a more profound effect upon thrombopoiesis than the smaller oncoprotein encoded by the e13a2 transcript. It has been suggested that $p210^{Bcr\text{-}Abl}$ might exert an influence on platelet formation through an interaction with the cytoskeleton of megakaryocytes [93]. Interestingly, the e14a2 transcript has also been associated with Ph+ essential thrombocythemia (ET) [96] although data from a more recent study indicate that CML that mimicks ET is not exclusively correlated with this particular *BCR-ABL* mRNA [97].

In addition to the N-terminal sequences encoded by exon e1 of *BCR*, which are present in $p190^{Bcr\text{-}Abl}$, $p210^{Bcr\text{-}Abl}$ also contains a *dbl*-like domain that encodes a guanine nucleotide exchange factor [59]. The contribution made by this domain to the phenotype of classical CML has yet to be determined. Significantly, the biological and clinical properties of adult Ph+ ALL are identical, regardless of whether $p190^{Bcr\text{-}Abl}$ or $p210^{Bcr\text{-}Abl}$ is expressed [27].

$P230^{Bcr\text{-}Abl}$, encoded by mRNA transcripts with an e19a2 junction, is the largest of the Bcr-Abl fusion proteins. In 1996, considerable interest was generated by a report by Pane et al. [30] which described three cases of CML with e19a2 fusion transcripts. All three patients had presented with a rare myeloproliferative disorder: neutrophilic-chronic myeloid leukaemia (CML-N), also known as chronic neutrophilic leukaemia (CNL). CML-N may be distinguished from CML by its milder clinical course. Patients have lower total white blood cell counts, less severe anaemia and less prominent splenomegaly. Disease progression to blastic transformation is significantly delayed and may not occur at all. Interestingly, mature granulocytes are the predominant leukaemic myeloid cells undergoing circulation in peripheral blood. $p230^{Bcr\text{-}Abl}$ would thus seem to be less disruptive of the normal process of granulocytic differentiation than $p210^{Bcr\text{-}Abl}$. The association of CML-N with patients having an e19a2 fusion transcript led those investigators to re-evaluate their earlier diagnosis of CML for the two patients in whom the e19a2 transcript was originally identified [98]. In retrospect, it was considered that these patients may also have been suffering from a milder myeloproliferative disorder [30]. It should be noted, however, that there are a small number of reports that describe patients having an e19a2 fusion transcript and a disease phenotype indistinguishable from classical CML [99, 100].

Elevated platelet counts have been reported in CML patients who have e19a2 transcripts [101, 102]. These findings prompted a re-examination of the haematological data from the available literature [102] and thrombocytosis was identified in patients described in earlier reports [30, 98, 103]. The validity of the term N-CML was questioned and an alternative classification of Ph+ ET was proposed as a more appropriate description of this disease phenotype [102]. Proponents of the N-CML classification regard this description as preferable as it conveys the notion that although the visible cytogenetic abnormality in these patients – the Ph chromosome – is the same as in CML, the molecular lesion – a breakpoint in the μ-*bcr* of the *BCR* gene – is different and may be the key determinant for the more benign phenotype [104]. Moreover, there are currently no known genetic or biological markers that are specific for ET and the diagnostic criteria for this disease specifically exclude patients with a Ph chromosome or a *BCR-ABL* fusion gene [105]. Although very high platelet counts do seem to be associated with expression of $p230^{Bcr\text{-}Abl}$ in some cases of N-CML, in the most recent survey of the literature [106], only five out of twenty-four patients with e19a2 transcripts were found to have excessive platelet counts (>1,000 × $10^9$/l).

$P230^{Bcr\text{-}Abl}$ contains over 90% of the peptide sequence of Bcr, and lacks only the last two thirds of the C-terminal $GAP^{rac}$ homology domain [107]. It is possible that the remaining N-terminal third is functionally active since mutational analysis revealed that this region is essential for GAP activity [107]. It has been postulated that CML progenitors expressing the $p230^{Bcr\text{-}Abl}$ oncoprotein retain the capacity to progress through the normal stages of granulocytic differentiation because both the normal and the translocated alleles of the *BCR* gene may express a protein with a physiologically 'normal' GAP activity [27]. This hypothesis is consistent with the notion that the additional Bcr sequence in the $p230^{Bcr\text{-}Abl}$ oncoprotein, relative to $p190^{Bcr\text{-}Abl}$ and $p210^{Bcr\text{-}Abl}$, contributes a 'gain of function' that ultimately results in a milder myeloproliferative phenotype.

The less aggressive clinical course of CML has also been attributed to a quantitative reduction in transcription from fusion genes having e19a2 junctions [30]. Support for this hypothesis comes from a recent report in which the number of e19a2 mRNA transcripts was determined by real-time quantitative PCR (RQ-PCR) in seven patients diagnosed with CML-N [106]. Minimal amounts of e19a2 mRNA were detected in the majority of these patients, suggesting that transcription from *BCR-ABL* fusion genes with e19a2 junctions occurs frequently at an extremely low level. Interestingly, the $p230^{Bcr\text{-}Abl}$ protein is consistently undetectable by immunoblotting [106] and in the only case in which the protein has been detected, the patient exhibited duplication of the Ph chromosome [108]. Transcription may be subject to silencing because of the presence of at least three transcriptional repressor sequences within the intron downstream of BCR exon e14 [109]. These sequences are not present in e1a2 fusion genes encoding $p190^{Bcr\text{-}Abl}$ and are incomplete in e14a2 fusion genes encoding $p210^{Bcr\text{-}Abl}$ [30].

A recent report of a novel e15a2 *BCR-ABL* fusion transcript detected by nested PCR in two patients with an atypical myeloproliferative disorder raises intriguing questions concerning the contribution made by *BCR* sequences to the disease phenotype [37]. This transcript was unusual in that the breakpoint in *BCR* occurred outside the three defined breakpoint cluster regions within exon 15. Moreover, the *ABL* breakpoint fell within exon a2. Curiously, both patients were diagnosed as having a mild myeloproliferative disorder with neutrophilic leucocytosis reminiscent of the neutrophilic chronic myeloid leukaemia associated with $p230^{Bcr\text{-}Abl}$. The translated oncoprotein would be expected to have a greater similarity to $p210^{Bcr\text{-}Abl}$ than to $p230^{Bcr\text{-}Abl}$. The only substantial difference between the e15a2 and e14a2 fusion genes, in terms of *BCR* content, is that the former have an intact intron downstream of *BCR* exon e14 and the first 86 bp of exon 15. A protein translated from the e15a2 transcript, in common with $p210^{Bcr\text{-}Abl}$, will lack the $GAP^{Rac}$ domain present within $p230^{Bcr\text{-}Abl}$. Hence, the extent of the *BCR* sequence that needs to be present in the *BCR-ABL* oncogene in order to obtain a less severe disease phenotype may be less than was previously supposed.

## Altered Cellular Adhesion

In normal haemopoieis, progenitor cells adhere to the stromal cells of the bone marrow and their associated extracellular matrix. The latter contains proteins, such as fibronectin, that are produced by stromal cells and which function as adhesive ligands for receptors expressed on the surface of haemopoietic progenitor cells. The process of adhesion is believed to be essential for the regulation of haemopoiesis since it provides a means of anchoring progenitors within the vicinity of cytokine-secreting cells (reviewed by Verfaillie et al. [110]). Haemopoietic progenitors are thus held in position to receive local signals communicated by cytokines. Progenitor cells are pluripotent and their fate is determined by the nature of these signals.

Depending on which cytokine messages they receive (or fail to receive), progenitors may survive or may be induced to undergo apoptosis. Surviving progenitors may be stimulated to proliferate or their growth may be arrested. Proliferation may involve replication and self-renewal of the pluripotent progenitors or it may involve differentiation and commitment to either the lymphoid or myeloid lineages.

In CML, Ph+ progenitors exhibit reduced adhesion to stromal cells and the extracellular matrix [111]. The failure of CML progenitors to adhere to these components of the bone marrow implies that they also lose the regulatory signals that are supplied to normal, adherent haemopoietic progenitors. Cell surface receptors of the integrin family have been shown to behave differently in CML progenitors than in normal haemopoietic progenitors. Integrins have intracellular, transmembrane and large extracellular domains that recognise and bind to adhesive ligands on stromal cells and in the extracellular matrix. Structurally, integrins are heterodimers consisting of α and β subunits held together by a non-covalent association. The intracellular domain of the β subunit is linked to the cytoskeleton. Binding of integrins to ligands is believed to initiate a signal transduction cascade that is propagated through the cytoskeleton. Hence, integrins have a putative role in transducing and transmitting signals that originate from outside the cell, in the local microenvironment of the bone marrow, to the inside of the cell (outside-in signalling). It has been demonstrated that the β1 integrin subunit expressed by CML progenitors has a lower affinity for adhesive ligands than the corresponding β1 integrin expressed by normal progenitors [110]. Since the cytoplasmic domain of β1 integrin is coupled to the cytoskeleton, reduced binding of its extracellular domain would be expected to result in a concomitant reduction in outside-in signalling. Although the precise nature of these signals is unknown, the increased proliferation of CML progenitors suggests that they are inhibitory. Significantly, the defective adhesion of CML progenitors may be reversed with interferon-α (IFN-α), a therapeutic agent with a history of efficacy in the treatment of CML [112].

Inevitably, *BCR-ABL* expression has been linked to the altered adhesion of CML progenitors. The relatively few proteins that are phosphorylated on tyrosine residues in CML progenitor cells have been identified as being particularly important substrates for Bcr-Abl. These include Cbl, Crkl and paxillin [113]. Crkl would appear to be of prime importance since it has been demonstrated to have roles in the regulation of cellular motility [114] and adhesion via its association with focal adhesion proteins such as Fak, p130[Cas] [115], Hef1 [116] and paxillin. Phosphorylation of focal adhesion molecules may impair or abrogate their ability to function as components of an integrin-mediated signalling pathway. In addition, transcriptional activation of integrin genes may be involved in altered cellular adhesion since Bcr-Abl has been shown to cause the upregulation of α6 integrin mRNA [117].

Studies of integrin signalling in cell lines have yielded conflicting results that are at odds with the findings obtained with primary progenitor cells. Adhesion is found to be enhanced, rather than inhibited, in cell lines that have been transfected with *BCR-ABL* [118, 119]. This apparently contradictory but reproducible property of transfectants may be an artefact caused by overexpression of $p210^{Bcr\text{-}Abl}$ at levels greatly in excess of those that occur in CML progenitor cells [120]. Further confusion surrounds the mechanism by which Bcr-Abl is proposed to disrupt integrin signalling in cell lines. Thus, β1 integrin expression [119] or function [118] has been reported as being enhanced in *BCR-ABL*-expressing cells when compared with the untransfected, parental cell lines.

### Activation of Mitogenic Signalling Pathways

Bcr-Abl is known to activate a number of important cell-signalling pathways that have mitogenic potential (reviewed by Deininger et al. [121]). It is important to remember that, in many cases, the available data come from experiments that make use of *BCR-ABL*+ cell lines and activation of some of these pathways in primary CML cells has yet to be verified.

#### *Ras and the MAP Kinase Pathways*

Bcr-Abl binds directly to proteins that activate the Ras signalling pathway [122]. Autophosphorylation of tyrosine 177 generates a binding site for the adapter molecule, Grb-2 [52]. Grb-2 associates with the Sos protein which stimulates the conversion of the inactive GDP-bound form of Ras to the active GTP-bound state [122]. Ras may also be activated by two other adapter molecules, Shc and Crkl, which are known to be substrates of Bcr-Abl [123, 124]. Although Crkl appears to be necessary for the transformation of fibroblasts by Bcr-Abl [125], direct binding of Crkl to Bcr-Abl is not required for the transformation of myeloid cells [126]. Activated Ras binds to the serine-threonine kinase, Raf-1, recruiting it to the plasma membrane where it is activated by tyrosine phosphorylation [127] and initiates a signalling cascade via the mitogen-activated protein kinase (MAPK) pathway. MAPK signalling is regulated by a three-tiered cascade consisting of a MAP kinase (MAPK), a MAPK kinase (MAPKK or MEK) and a MAPKK kinase (MAPKKK or MEKK). Four distinct classes of mammalian MAPK have been identified: extracellular signal-related kinases (ERK)-1/2, Jun amino-terminal kinases (JNK1/2/3), p38 proteins (p38α/β/γ/δ) and ERK5 [128]. Raf-1 signals via a pathway involving Mek1/2 (MAPK kinases) and Erk1/2 (MAP kinases) [127]. Ultimately, this pathway culminates in gene transcription [129]. Although Erk2 activation could not be detected in Ba/F3 cells transfected with a temperature sensitive p210$^{Bcr\text{-}Abl}$-mutant [130], a subsequent study has demonstrated that Bcr-Abl can activate the Ras-Raf-Mek-Erk signalling pathway in haemopoietic cells [131]. In addition, activation of the JNK or stress-activated protein kinase (SAPK) pathway by Bcr-Abl has been shown to be required for transformation of fibroblasts [132]. Signalling is Ras-dependent and occurs via the serine-threonine protein kinase GCKR (germinal centre kinase related) [133]. Interestingly, the p38 MAPK pathway has recently been shown to be activated by IFN-α in *BCR-ABL*+ cell lines and in primary cells from CML patients [134]. This finding led the authors to speculate that Bcr-Abl may negatively regulate this pathway. Activation of p38 MAPK by IFN-α-dependent tyrosine kinases was proposed to overcome this inhibition and to account for the antiproliferative effect of IFN-α in leukaemic progenitors.

*Jak-Stat Pathway*

Phosphorylation of members of the signal transducer and activator of transcription (Stat) family of transcription factors has been reported in *BCR-ABL*+ cell lines [135] and in primary CML cells [136]. In the normal physiological sequence of events, phosphorylation of Stat proteins occurs downstream of activation of Janus kinases (Jak) by cytokine and growth factor receptors [137]. In contrast, Bcr-Abl appears to phosphorylate Stat proteins directly, without prior activation of Jak proteins. A high level of Stat5 and a lower level of Stat1 and Stat3 activation have been demonstrated in cell lines transformed with p210$^{Bcr\text{-}Abl}$ and p190$^{Bcr\text{-}Abl}$ [135]. Activation of Stat5 by p210$^{Bcr\text{-}Abl}$ has been shown to contribute to the malignant transformation of K562 cells [138]. The functions of Stat5 are pleiotropic [139] but in *BCR-ABL*+ cell lines this protein appears to direct enhanced transcription of Bcl-$x_L$ leading to inhibition of apoptosis [140–142]. Interestingly, p190$^{Bcr\text{-}Abl}$ differs from p210$^{Bcr\text{-}Abl}$ in that it is also able to activate Stat6 [135]. It remains to be seen whether the predominantly lymphoblastic phenotype associated with p190$^{Bcr\text{-}Abl}$ is related to this property of the shorter form of the oncoprotein.

*PI-3 Kinase Pathway*

Proliferation of *BCR-ABL*+ cell lines and primary cells is dependent upon phosphatidylinositol-3 (PI-3) kinase activity [143]. PI-3 kinase catalyses the phosphorylation of phosphoinositols on the D-3 position of the inositol ring. The resulting 3′-phosphorylated inositides function as cell-signalling molecules. Bcr-Abl apparently activates this pathway by forming a multimeric complex with PI-3 kinase, P120$^{Cbl}$ and the adaptor molecules Crk and Crkl [65]. In *BCR-ABL*-expressing cells, activated PI-3 kinase stimulates the serine-threonine kinase, Akt, which functions as a downstream effector [144]. Activation of Akt is pro-mitogenic and causes downregulation of the key cell cycle inhibitor, p27$^{Kip1}$ [145]. This finding is consistent with a model in which activation of the PI-3 kinase/Akt pathway by Bcr-Abl leads to suppression of p27$^{Kip1}$ that, in turn, causes accelerated entry into the S phase of the cell cycle. In addition, activated Akt may function in an anti-apoptotic capacity [146]. Survival signals provided by binding of IL-3 to its receptor are transduced via the PI-3 kinase/Akt pathway. A key substrate of Akt is the pro-apoptotic protein or 'death agonist', Bad [147]. Bad promotes cell death by binding to, and thereby inactivating, the anti-apoptotic and membrane-associated proteins Bcl-2 and Bcl-$x_L$. Phosphorylation of Bad by Akt prevents it from binding to these proteins. In its phosphorylated state, Bad is sequestered in the cytoplasm through an association with 14–3-3 proteins [148]. Taken together, these studies provide evidence that Bcr-Abl is able to substitute for the survival signal provided by IL-3 by activation of the PI-3 kinase/Akt pathway.

*Myc Pathway*

The first evidence that Myc could be activated downstream of Bcr-Abl and was involved in cellular transformation came from experiments with rat fibroblasts [149]. Fibroblasts were stably transfected with either the *MYC* gene or a dominant negative mutant of *MYC*. These transfectants, together with wild-type cells, were then infected with a retrovirus bearing p190$^{Bcr\text{-}Abl}$. Transformation by p190$^{Bcr\text{-}Abl}$, as assessed by colony formation, was found to be enhanced 3-fold in cells transfected with *MYC* when compared with wild-type cells. Conversely, transformation was inhibited 10-fold in cells transfected with a dominant negative mutant of *MYC* when compared with the wild-type. Moreover, overexpression of a dominant negative mutant of *MYC* blocked the transformation of mouse bone marrow cells by *BCR-ABL* [149]. The pathway linking Bcr-Abl to the induction of *MYC* expression is currently unknown, although a model has been proposed to describe the induction of *MYC* by v-Abl [150]. According to this model, which was proposed on the basis of data obtained from v-Abl-transformed cells, v-Abl initiates a phosphorylation cascade involving, Ras, Raf, cyclin-dependent kinases (cdks) and E2F transcription factors. The signalling cascade culminates with activation of the *MYC* promoter by binding of E2F transcription factors. Myc is classed as a proto-oncogene because it is overexpressed in many human malignancies. Under normal physiological conditions Myc functions as a transcription factor. In a recent report [151], 53 target genes of Myc were identified by microarray analysis. A small subset of these candidate genes also exhibited in vivo promoter occupancy by Myc indicating that they were direct targets. These targets had multiple functions and included genes involved in the cell cycle [cyclin-dependent kinase 4 *(CDK4)*, cyclin B1 *(CCNB1)*, cyclin E binding protein 1*(CEB1)*] and genes involved in apoptosis [prohibitin *(PHB)*]. *MNT*, an antagonist of Myc function, was also upregulated, implying the existence of a negative feedback mechanism.

### Inhibition of Apoptosis

A long-standing controversy is concerned with whether Bcr-Abl functions as an inhibitor of apoptosis in vivo. Apoptosis, defined as genetically programmed cell death, limits the expansion of a clone since the size of a cell population depends upon the balance between the rate of proliferation and the rate of cell death. *BCR-ABL* expression is presumed to cause an expansion of the leukaemic clone – an absolute requirement for the development of CML. Inhibition of apoptosis has been proposed as the mechanism by which Bcr-Abl is able to induce clonal expansion in CML [152].

Apoptosis caused by growth factor withdrawal is eliminated when factor-dependent murine [153] or human [154] cell lines are transfected with exogenous *BCR-ABL*. Cell lines that have been transfected with *BCR-ABL* have also been reported as exhibiting increased resistance to apoptosis induced by DNA-damaging agents [152, 155]. The extent to which these cell lines model CML progenitors is debatable since transfection results in overexpression of *BCR-ABL*. In addition, the complete removal of growth factors and exposure to high doses of ionising radiation can scarcely be regarded as having much relevance to the situation that pertains in vivo. Moreover, the importance of apoptosis in normal haemopoiesis is questionable. It has yet to be demonstrated that CML progenitors undergo apoptosis to any significant degree in the environment of the bone marrow. Conflicting results have come from in vitro studies of primary progenitors in growth-factor-deprived conditions. In one report [152], CML progenitors appeared to be more resistant to apoptosis than normal progenitors, while in another[156], normal and CML progenitors were equally susceptible to apoptosis induced by growth factor deprivation.

The mechanisms by which Bcr-Abl inhibits apoptosis in cell lines are not well understood. The release of cytochrome C from mitochondria, a pre-requisite for caspase-3 activation, is apparently blocked in *BCR-ABL*-expressing cell lines [157]. In addition, Bcr-Abl can delay apoptosis through interaction with a cytosolic pathway that is upstream of procaspase-3 activation [158]. Members of the Bcl-2 family of proteins may be involved in mediating the anti-apoptotic effect of Bcr-Abl. Up-regulation of Bcl-2 by Bcr-Abl has been demonstrated in two different cellular contexts, one of which was dependent upon the Ras pathway [159] and the other upon the PI-3 kinase pathway [144]. Bcl-2 has a critical role in targeting the protein kinase Raf-1 to mitochondria [160]. Raf-1, in turn, inactivates the pro-apoptotic protein, Bad, by phosphorylating it on serine residues [148]. A recent study [161], has shown that a Bad-dependent pathway mediates part of the survival signal in *BCR-ABL*-expressing cell lines. Downregulation of interferon consensus sequence binding protein (ICSBP) by Bcr-Abl has also been implicated as a key anti-apoptotic event [162, 163].

Surprisingly, a recent report has demonstrated that Bcr-Abl can actively induce apoptosis when trapped in the nucleus [164]. Treatment of human and murine Bcr-Abl+ cell lines with imatinib mesylate (STI571) stimulated entry of the oncoprotein into the nucleus. A 'pool' of Bcr-Abl molecules was trapped in the nucleus by co-incubating the cells with leptomycin B, a substance that blocks the nuclear export signal receptor, CRM1/exportin-1. The cells were then washed in order to restore tyrosine kinase activity to the trapped nuclear pool of Bcr-Abl by removing imatinib mesylate. Upon recovery of its tyrosine kinase activity, the nuclear Bcr-Abl caused a rapid induction of apoptosis. This finding is particularly significant because it demonstrates that the cellular compartment in which Bcr-Abl is localised is all-important in determining whether the outcome of its deregulated tyrosine kinase activity is pro- or anti-apoptotic [165].

### Proteasomal Degradation

Recent reports have demonstrated that Bcr-Abl can promote the degradation of proteins via the ubiquitin-proteasomal pathway. This phenomenon was first identified when the expression of the Abl-interactor proteins, Abi 1 and Abi 2, was studied in haemopoietic cell lines transduced with *BCR-ABL* [84]. Surprisingly, the introduction of exogenous *BCR-ABL* into cell lines resulted in an almost complete abrogation of the expression of Abi proteins. This effect was dependent upon the tyrosine kinase activity of Bcr-Abl and was evidently a post-translational event since mRNA levels for Abi 1 and Abi 2 were the same regardless of whether the cells had been transduced or not. Downregulation of the Abi proteins could be reversed by incubating the cells with a proteasome inhibitor. Importantly, Abi protein expression could not be detected in Ph+ bone marrow cells from a CML patient in blast crisis but Abi proteins were present in Ph– bone marrow cells from a patient who had AML. This finding suggests that proteasomal degradation of inhibitory proteins elicited by Bcr-Abl may be partly responsible for disease progression and blastic transformation in CML.

Particularly relevant to a consideration of the mechanisms underlying the transition from chronic phase to blast crisis in CML is the finding that Bcr-Abl downregulates the DNA repair protein DNA-PKcs [166]. DNA-PKcs is the catalytic subunit of a major DNA repair complex, DNA-dependent protein kinase. Downregulation of DNA-PKcs was dependent upon the tyrosine kinase activity of Bcr-Abl and could be reversed by treatment of cells with proteasome inhibitors. Cell lines in which Bcr-Abl was shown to downregulate DNA-PKcs had a DNA repair deficiency and were more sensitive to ionizing radiation than the parental cells. Interestingly, and in contrast to the observation that downregulation of Abi proteins was associated with blast crisis, DNA-PKcs expression was negligible or undetectable in CD34+ cells from CML patients in chronic phase. High levels of DNA-PKcs protein were detected in the control sample: Ph– CD34+ cells from a patient with a neuroblastoma. This result is the first indication that the proteasomal destruction of physiologically important proteins may be occurring in Ph+ myeloid cells prior to the acceleration of CML to its aggressive, acute phase. In addition, Bcr-Abl has been shown to promote the ubiquitin-proteasomal degradation of the cyclin-dependent kinase inhibitor, p27, in human and murine cell lines [167]. Furthermore, Bcr-Abl can stabilise and maintain the expression of Mdm2, a protein that catalyses the ubiquitination of the tumour suppressor, p53 [168]. This interaction, which has been demonstrated in *BCR-ABL*-expressing cell lines, causes a profound downregulation of p53 via the ubiquitin-proteasomal pathway. The current state of knowledge regarding the ability of Bcr-Abl to enhance the degradation of target proteins is somewhat limited, but it seems likely that this novel mechanism contributes to the overall transforming potential of the oncoprotein.

### Prospects for Therapy

Currently, the only treatment for CML that can unequivocally eradicate the leukaemic clone is an allogeneic stem cell transplant [169]. Unfortunately, this procedure can only be offered to a relatively limited number of young patients who have HLA-matched donors. In addition, there is an attendant and appreciable risk of morbidity or mortality associated with this approach. Until very recently, the preferred treatment for newly diagnosed CML patients was IFN-α, either administered alone or in combination with cytarabine. With the advent of imatinib mesylate, clinicians are now able to prescribe a compound that has a high degree of efficacy and specificity as a consequence of being the product of a rational approach to drug design.

The development of imatinib mesylate was stimulated by the realisation that the major part of the transforming potential of Bcr-Abl was due to its deregulated tyrosine kinase activity. In the early 1990s, scientists at Ciba-Geigy embarked on a programme to synthesise small molecules capable of inhibiting the tyrosine kinase activity of Bcr-Abl by competing with adenosine triphosphate (ATP) for occupancy of its kinase pocket (fig. 5). Development of these compounds was informed and guided by knowledge of the structure of the ATP binding site of protein kinases [7]. The most promising compound to emerge from this programme was a 2-phenylaminopyrimidine, CGP57418B [170]. Preclinical studies showed that micromolar concentrations of CGP57418B, later renamed STI571, were sufficient to inhibit autophosphorylation of the Abl- and platelet-derived growth factor receptor (PDGFR) protein tyrosine kinases. Moreover, the effect on other tyrosine kinases was negligible. STI571 selectively inhibited the growth of CML primary cells and *BCR-ABL*+ cell lines in vitro [171, 172] and in mice [171]. The exceptional specificity of this compound suggested that it might have value as a therapeutic agent. Phase I clinical trials of STI571 began in the United States in 1998. The

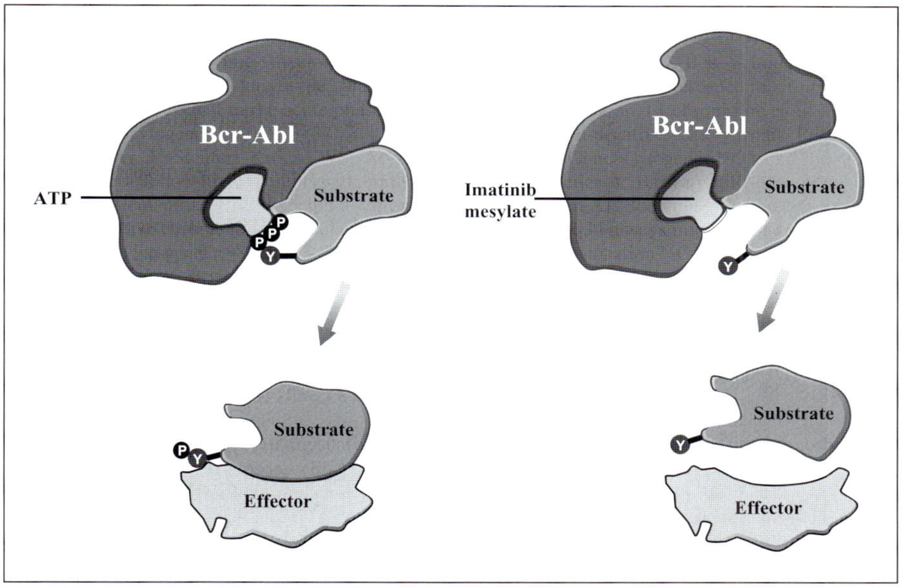


**Fig. 5.** Mechanism of action of imatinib mesylate. Imatinib mesylate competes with ATP for occupancy of its specific binding pocket within the kinase domain of Bcr-Abl. When the binding pocket is occupied by ATP, Bcr-Abl can catalyse the phosphorylation of tyrosine residues on substrate proteins (left). Imatinib mesylate is structurally similar to ATP and can bind to the ATP-binding pocket within the kinase domain of Bcr-Abl (right). Inhibition of tyrosine kinase activity occurs because imatinib mesylate lacks the essential phosphate groups that are provided by ATP. Thus, phosphorylation of tyrosine residues on substrate proteins cannot take place. Substrate proteins with unphosphorylated tyrosine residues fail to adopt the necessary conformation for binding to effector molecules, and downstream signalling is inhibited as a consequence.

results of these trials [173] were encouraging. The drug showed little toxicity and all patients receiving more than 300 mg/day entered a complete haematological remission. Furthermore, cytogenetic responses were noted in many of these patients. The results were sufficient to justify phase II trials for patients in accelerated phase [174] and in myeloid blast crisis [175]. In the accelerated-phase trial 60% of patients had a major cytogenetic response with 41% having a complete response, whereas in the blast crisis study, only 16% of patients had a major cytogenetic response with 7% achieving a complete response.

In May 2001, STI571 was approved by the US Food and Drug Administration for the treatment of CML that is refractory to interferon therapy [8]. The introduction of the drug into clinical practice necessitated another change of name and STI571 was re-christened imatinib mesylate and marketed under the proprietary name Glivec (Gleevec in the US). In the United Kingdom, the National Institute of Clinical Excellence (NICE) is currently appraising the clinical and cost effectiveness of imatinib mesylate for the treatment of CML.

Undoubtedly, imatinib mesylate offers a number of advantages over IFN-α. Imatinib mesylate can be given orally whereas IFN-α must be injected. In addition, the toxicity profile of imatinib mesylate is very mild for a drug with such a high degree of potency, although some rare, severe adverse effects have been reported [176]. Perhaps the most significant benefit of imatinib mesylate when compared with IFN-α, is the higher rate of cytogenetic response. Nevertheless, there is as yet no direct evidence that imatinib prolongs life [169]. A serious problem with this inhibitor is the phenomenon of acquired resistance. Since this severely curtails the benefit that the patient is able to derive from the drug, it is vital to determine the underlying cause of this process. Experiments in which *BCR-ABL*+ cell lines have been made resistant to imatinib mesylate have shown that a number of different mechanisms may allow cells to evade the inhibitory effect of the drug. Thus, upregulation of the Bcr-Abl protein associated with amplification of the *BCR-ABL* gene and overexpression of the multidrug resistance P-glycoprotein have both been demonstrated in resistant cell lines [177]. Individual point mutations within the kinase domain encoding sequences of *BCR-ABL* have been detected by several groups in patients who developed resistance to imatinib mesylate [178–182]. It is not yet clear whether resistance to imatinib will also be a significant problem in newly diagnosed chronic-phase patients and whether the mechanisms identified so far as causes of relapse in advanced disease will also apply to them.

Prospective studies are ongoing to determine whether the efficacy of imatinib mesylate can be enhanced if it is administered in combination with other drugs such as IFN-α or cytarabine. It remains to be seen whether treatment with imatinib mesylate, either alone or in combination with other therapeutic agents, is capable of completely eliminating the leukaemic clone. Recent reports suggest that this outcome is unlikely since treatment of primitive

CML progenitors grown in vitro with imatinib mesylate failed to produce a significant increase in the proportion of cells undergoing apoptosis [183, 184]. Despite its shortcomings, the success of imatinib mesylate demonstrates the effectiveness of a rational, targeted approach to drug design. In all probability, imatinib mesylate represents the first of an emerging class of therapeutic agents for CML that target particular molecular determinants of the Bcr-Abl oncoprotein.

## Conclusion

Considerable progress has been made in understanding the cytogenetic and molecular genetic aspects of CML since the discovery of the Ph chromosome in 1960 [1]. Nevertheless, large gaps in our understanding remain to be filled and these represent challenges for the future. A greater knowledge of the normal physiological roles of Bcr and Abl will be required, particularly with regard to their functional domains. This is particularly important for Bcr since the absence or inclusion of domains from this protein in the Bcr-Abl fusion seems to significantly affect the disease phenotype. Recent contributions to this are the publications on the crystal structure of the oligomerisation domain of Bcr [55] and a re-evaluation of the mechanism of regulation of the Abl kinase [79].

A number of fundamental questions remain to be answered. In particular, if CML is caused by a single genetic lesion – the t(9;22) chromosomal translocation – one may ask how this can be reconciled with the extreme variability that is observed in the clinical course of the disease? Moreover, what is the molecular basis for the genomic instability that seemingly drives the progression of CML towards blast crisis? This latter question is vital and the answer (or answers) to this conundrum could have profound implications for improving the life expectancy and quality of life for CML patients.

## References


1 Nowell PC, Hungerford DA: A minute chromosome in human chronic granulocytic leukemia. Science 1960;132:1497.

2 Rowley JD: Letter: A new consistent chromosomal abnormality in chronic myelogenous leukaemia identified by quinacrine fluorescence and Giemsa staining. Nature 1973;243:290–293.

3 Bartram CR, de Klein A, Hagemeijer A, van Agthoven T, Geurts vK, Bootsma D, Grosveld G, Ferguson-Smith MA, Davies T, Stone M: Translocation of c-abl oncogene correlates with the presence of a Philadelphia chromosome in chronic myelocytic leukaemia. Nature 1983;306:277–280.

4 Groffen J, Stephenson JR, Heisterkamp N, de Klein A, Bartram CR, Grosveld G: Philadelphia chromosomal breakpoints are clustered within a limited region, bcr, on chromosome 22. Cell 1984;36:93–99.

5 Lugo TG, Pendergast AM, Muller AJ, Witte ON: Tyrosine kinase activity and transformation potency of bcr-abl oncogene products. Science 1990;247:1079–1082.

6 Daley GQ, Van Etten RA, Baltimore D: Induction of chronic myelogenous leukemia in mice by the P210bcr/abl gene of the Philadelphia chromosome. Science 1990;247:824–830.

7 Druker BJ, Lydon NB: Lessons learned from the development of an abl tyrosine kinase inhibitor for chronic myelogenous leukemia. J Clin Invest 2000;105:3–7.

8 Savage DG, Antman KH: Imatinib mesylate – a new oral targeted therapy. N Engl J Med 2002;346:683–693.

9 Richardson C, Jasin M: Frequent chromosomal translocations induced by DNA double-strand breaks. Nature 2000;405:697–700.

10 Khanna KK, Jackson SP: DNA double-strand breaks: Signaling, repair and the cancer connection. Nat Genet 2001;27:247–254.

11 Surrallés J, Puerto S, Ramirez MJ, Creus A, Marcos R, Mullenders LH, Natarajan AT: Links between chromatin structure, DNA repair and chromosome fragility. Mutat Res 1998;404:39–44.

12 Tanaka K, Takechi M, Hong J, Shigeta C, Oguma N, Kamada N, Takimoto Y, Kuramoto A, Dohy H, Kyo T: 9;22 translocation and bcr rearrangements in chronic myelocytic leukemia patients among atomic bomb survivors. J Radiat Res (Tokyo) 1989;30:352–358.

13 Corso A, Lazzarino M, Morra E, Merante S, Astori C, Bernasconi P, Boni M, Bernasconi C: Chronic myelogenous leukemia and exposure to ionizing radiation – a retrospective study of 443 patients. Ann Hematol 1995;70:79–82.

14 Preston DL, Kusumi S, Tomonaga M, Izumi S, Ron E, Kuramoto A, Kamada N, Dohy H, Matsuo T, Matsui T: Cancer incidence in atomic bomb survivors. III. Leukemia, lymphoma and multiple myeloma, 1950–1987. Radiat Res 1994;137(suppl 2):S68–S97.

15 Berridge MV, Tan AS: Characterization of the cellular reduction of 3-(4,5-dimethylthiazol-2-yl)-2,5-diphenyltetrazolium bromide (MTT): Subcellular localization, substrate dependence, and involvement of mitochondrial electron transport in MTT reduction. Arch Biochem Biophys 1993;303:474–482.

16 Deininger MW, Bose S, Gora-Tybor J, Yan XH, Goldman JM, Melo JV: Selective induction of leukemia-associated fusion genes by high-dose ionizing radiation. Cancer Res 1998;58:421–425.

17 Kozubek S, Lukasova E, Mareckova A, Skalnikova M, Kozubek M, Bartova E, Kroha V, Krahulcova E, Slotova J: The topological organization of chromosomes 9 and 22 in cell nuclei has a determinative role in the induction of t(9,22) translocations and in the pathogenesis of t(9,22) leukemias. Chromosoma 1999;108:426–435.

18 Neves H, Ramos C, da Silva MG, Parreira A, Parreira L: The nuclear topography of ABL, BCR, PML, and RARalpha genes: Evidence for gene proximity in specific phases of the cell cycle and stages of hematopoietic differentiation. Blood 1999;93:1197–1207.

19 Anastasi J, Moinuddin R, Daugherty C: The juxtaposition of ABL with BCR and risk for fusion may come at the time of BCR replication in late S-phase. Blood 1999;94:1137–1138.

20 Chissoe SL, Bodenteich A, Wang YF, Wang YP, Burian D, Clifton SW, Crabtree J, Freeman A, Iyer K, Jian L: Sequence and analysis of the human ABL gene, the BCR gene, and regions involved in the Philadelphia chromosomal translocation. Genomics 1995;27:67–82.

21 Saglio G, Storlazzi CT, Giugliano E, Surace C, Anelli L, Rege-Cambrin G, Zagaria A, Jimenez VA, Heiniger A, Scaravaglio P, Torres GA, Roman GJ, Archiciacono N, Banfi S, Rocchi M: A 76-kb duplicon maps close to the BCR gene on chromosome 22 and the ABL gene on chromosome 9: Possible involvement in the genesis of the Philadelphia chromosome translocation. Proc Natl Acad Sci USA 2002;99:9882–9887.

22 Sinclair PB, Nacheva EP, Leversha M, Telford N, Chang J, Reid A, Bench A, Champion K, Huntly B, Green AR: Large deletions at the t(9;22) breakpoint are common and may identify a poor-prognosis subgroup of patients with chronic myeloid leukemia. Blood 2000;95:738–743.

23 Huntly BJ, Reid AG, Bench AJ, Campbell LJ, Telford N, Shepherd P, Szer J, Prince HM, Turner P, Grace C, Nacheva EP, Green AR: Deletions of the derivative chromosome 9 occur at the time of the Philadelphia translocation and provide a powerful and independent prognostic indicator in chronic myeloid leukemia. Blood 2001;98:1732–1738.

24 de la FJ, Merx K, Steer EJ, Muller M, Szydlo RM, Maywald O, Berger U, Hehlmann R, Goldman JM, Cross NC, Melo JV, Hochhaus A: ABL-BCR expression does not correlate with deletions on the derivative chromosome 9 or survival in chronic myeloid leukemia. Blood 2001;98:2879–2880

25 Melo JV, Gordon DE, Tuszynski A, Dhut S, Young BD, Goldman JM: Expression of the ABL-BCR fusion gene in Philadelphia-positive acute lymphoblastic leukemia. Blood 1993;81:2488–2491.

26 Shepherd P, Suffolk R, Halsey J, Allan N: Analysis of molecular breakpoint and m-RNA transcripts in a prospective randomized trial of interferon in chronic myeloid leukaemia: No correlation with clinical features, cytogenetic response, duration of chronic phase, or survival. Br J Haematol 1995;89:546–554.

27 Melo JV: The diversity of BCR-ABL fusion proteins and their relationship to leukemia phenotype. Blood 1996;88:2375–2384.

28 Ravandi F, Cortes J, Albitar M, Arlinghaus R, Qiang GJ, Talpaz M, Kantarjian HM: Chronic myelogenous leukaemia with p185(BCR/ABL) expression: Characteristics and clinical significance. Br J Haematol 1999;107:581–586.

29 Dhingra K, Talpaz M, Kantarjian H, Ku S, Rothberg J, Gutterman JU, Kurzrock R: Appearance of acute leukemia-associated P190BCR-ABL in chronic myelogenous leukemia may correlate with disease progression. Leukemia 1991;5:191–195.

30 Pane F, Frigeri F, Sindona M, Luciano L, Ferrara F, Cimino R, Meloni G, Saglio G, Salvatore F, Rotoli B: Neutrophilic-chronic myeloid leukemia: A distinct disease with a specific molecular marker (BCR/ABL with C3/A2 junction). Blood 1996;88:2410–2414.

31 van Rhee F, Hochhaus A, Lin F, Melo JV, Goldman JM, Cross NC: p190 BCR-ABL mRNA is expressed at low levels in p210-positive chronic myeloid and acute lymphoblastic leukemias. Blood 1996;87:5213–5217.

32 Lichty BD, Keating A, Callum J, Yee K, Croxford R, Corpus G, Nwachukwu B, Kim P, Guo J, Kamel-Reid S: Expression of p210 and p190 BCR-ABL due to alternative splicing in chronic myelogenous leukaemia. Br J Haematol 1998;103:711–715.

33 Lichty BD, Kamel-Reid S: Exon-skipping in BCR/ABL is induced by ABL exon 2. Biochem J 2000;348(pt 1):63–69.

34 Hochhaus A, Reiter A, Skladny H, Melo JV, Sick C, Berger U, Guo JQ, Arlinghaus RB, Hehlmann R, Goldman JM, Cross NC: A novel BCR-ABL fusion gene (e6a2) in a patient with Philadelphia chromosome-negative chronic myelogenous leukemia. Blood 1996;88:2236–2240.

35 Leibundgut EO, Jotterand M, Rigamonti V, Parlier V, Muhlematter D, Tobler A, Solenthaler M: A novel BCR-ABL transcript e2a2 in a chronic myelogenous leukaemia patient with a duplicated Ph-chromosome and monosomy 7. Br J Haematol 1999;106:1041–1044.

36 How GF, Lim LC, Kulkarni S, Tan LT, Tan P, Cross NC: Two patients with novel BCR/ABL fusion transcripts (e8/a2 and e13/a2) resulting from translocation breakpoints within BCR exons. Br J Haematol 1999;105:434–436.

37 Moreno MP, Cortinas MN, Bonomi R, Cardeza A, Uriarte MR: A novel BCR-ABL fusion transcript (e15a2) in 2 patients with atypical chronic myeloproliferative syndrome. Blood 2001;97:3668–3669.

38 Byrne JL, Carter GI, Davies JM, Haynes AP, Russell NH, Cross NC: A novel BCR-ABL fusion gene (e2/1a) in a patient with Philadelphia-positive chronic myelogenous leukaemia and an aggressive clinical course. Br J Haematol 1998;103:791–794.

39 Soekarman D, van Denderen J, Hoefsloot L, Moret M, Meeuwsen T, van Baal J, Hagemeijer A, Grosveld G: A novel variant of the bcr-abl fusion product in Philadelphia chromosome-positive acute lymphoblastic leukemia. Leukemia 1990;4:397–403.

40 Wilson GA, Vandenberghe EA, Pollitt RC, Rees DC, Goodeve AC, Peake IR, Reilly JT: Are aberrant BCR-ABL transcripts more common than previously thought? Br J Haematol 2000;111:1109–1111.

41 Al Ali H-K, Leiblein S, Kovacs I, Hennig E, Niederwieser D, Deininger MWN: CML with an e1a3 BCR-ABL fusion: Rare, benign, and a potential diagnostic pitfall. Blood 2002;100:1092–1093.

42 Inukai T, Sugita K, Suzuki T, Ijima K, Goi K, Tezuka T, Kojika S, Hatakeyama K, Kagami K, Mori T: A novel 203 kD aberrant BCR-ABL product in a girl with Philadelphia chromosome positive acute lymphoblastic leukaemia. Br J Haematol 1993;85:823–825.

43 Cross NC, Melo JV, Feng L, Goldman JM: An optimized multiplex polymerase chain reaction (PCR) for detection of BCR-ABL fusion mRNAs in haematological disorders. Leukemia 1994;8:186–189.

44 Branford S, Rudzki Z, Hughes TP: A novel BCR-ABL transcript (e8a2) with the insertion of an inverted sequence of ABL intron 1b in a patient with Philadelphia-positive chronic myeloid leukaemia. Br J Haematol 2000;109:635–637.

45 Roman J, Parziale A, Gottardi E, De Micheli D, Cilloni D, Tiribelli M, Gonzalez MG, del Carmen RM, Torres A, Saglio G: Novel type of BCR-ABL transcript in a chronic myelogenous leukaemia patient relapsed after bone marrow transplantation. Br J Haematol 2000;111:644–646.

46 Shiratsuchi M, Muta K, Minami R, Motomura S, Suehiro Y, Abe Y, Shiokawa S, Umemura T, Fukui T, Nishimura J, Nawata H: Aberrant BCR-ABL transcript with intronic insertion in a patient with philadelphia chromosome-positive chronic myeloid leukemia: Implications for disease progression. Leuk Lymphoma 2001;41:411–415.

47 Biernaux C, Sels A, Huez G, Stryckmans P: Very low level of major BCR-ABL expression in blood of some healthy individuals. Bone Marrow Transplant 1996;17(suppl 3):S45–S47.

48 Bose S, Deininger M, Gora-Tybor J, Goldman JM, Melo JV: The presence of typical and atypical BCR-ABL fusion genes in leukocytes of normal individuals: Biologic significance and implications for the assessment of minimal residual disease. Blood 1998;92:3362–3367.

49 Laurent E, Talpaz M, Kantarjian H, Kurzrock R: The BCR gene and philadelphia chromosome-positive leukemogenesis. Cancer Res 2001;61:2343–2355.

50 Maru Y, Witte ON: The BCR gene encodes a novel serine/threonine kinase activity within a single exon. Cell 1991;67:459–468.

51 Reuther GW, Fu H, Cripe LD, Collier RJ, Pendergast AM: Association of the protein kinases c-Bcr and Bcr-Abl with proteins of the 14–3-3 family. Science 1994;266:129–133.

52 Pendergast AM, Quilliam LA, Cripe LD, Bassing CH, Dai Z, Li N, Batzer A, Rabun KM, Der CJ, Schlessinger J: BCR-ABL-induced oncogenesis is mediated by direct interaction with the SH2 domain of the GRB-2 adaptor protein. Cell 1993;75:175–185.

53 Pendergast AM, Muller AJ, Havlik MH, Maru Y, Witte ON: BCR sequences essential for transformation by the BCR-ABL oncogene bind to the ABL SH2 regulatory domain in a non-phosphotyrosine-dependent manner. Cell 1991;66:161–171.

54 McWhirter JR, Galasso DL, Wang JY: A coiled-coil oligomerization domain of Bcr is essential for the transforming function of Bcr-Abl oncoproteins. Mol Cell Biol 1993;13:7587–7595.

55 Zhao X, Ghaffari S, Lodish H, Malashkevich VN, Kim PS: Structure of the Bcr-Abl oncoprotein oligomerization domain. Nat Struct Biol 2002;9:117–120.

56 Janssen JW, Ridge SA, Papadopoulos P, Cotter F, Ludwig WD, Fonatsch C, Rieder H, Ostertag W, Bartram CR, Wiedemann LM: The fusion of TEL and ABL in human acute lymphoblastic leukaemia is a rare event. Br J Haematol 1995;90:222–224.

57 Golub TR, Goga A, Barker GF, Afar DE, McLaughlin J, Bohlander SK, Rowley JD, Witte ON, Gilliland DG: Oligomerization of the ABL tyrosine kinase by the Ets protein TEL in human leukemia. Mol Cell Biol 1996;16:4107–4116.
58 He Y, Wertheim JA, Xu L, Miller JP, Karnell FG, Choi JK, Ren R, Pear WS: The coiled-coil domain and Tyr177 of bcr are required to induce a murine chronic myelogenous leukemia-like disease by bcr/abl. Blood 2002;99:2957–2968.
59 Hart MJ, Eva A, Evans T, Aaronson SA, Cerione RA: Catalysis of guanine nucleotide exchange on the CDC42Hs protein by the dbl oncogene product. Nature 1991;354:311–314.
60 Cen H, Papageorge AG, Zippel R, Lowy DR, Zhang K: Isolation of multiple mouse cDNAs with coding homology to *Saccharomyces cerevisiae* CDC25: Identification of a region related to Bcr, Vav, Dbl and CDC24. EMBO J 1992;11:4007–4015.
61 Diekmann D, Brill S, Garrett MD, Totty N, Hsuan J, Monfries C, Hall C, Lim L, Hall A: Bcr encodes a GTPase-activating protein for p21rac. Nature 1991;351:400–402.
62 Voncken JW, van Schaick H, Kaartinen V, Deemer K, Coates T, Landing B, Pattengale P, Dorseuil O, Bokoch GM, Groffen J: Increased neutrophil respiratory burst in bcr-null mutants. Cell 1995;80:719–728.
63 Abelson HT, Rabstein LS: Lymphosarcoma: Virus-induced thymic-independent disease in mice. Cancer Res 1970;30:2213–2222.
64 Feller SM, Knudsen B, Hanafusa H: c-Abl kinase regulates the protein binding activity of c-Crk. EMBO J 1994;13:2341–2351.
65 Sattler M, Salgia R, Okuda K, Uemura N, Durstin MA, Pisick E, Xu G, Li JL, Prasad KV, Griffin JD: The proto-oncogene product p120CBL and the adaptor proteins CRKL and c-CRK link c-ABL, p190BCR/ABL and p210BCR/ABL to the phosphatidylinositol-3′ kinase pathway. Oncogene 1996;12:839–846.
66 McWhirter JR, Wang JY: An actin-binding function contributes to transformation by the Bcr-Abl oncoprotein of Philadelphia chromosome-positive human leukemias. EMBO J 1993;12:1533–1546.
67 Kipreos ET, Wang JY: Cell cycle-regulated binding of c-Abl tyrosine kinase to DNA. Science 1992;256:382–385.
68 Wen ST, Jackson PK, Van Etten RA: The cytostatic function of c-Abl is controlled by multiple nuclear localization signals and requires the p53 and Rb tumor suppressor gene products. EMBO J 1996;15:1583–1595.
69 Taagepera S, McDonald D, Loeb JE, Whitaker LL, McElroy AK, Wang JY, Hope TJ: Nuclear-cytoplasmic shuttling of C-ABL tyrosine kinase. Proc Natl Acad Sci USA 1998;95:7457–7462.
70 Van Etten RA, Jackson P, Baltimore D: The mouse type IV c-abl gene product is a nuclear protein, and activation of transforming ability is associated with cytoplasmic localization. Cell 1989;58:669–678.
71 Van Etten RA: Cycling, stressed-out and nervous: Cellular functions of c-Abl. Trends Cell Biol 1999;9:179–186.
72 Sawyers CL, McLaughlin J, Goga A, Havlik M, Witte O: The nuclear tyrosine kinase c-Abl negatively regulates cell growth. Cell 1994;77:121–131.
73 Yuan ZM, Shioya H, Ishiko T, Sun X, Gu J, Huang YY, Lu H, Kharbanda S, Weichselbaum R, Kufe D: p73 is regulated by tyrosine kinase c-Abl in the apoptotic response to DNA damage. Nature 1999;399:814–817.
74 Lewis JM, Schwartz MA: Integrins regulate the association and phosphorylation of paxillin by c-Abl. J Biol Chem 1998;273:14225–14230.
75 Cooper JA, Gould KL, Cartwright CA, Hunter T: Tyr527 is phosphorylated in pp60c-src: Implications for regulation. Science 1986;231:1431–1434.
76 Xu W, Harrison SC, Eck MJ: Three-dimensional structure of the tyrosine kinase c-Src. Nature 1997;385:595–602.
77 Pendergast AM, Muller AJ, Havlik MH, Clark R, McCormick F, Witte ON: Evidence for regulation of the human ABL tyrosine kinase by a cellular inhibitor. Proc Natl Acad Sci USA 1991;88:5927–5931.
78 Goga A, McLaughlin J, Pendergast AM, Parmar K, Muller A, Rosenberg N, Witte ON: Oncogenic activation of c-ABL by mutation within its last exon. Mol Cell Biol 1993;13:4967–4975.
79 Pluk H, Dorey K, Superti-Furga G: Autoinhibition of c-Abl. Cell 2002;108:247–259.
80 Mayer BJ, Baltimore D: Mutagenic analysis of the roles of SH2 and SH3 domains in regulation of the Abl tyrosine kinase. Mol Cell Biol 1994;14:2883–2894.
81 Ilaria RL Jr, Van Etten RA: The SH2 domain of P210BCR/ABL is not required for the transformation of hematopoietic factor-dependent cells. Blood 1995;86:3897–3904.
82 Shi Y, Alin K, Goff SP: Abl-interactor-1, a novel SH3 protein binding to the carboxy-terminal portion of the Abl protein, suppresses v-abl transforming activity. Genes Dev 1995;9:2583–2597.
83 Dai Z, Pendergast AM: Abi-2, a novel SH3-containing protein interacts with the c-Abl tyrosine kinase and modulates c-Abl transforming activity. Genes Dev 1995;9:2569–2582.
84 Dai Z, Quackenbush RC, Courtney KD, Grove M, Cortez D, Reuther GW, Pendergast AM: Oncogenic Abl and Src tyrosine kinases elicit the ubiquitin-dependent degradation of target proteins through a Ras-independent pathway. Genes Dev 1998;12:1415–1424.
85 Wen ST, Van Etten RA: The PAG gene product, a stress-induced protein with antioxidant properties, is an Abl SH3-binding protein and a physiological inhibitor of c-Abl tyrosine kinase activity. Genes Dev 1997;11:2456–2467.
86 Mills KI, MacKenzie ED, Birnie GD: The site of the breakpoint within the bcr is a prognostic factor in Philadelphia-positive CML patients. Blood 1988;72:1237–1241.
87 Shtalrid M, Talpaz M, Kurzrock R, Kantarjian H, Trujillo J, Gutterman J, Yoffe G, Blick M: Analysis of breakpoints within the bcr gene and their correlation with the clinical course of Philadelphia-positive chronic myelogenous leukemia. Blood 1988;72:485–490.
88 Mills KI, Sproul AM, Leibowitz D, Burnett AK: Mapping of breakpoints, and relationship to BCR-ABL RNA expression, in Philadelphia-chromosome-positive chronic myeloid leukaemia patients with a breakpoint around exon 14 (b3) of the BCR gene. Leukemia 1991;5:937–941.
89 Morris SW, Daniel L, Ahmed CM, Elias A, Lebowitz P: Relationship of bcr breakpoint to chronic phase duration, survival, and blast crisis lineage in chronic myelogenous leukemia patients presenting in early chronic phase. Blood 1990;75:2035–2041.
90 Jaubert J, Martiat P, Dowding C, Ifrah N, Goldman JM: The position of the M-BCR breakpoint does not predict the duration of chronic phase or survival in chronic myeloid leukemia. Br J Haematol 1990;74:30–35.
91 Rozman C, Urbano-Ispizua A, Cervantes F, Rozman M, Colomer D, Feliz P, Pujades A, Vives Corrons JL: Analysis of the clinical relevance of the breakpoint location within M-BCR and the type of chimeric mRNA in chronic myelogenous leukemia. Leukemia 1995;9:1104–1107.
92 Inokuchi K, Inoue T, Tojo A, Futaki M, Miyake K, Yamada T, Tanabe Y, Ohki I, Dan K, Ozawa K: A possible correlation between the type of bcr-abl hybrid messenger RNA and platelet count in Philadelphia-positive chronic myelogenous leukemia. Blood 1991;78:3125–3127.
93 Perego RA, Costantini M, Cornacchini G, Gargantini L, Bianci C, Pungolino E, Rovida E, Morra E: The possible influences of B2A2 and B3A2 BCR/ABL protein structure on thrombopoiesis in chronic myeloid leukaemia. European Journal of Cancer 2000;36:1395–1401.
94 Ardern JC, Speak J, Hyde K, Hunt LP, Lawson R, Gorst DW, Geary CG, Lucas GS: Molecular analysis in chronic granulocytic leukaemia: Location of breakpoints within M-BCR and relationship with presentation platelet counts. Clin Lab Haematol 1993;15:253–258.
95 Opalka B, Wandl UB, Stutenkemper R, Kloke O, Seeber S, Niederle N: No correlation between the type of bcr-abl hybrid messenger RNA and platelet counts in chronic myelogenous leukemia. Blood 1992;80:1854–1855.
96 Martiat P, Ifrah N, Rassool F, Morgan G, Giles F, Gow J, Goldman JM: Molecular analysis of Philadelphia positive essential thrombocythemia. Leukemia 1989;3:563–565.
97 Emilia G, Luppi M, Ferrari MG, Temperani P, Marasca R, Giacobbi F, Vaccari P, Bandieri E, Di Donato C, Carapezzi C, Torelli G: Chronic myeloid leukemia with thrombocythemic onset may be associated with different BCR/ABL variant transcripts. Cancer Genet Cytogenet 1998;101:75–77.

98 Saglio G, Guerrasio A, Rosso C, Zaccaria A, Tassinari A, Serra A, Rege-Cambrin G, Mazza U, Gavosto F: New type of Bcr/Abl junction in Philadelphia chromosome-positive chronic myelogencus leukemia. Blood 1990; 76:1819–1824.
99 Briz M, Vilches C, Cabrera R, Fores R, Fernandez MN: Typical chronic myelogenous leukemia with e19a2 junction BCR/ABL transcript. Blood 1997;90:5024–5025.
100 Wilson G, Frost L, Goodeve A, Vandenberghe E, Peake I, Reilly J: BCR-ABL transcript with an e19a2 (c3a2) junction in classical chronic myeloic leukemia. Blood 1997;89: 3064.
101 Yamagata T, Mitani K, Kanda Y, Yazaki Y, Hirai H: Elevated platelet count features the variant type of BCR/ABL junction in chronic myelogenous leukaemia. Br J Haematol 1996; 94:370–372.
102 Mittre H, Leymarie P, Macro M, Leporrier M: A new case of chronic myeloid leukemia with c3/a2 BCR/ABL junction. Is it really a distinct disease? Blood 1997;89:4239–4241.
103 Wada H, Mizutani S, Nishimura J, Usuki Y, Kohsaki M, Komai M, Kaneko H, Sakamoto S, Delia D, Kanamaru A: Establishment and molecular characterization of a novel leukemic cell line with Philadelphia chromosome expressing p230 BCR/ABL fusion protein. Cancer Res 1995;55:3192–3196.
104 Rotoli B, Luciano L, Pane F, Salvatore F, Saglio G: A new case of chronic myeloid leukemia with c3/a2 BCR/ABL junction. Is it really a distinct disease? Blood 1997;89: 4239–4241.
105 Imbert M, Pierre R, Thiele J, Vardiman J, Brunning R, Flandrin G: Essential thrombocythaemia. Tumours of Haematopoietic and Lymphoid Tissues. Lyon, IARC Press, 2001, pp39–41.
106 Verstovsek S, Lin H, Kantarjian H, Saglio G, De Micheli D, Pane F, Garcia-Manero G, Intrieri M, Rotoli B, Salvatore F, Guo JQ, Talpaz M, Specchia G, Pizzolo G, Liberati AM, Cortes J, Quackenbush RC, Arlinghaus RB: Neutrophilic-chronic myeloid leukemia. Cancer 2002;94:2416–2425.
107 Ahmed S, Lee J, Wen LP, Zhao Z, Ho J, Best A, Kozma R, Lim L: Breakpoint cluster region gene product-related domain of n-chimaerin. Discrimination between Rac-binding and GTPase-activating residues by mutational analysis. J Biol Chem 1994;269:17642–17648.
108 Haskovec C, Ponzetto C, Polak J, Maritano D, Zemanova Z, Serra A, Michalova K, Klamova H, Cermak J, Saglio G: P230 BCR/ABL protein may be associated with an acute leukaemia phenotype. Br J Haematol 1998;103: 1104–1108.
109 Stewart MJ, Cox G, Reifel-Miller A, Kim SY, Westbrook CA, Leibowitz DS: A novel transcriptional suppressor located within a downstream intron of the BCR gene. J Biol Chem 1994;269:10820–10829.
110 Verfaillie CM, Hurley R, Lundell BI, Zhao C, Bhatia R: Integrin-mediated regulation of hematopoiesis: Do BCR/ABL-induced defects in integrin function underlie the abnormal circulation and proliferation of CML progenitors? Acta Haematol 1997;97:40–52.
111 Gordon MY, Dowding CR, Riley GP, Goldman JM, Greaves MF: Altered adhesive interactions with marrow stroma of haematopoietic progenitor cells in chronic myeloid leukaemia. Nature 1987;328:342–344.
112 Bhatia R, Wayner EA, McGlave PB, Verfaillie CM: Interferon-alpha restores normal adhesion of chronic myelogenous leukemia hematopoietic progenitors to bone marrow stroma by correcting impaired beta 1 integrin receptor function. J Clin Invest 1994;94:384–391.
113 Salgia R, Li JL, Lo SH, Brunkhorst B, Kansas GS, Sobhany ES, Sun Y, Pisick E, Hallek M, Ernst T: Molecular cloning of human paxillin, a focal adhesion protein phosphorylated by P210BCR/ABL. J Biol Chem 1995;270: 5039–5047.
114 Uemura N, Griffin JD: The adapter protein Crkl links Cbl to C3G after integrin ligation and enhances cell migration. J Biol Chem 1999;274:37525–37532.
115 Salgia R, Pisick E, Sattler M, Li JL, Uemura N, Wong WK, Burky SA, Hirai H, Chen LB, Griffin JD: p130CAS forms a signaling complex with the adapter protein CRKL in hematopoietic cells transformed by the BCR/ABL oncogene. J Biol Chem 1996;271:25198–25203.
116 Sattler M, Salgia R, Shrikhande G, Verma S, Uemura N, Law SF, Golemis EA, Griffin JD: Differential signaling after beta1 integrin ligation is mediated through binding of CRKL to p120(CBL) and p110(HEF1). J Biol Chem 1997;272:14320–14326.
117 Deininger MW, Vieira S, Mendiola R, Schultheis B, Goldman JM, Melo JV: BCR-ABL tyrosine kinase activity regulates the expression of multiple genes implicated in the pathogenesis of chronic myeloid leukemia. Cancer Res 2000;60:2049–2055.
118 Bazzoni G, Carlesso N, Griffin JD, Hemler ME: Bcr/Abl expression stimulates integrin function in hematopoietic cell lines. J Clin Invest 1996;98:521–528.
119 Krämer A, Hörner S, Willer A, Fruehauf S, Hochhaus A, Hallek M, Hehlmann R: Adhesion to fibronectin stimulates proliferation of wild-type and bcr/abl-transfected murine hematopoietic cells. Proc Natl Acad Sci USA 1999;96:2087–2092.
120 Barnes DJ, Goldman JM, Melo JV: Dose dependent effects of Bcr-Abl on the phenotype of murine haematopoietic cells (abstract). Blood 2001;98:145a.
121 Deininger MW, Goldman JM, Melo JV: The molecular biology of chronic myeloid leukemia. Blood 2000;96:3343–3356.
122 Puil L, Liu J, Gish G, Mbamalu G, Bowtell D, Pelicci PG, Arlinghaus R, Pawson T: Bcr-Abl oncoproteins bind directly to activators of the Ras signalling pathway. EMBO J 1994;13: 764–773.
123 Pelicci G, Lanfrancone L, Salcini AE, Romano A, Mele S, Grazia BM, Segatto O, Di Fiore PP, Pelicci PG: Constitutive phosphorylation of Shc proteins in human tumors. Oncogene 1995;11:899–907.
124 Oda T, Heaney C, Hagopian JR, Okuda K, Griffin JD, Druker BJ: Crkl is the major tyrosine-phosphorylated protein in neutrophils from patients with chronic myelogenous leukemia. J Biol Chem 1994;269:22925–22928.
125 Senechal K, Halpern J, Sawyers CL: The CRKL adaptor protein transforms fibroblasts and functions in transformation by the BCR-ABL oncogene. J Biol Chem 1996;271: 23255–23261.
126 Heaney C, Kolibaba K, Bhat A, Oda T, Ohno S, Fanning S, Druker BJ: Direct binding of CRKL to BCR-ABL is not required for BCR-ABL transformation. Blood 1997;89:297–306.
127 Marais R, Light Y, Paterson HF, Marshall CJ: Ras recruits Raf-1 to the plasma membrane for activation by tyrosine phosphorylation. EMBO J 1995;14:3136–3145.
128 Chang L, Karin M: Mammalian MAP kinase signalling cascades. Nature 2001;410:37–40.
129 Cahill MA, Janknecht R, Nordheim A: Signalling pathways: Jack of all cascades. Curr Biol 1996;6:16–19.
130 Kabarowski JH, Allen PB, Wiedemann LM: A temperature sensitive p210 BCR-ABL mutant defines the primary consequences of BCR-ABL tyrosine kinase expression in growth factor dependent cells. EMBO J 1994; 13:5887–5895.
131 Cortez D, Reuther G, Pendergast AM: The Bcr-Abl tyrosine kinase activates mitogenic signaling pathways and stimulates G1-to-S phase transition in hematopoietic cells. Oncogene 1997;15:2333–2342.
132 Raitano AB, Halpern JR, Hambuch TM, Sawyers CL: The Bcr-Abl leukemia oncogene activates Jun kinase and requires Jun for transformation. Proc Natl Acad Sci USA 1995;92:11746–11750.
133 Shi CS, Tuscano JM, Witte ON, Kehrl JH: GCKR links the Bcr-Abl oncogene and Ras to the stress-activated protein kinase pathway. Blood 1999;93:1338–1345.
134 Mayer IA, Verma A, Grumbach IM, Uddin S, Lekmine F, Ravandi F, Majchrzak B, Fujita S, Fish EN, Platanias LC: The p38 MAPK pathway mediates the growth inhibitory effects of interferon-alpha in BCR-ABL-expressing cells. J Biol Chem 2001;276:28570–28577.
135 Ilaria RL, Jr., Van Etten RA: P210 and P190(BCR/ABL) induce the tyrosine phosphorylation and DNA binding activity of multiple specific STAT family members. J Biol Chem 1996;271:31704–31710.
136 Chai SK, Nichols GL, Rothman P: Constitutive activation of JAKs and STATs in BCR-Abl-expressing cell lines and peripheral blood cells derived from leukemic patients. J Immunol 1997;159:4720–4728.

137 Horvath CM: STAT proteins and transcriptional responses to extracellular signals. Trends Biochem Sci 2000;25:496–502.
138 de Groot RP, Raaijmakers JA, Lammers JW, Jove R, Koenderman L: STAT5 activation by BCR-Abl contributes to transformation of K562 leukemia cells. Blood 1999;94:1108–1112.
139 Nosaka T, Kawashima T, Misawa K, Ikuta K, Mui AL, Kitamura T: STAT5 as a molecular regulator of proliferation, differentiation and apoptosis in hematopoietic cells. EMBO J 1999;18:4754–4765.
140 Horita M, Andreu EJ, Benito A, Arbona C, Sanz C, Benet I, Prosper F, Fernandez-Luna JL: Blockade of the Bcr-Abl kinase activity induces apoptosis of chronic myelogenous leukemia cells by suppressing signal transducer and activator of transcription 5-dependent expression of Bcl-xL. J Exp Med 2000;191: 977–984.
141 Sillaber C, Gesbert F, Frank DA, Sattler M, Griffin JD: STAT5 activation contributes to growth and viability in Bcr/Abl- transformed cells. Blood 2000;95:2118–2125.
142 Gesbert F, Griffin JD: Bcr/Abl activates transcription of the Bcl-X gene through STAT5. Blood 2000;96:2269–2276.
143 Skorski T, Kanakaraj P, Nieborowska-Skorska M, Ratajczak MZ, Wen SC, Zon G, Gewirtz AM, Perussia B, Calabretta B: Phosphatidylinositol-3 kinase activity is regulated by BCR/ABL and is required for the growth of Philadelphia chromosome-positive cells. Blood 1995;86:726–736.
144 Skorski T, Bellacosa A, Nieborowska-Skorska M, Majewski M, Martinez R, Choi JK, Trotta R, Wlodarski P, Perrotti D, Chan TO, Wasik MA, Tsichlis PN, Calabretta B: Transformation of hematopoietic cells by BCR/ABL requires activation of a PI-3k/Akt-dependent pathway. EMBO J 1997;16:6151–6161.
145 Gesbert F, Sellers WR, Signoretti S, Loda M, Griffin JD: BCR/ABL regulates expression of the cyclin-dependent kinase inhibitor p27Kip1 through the phosphatidylinositol 3-Kinase/AKT pathway. J Biol Chem 2000; 275:39223–39230.
146 Franke TF, Kaplan DR, Cantley LC: PI3K: Downstream AKTion blocks apoptosis. Cell 1997;88:435–437.
147 del Peso L, Gonzalez-Garcia M, Page C, Herrera R, Nunez G: Interleukin-3-induced phosphorylation of BAD through the protein kinase Akt. Science 1997;278:687–689.
148 Zha J, Harada H, Yang E, Jockel J, Korsmeyer SJ: Serine phosphorylation of death agonist BAD in response to survival factor results in binding to 14-3-3 not BCL-X(L). Cell 1996;87:619–628.
149 Sawyers CL, Callahan W, Witte ON: Dominant negative MYC blocks transformation by ABL oncogenes. Cell 1992;70:901–910.
150 Zou X, Rudchenko S, Wong K, Calame K: Induction of c-myc transcription by the v-Abl tyrosine kinase requires Ras, Raf1, and cyclin-dependent kinases. Genes Dev 1997;11: 654–662.
151 Menssen A, Hermeking H: From the Cover: Characterization of the c-MYC-regulated transcriptome by SAGE: Identification and analysis of c-MYC target genes. Proc Natl Acad Sci USA 2002;99:6274–6279.
152 Bedi A, Zehnbauer BA, Barber JP, Sharkis SJ, Jones RJ: Inhibition of apoptosis by BCR-ABL in chronic myeloid leukemia. Blood 1994;83:2038–2044.
153 Daley GQ, Baltimore D: Transformation of an interleukin 3-dependent hematopoietic cell line by the chronic myelogenous leukemia-specific P210bcr/abl protein. Proc Natl Acad Sci USA 1988;85:9312–9316.
154 Sirard C, Laneuville P, Dick JE: Expression of bcr-abl abrogates factor-dependent growth of human hematopoietic M07E cells by an autocrine mechanism. Blood 1994;83:1575–1585.
155 Bedi A, Barber JP, Bedi GC, el Deiry WS, Sidransky D, Vala MS, Akhtar AJ, Hilton J, Jones RJ: BCR-ABL-mediated inhibition of apoptosis with delay of G2/M transition after DNA damage: A mechanism of resistance to multiple anticancer agents. Blood 1995;86: 1148–1158.
156 Amos TA, Lewis JL, Grand FH, Gooding RP, Goldman JM, Gordon MY: Apoptosis in chronic myeloid leukaemia: Normal responses by progenitor cells to growth factor deprivation, X-irradiation and glucocorticoids. Br J Haematol 1995;91:387–393.
157 Amarante-Mendes GP, Naekyung KC, Liu L, Huang Y, Perkins CL, Green DR, Bhalla K: Bcr-Abl exerts its antiapoptotic effect against diverse apoptotic stimuli through blockage of mitochondrial release of cytochrome C and activation of caspase-3. Blood 1998;91:1700–1705.
158 Dubrez L, Eymin B, Sordet O, Droin N, Turhan AG, Solary E: BCR-ABL delays apoptosis upstream of procaspase-3 activation. Blood 1998;91:2415–2422.
159 Sanchez-Garcia I, Martin-Zanca D: Regulation of Bcl-2 gene expression by BCR-ABL is mediated by Ras. J Mol Biol 1997;267:225–228.
160 Wang HG, Rapp UR, Reed JC: Bcl-2 targets the protein kinase Raf-1 to mitochondria. Cell 1996;87:629–638.
161 Neshat MS, Raitano AB, Wang HG, Reed JC, Sawyers CL: The survival function of the Bcr-Abl oncogene is mediated by Bad-dependent and -independent pathways: Roles for phosphatidylinositol 3-kinase and Raf. Mol Cell Biol 2000;20:1179–1186.
162 Gabriele L, Phung J, Fukumoto J, Segal D, Wang IM, Giannakakou P, Giese NA, Ozato K, Morse HC, III: Regulation of apoptosis in myeloid cells by interferon consensus sequence-binding protein. J Exp Med 1999; 190:411–421.
163 Hao SX, Ren R: Expression of interferon consensus sequence binding protein (ICSBP) is downregulated in Bcr-Abl-induced murine chronic myelogenous leukemia-like disease, and forced coexpression of ICSBP inhibits Bcr-Abl-induced myeloproliferative disorder. Mol Cell Biol 2000;20:1149–1161.
164 Vigneri P, Wang JY: Induction of apoptosis in chronic myelogenous leukemia cells through nuclear entrapment of BCR-ABL tyrosine kinase. Nat Med 2001;7:228–234.
165 Melo JV: Inviting leukaemic cells to waltz with the devil. Nature Medicine 2001;7:156–157.
166 Deutsch E, Dugray A, AbdulKarim B, Marangoni E, Maggiorella L, Vaganay S, M'Kacher R, Rasy SD, Eschwege F, Vainchenker W, Turhan AG, Bourhis J: BCR-ABL down-regulates the DNA repair protein DNA-PKcs. Blood 2001;97:2084–2090.
167 Jonuleit T, van der KH, Miething C, Michels H, Hallek M, Duyster J, Aulitzky WE: Bcr-Abl kinase down-regulates cyclin-dependent kinase inhibitor p27 in human and murine cell lines. Blood 2000;96:1933–1939.
168 Goetz AW, van der KH, Maya R, Oren M, Aulitzky WE: Requirement for Mdm2 in the survival effects of Bcr-Abl and interleukin 3 in hematopoietic cells. Cancer Res 2001;61: 7635–7641.
169 Goldman JM, Melo JV: Targeting the BCR-ABL tyrosine kinase in chronic myeloid leukemia. N Engl J Med 2001;344:1084–1086.
170 Buchdunger E, Zimmermann J, Mett H, Meyer T, Muller M, Druker BJ, Lydon NB: Inhibition of the Abl protein-tyrosine kinase in vitro and in vivo by a 2-phenylaminopyrimidine derivative. Cancer Res 1996;56: 100–104.
171 Druker BJ, Tamura S, Buchdunger E, Ohno S, Segal GM, Fanning S, Zimmermann J, Lydon NB: Effects of a selective inhibitor of the Abl tyrosine kinase on the growth of Bcr-Abl positive cells. Nat Med 1996;2:561–566.
172 Deininger MW, Goldman JM, Lydon N, Melo JV: The tyrosine kinase inhibitor CGP57148B selectively inhibits the growth of BCR-ABL-positive cells. Blood 1997;90: 3691–3698.
173 Druker BJ, Talpaz M, Resta DJ, Peng B, Buchdunger E, Ford JM, Lydon NB, Kantarjian H, Capdeville R, Ohno-Jones S, Sawyers CL: Efficacy and safety of a specific inhibitor of the BCR-ABL tyrosine kinase in chronic myeloid leukemia. N Engl J Med 2001;344: 1031–1037.
174 Kantarjian H, Sawyers C, Hochhaus A, Guilhot F, Schiffer C, Gambacorti-Passerini C, Niederwieser D, Resta D, Capdeville R, Zoellner U, Talpaz M, Druker B: Hematologic and cytogenetic responses to imatinib mesylate in chronic myelogenous leukemia. N Engl J Med 2002;346:645–652.
175 Sawyers CL, Hochhaus A, Feldman E, Goldman JM, Miller CB, Ottmann OG, Schiffer CA, Talpaz M, Guilhot F, Deininger MW, Fischer T, O'Brien SG, Stone RM, Gambacorti-Passerini CB, Russell NH, Reiffers JJ, Shea TC, Chapuis B, Coutre S, Tura S, Morra E, Larson RA, Saven A, Peschel C, Gratwohl A, Mandelli F, Ben Am M, Gathmann I, Capdeville R, Paquette RL, Druker BJ: Imatinib induces hematologic and cytogenetic responses in patients with chronic myelogenous leukemia in myeloid blast crisis: Results of a phase II study. Blood 2002;99:3530–3539.

176 Elliott MA, Mesa RA, Tefferi A: Adverse events after imatinib mesylate therapy. N Engl J Med 2002;346:712–713.
177 Mahon FX, Deininger MW, Schultheis B, Chabrol J, Reiffers J, Goldman JM, Melo JV: Selection and characterization of BCR-ABL positive cell lines with differential sensitivity to the tyrosine kinase inhibitor STI571: Diverse mechanisms of resistance. Blood 2000; 96:1070–1079.
178 Gorre ME, Mohammed M, Ellwood K, Hsu N, Paquette R, Rao PN, Sawyers CL: Clinical resistance to STI-571 cancer therapy caused by BCR-ABL gene mutation or amplification. Science 2001;293:876–880.
179 Barthe C, Cony-Makhoul P, Melo JV, Mahon JR: Roots of clinical resistance to STI-571 cancer therapy. Science 2001;293:2163.
180 Kreil S, Müller MC, Lahaye T, La Rosée P, Corbin AS, Schoch C, Cross NCP, Berger U, Rieder H, Druker BJ, Gschaidmeier H, Hehlmann R, Hochhaus A: Molecular and chromosomal mechanisms of resistance in CML patients after STI571 (Glivec) therapy. Blood 2001;98:435a.
181 von Bubnoff N, Schneller F, Peschel C, Duyster J: BCR-ABL gene mutations in relation to clinical resistance of Philadelphia-chromosome-positive leukaemia to STI571: A prospective study. Lancet 2002;359:487–491.
182 Branford S, Rudzki Z, Walsh S, Grigg A, Arthur C, Taylor K, Herrmann R, Lynch KP, Hughes TP: High frequency of point mutations clustered within the adenosine triphosphate-binding region of BCR/ABL in patients with chronic myeloid leukemia or Ph-positive acute lymphoblastic leukemia who develop imatinib (STI571) resistance. Blood 2002;99: 3472–3475.
183 Graham SM, Jorgensen HG, Allan E, Pearson C, Alcorn MJ, Richmond L, Holyoake TL: Primitive, quiescent, Philadelphia-positive stem cells from patients with chronic myeloid leukemia are insensitive to STI571 in vitro. Blood 2002;99:319–325.
184 Holtz MS, Slovak ML, Zhang F, Sawyers CL, Forman SJ, Bhatia R: Imatinib mesylate (STI571) inhibits growth of primitive malignant progenitors in chronic myelogenous leukemia through reversal of abnormally increased proliferation. Blood 2002;99:3792–3800.

Prof. Junia V. Melo
Department of Haematology – ICSTM
Royal Postgraduate Medical School
Hammersmith Hospital, Du Cane Road
London W12 0NN (UK)
E-Mail j.melo@ic.ac.uk

Bain BJ (ed): Chronic Myeloproliferative Disorders.
Basel, Karger, 2003, pp 32–43 / Acta Haematol 2002;107:64–75

# Cytogenetic and Molecular Monitoring of Residual Disease in Chronic Myeloid Leukaemia

Jaspal Kaeda[a] Andrew Chase[b] John M. Goldman[a]

[a]Department of Haematology, Faculty of Medicine, Imperial College of Science, Technology and Medicine, Hammersmith Hospital, London, [b]Leukaemia Research Unit, Wessex Regional Genetics Laboratory, Salisbury District Hospital, Salisbury, UK

**Key Words**

Chronic myeloid leukaemia · Ph chromosome · *BCR-ABL* · Fluorescence in situ hybridisation · Polymerase chain reaction · Reverse-transcriptase polymerase chain reaction · Real time polymerase chain reaction

**Abstract**

For patients with chronic myeloid leukaemia, methods for monitoring response to treatment have changed considerably in recent years. In the 1980s, the principal approach was repeated examination of bone marrow metaphases for the presence of the Ph chromosome in patients treated by interferon-α (IFN-α) or allogeneic stem cell transplantation. The use of fluorescence in situ hybridisation (FISH) techniques to detect the *BCR-ABL* fusion gene in Ph-positive leukaemia cells increased the sensitivity of cytogenetic studies to some degree. In the last 10 years, the reverse-transcriptase polymerase chain reaction (RT-PCR) has proved extremely valuable for assessing and monitoring minimal residual disease in patients who achieve Ph negativity after treatment with IFN-α or with the new Abl tyrosine kinase inhibitor imatinib mesylate or after allogeneic stem cell transplantation (SCT). Results are consistent with the notion that the majority of long-term survivors after allogeneic SCT are probably 'cured'; for other patients monitored serially in complete cytogenetic remission, rising numbers of BCR-ABL transcripts detected by RT-PCR can indicate the need for further therapy.

## Introduction

The optimal management of malignant haematological disorders must depend on accurate and long-term monitoring of a patient's response to therapy. For some decades clinicians relied on examination of cellular morphology in peripheral blood and marrow specimens collected at regular intervals from asymptomatic patients. These studies were and remain valuable but they lack sensitivity. As a consequence the patients not infrequently had advanced disease that was difficult to treat by the time that relapse was detected. With increasing knowledge of the chromosomal aberrations characteristic of malignant disorders other analyses such as cytogenetics and fluorescence in situ hybridisation (FISH) were applied, but even these were relatively insensitive, time consuming and invasive. In the last decade our knowledge of the molecular basis of chronic myeloid leukaemia (CML) has been exploited to monitor a patient's response to the various therapeutic agents employed in this disease. The main objective therefore is to monitor 'minimal residual disease' (MRD) using the most sensitive available assay in order to assess a patient's response to treatment and to recognise early signs of relapse. Cytogenetic studies however remain important to confirm molecular findings and to help to identify disease progression.

## Cytogenetics

*Cytogenetics at Diagnosis*

In around 90% of cases of CML chromosomal material is reciprocally exchanged between the long arm of one chromosome 9 and the long arm of one chromosome 22, a translocation event referred to as t(9;22)(q34;q11). The derivative 22q– is the Philadelphia (Ph) chromosome. The translocation results in the formation of the *BCR-ABL* fusion gene, the protein product of which has constitutive tyrosine kinase activity and is believed to be the major 'cause' of the chronic phase of CML. In a further 5% of patients with a haematological disorder classifiable as CML, a variant Ph translocation occurs that involves chromosomes 9, 22 and one other chromosome. In a small number of CML patients the karyotype is normal, i.e. the patients are described as Ph negative, although in about half of these cases the *BCR-ABL* fusion can be detected at the molecular level, indicating a submicroscopic insertion of *ABL* into *BCR* or vice versa. The remaining cases of Ph-negative CML have no *BCR-ABL* fusion gene and must therefore have a different molecular basis.

FISH with *BCR* and *ABL* probes has helped to clarify the nature of complex rearrangements and can identify Ph-negative, *BCR-ABL*-positive CML. Together with other molecular techniques, FISH has been used to demonstrate the presence of the *BCR-ABL* fusion gene in the variant forms described above.

*Methods*

*Metaphase Cytogenetics*

Although metaphase cytogenetics has been the mainstay for detection of MRD for many years, it has been increasingly replaced by reverse transcriptase-polymerase chain reaction (RT-PCR) and FISH technologies. One principal drawback of standard cytogenetics in comparison with other techniques is its lack of sensitivity. Usually only 20–50 metaphases are scored and thus a given cytogenetic abnormality may not be detected unless it is present in at least 2–5% of cells.

Another disadvantage of conventional cytogenetic analysis is that it requires dividing cells. Because the peripheral blood, at least in patients in haematological remission, contains very few cycling myeloid cells, bone marrow aspirates are required to provide a suitable sample. In addition, bone marrow specimens from patients early after stem cell transplantation (SCT) and from some patients treated with interferon-α (IFN-α) or imatinib mesylate (STI571, Glivec) either have low cell counts or fail to grow well in culture.

One advantage, not afforded by other techniques, is that the appearance of additional karyotypic abnormalities will be detected, and these may alert the clinician to impending disease progression.

*FISH on Interphase Cells*

The initial use of FISH technology to study interphase cells employed a 3′ *ABL* probe and 5′ *BCR* probe which were labelled differentially so as to give red and green fluorescent signals, respectively. In normal cells, two red and two green signals indicated the location of the normal *ABL* and *BCR* genes. In *BCR-ABL*-positive cells single red and single green signals indicated the normal *ABL* and *BCR* genes whilst a fused red-green signal, sometimes seen as yellow, indicated the *BCR-ABL* fusion gene. However, random colocalisation of red and green signals could also occur in normal cells and could give a high false-positive rate of 10–15% [1]. In addition, depending on the probes used and the precise location of the *ABL* breakpoint, the separation of the red and green portions of the 'fusion' signal could also lead to a high false-negative rate [1].

The probe systems currently in use have overcome these problems by utilising differentially labelled probes which flank the breakpoints of the *BCR* and *ABL* genes [2, 3]. In a normal cell, two red and two green signals indicate the normal *BCR* and *ABL* genes. In a CML cell, in most patients, single red and green signals indicate the normal *BCR* and *ABL* genes, and two red-green fusion signals indicate the presence of the *BCR-ABL* and *ABL-BCR* fusion genes. The requirement to identify two fusion signals in a *BCR-ABL*-positive cell substantially reduces the false-positive rate – to around 0.2 %.

In practice, it is essential that the actual rate of false positives is established in each laboratory by 'blind' scoring of known Ph-negative samples together with samples with low Ph positivity. Some groups have calculated a cut-off value for the level below which a sample is regarded as *BCR-ABL* negative, for example the mean false-positive rate plus three standard deviations.The advantages of FISH for *BCR-ABL* are that a large number of cells can be scored, usually 100–500, and peripheral blood cells can be examined avoiding the need for a bone-marrow aspirate. Despite these advantages, there are several issues which can cause problems with interpretation:

(1) The recent discovery of large deletions of the *ABL-BCR* gene and flanking regions on the derivative chromosome 9 in a minority of patients [4–6] has complicated the

interpretation of *BCR-ABL* FISH since, in these patients, the *ABL-BCR* fusion signal is missing. Deletions have been seen in 10–15% of patients and are more frequent in patients with variant Ph translocations in whom as many as 30% may show a deletion. For these patients, the false-positive rate will be around 10–15%, i.e. the same as that in the earlier systems described above. It is therefore crucial that the deletion status of patients being screened by *BCR-ABL* FISH is known so that the signal pattern can be interpreted appropriately [7].

(2) Although conventional cytogenetics and FISH usually give comparable results [8] some groups have noticed differences between the two techniques. First, peripheral blood granulocytes may give lower Ph-positive values than bone marrow cells in some patients [9, 10]. Experimental confirmation of this has been afforded by one study which demonstrated a higher proportion of Ph-negative myeloid progenitors in the peripheral blood of CML patients than was found by contemporary bone marrow cytogenetics [11]. Secondly, in a minority of patients treated with imatinib mesylate or IFN-α, bone marrow FISH gave consistently lower values for the proportions of Ph-positive cells than did cytogenetic studies [12 and pers. obs.].

It must be remembered that T lymphocytes are predominantly Ph negative [13], and this must be taken into account when scoring peripheral blood, preferably by scoring only the granulocyte fraction rather than total white cells.

(3) Interpretation of signal patterns may be problematic in non-expert hands. In a multicentre study comparing the scoring of CML samples by *BCR-ABL* FISH in a number of laboratories in the USA [14], 10.7% of Ph-negative samples, scored blind along with a range of Ph-positive samples, were scored as Ph positive.

In summary, *BCR-ABL* FISH is likely to play an increasing role in monitoring treatment of CML but is technically demanding and requires care in a number of areas as described above.

*Hypermetaphase FISH*

Finally, the application of FISH probes for *BCR* and *ABL* to 'hypermetaphases' has been developed by the MD Anderson group in Texas, USA [15]. Hypermetaphases are produced by exposing bone marrow cells in culture to colcemid for 24 h. Although the metaphases produced by this method are highly contracted and unsuitable for cytogenetic analysis, they are suitable for FISH and it is possible to routinely screen 400–500 metaphases. The issue of false positivity is likely to be minimised by this method since metaphases, rather than interphases, are being examined.

## *Clinical Results*

*Allogeneic SCT*

Although in early studies of patients treated by allogeneic SCT, cytogenetics and Southern blotting were the only techniques available to assess MRD, quantitative RT-PCR is now regarded as the method of choice. RT-PCR is able to detect one *BCR-ABL*-positive cell in $10^4$–$10^6$ normal cells, which is several orders of magnitude lower than that obtained by either cytogenetics or FISH. In addition, the ability to detect changes in the level of *BCR-ABL*-positive cells at these lower limits is of clinical importance, as will be discussed later.

Although cytogenetics is of limited value in measuring MRD, cytogenetic analysis can be useful to confirm cytogenetic relapse when this is already suspected as a result of RT-PCR findings.

*IFN-α*

For assessing response to IFN-α treatment cytogenetic analysis is still the standard method although FISH is increasingly replacing cytogenetics or is used to provide supplementary information. The standard scheme for classifying cytogenetic response in the bone marrow of CML patients is the following: no response: 100% Ph-positive metaphases; minor response: 35–95% Ph-positive metaphases; partial response: 1–34% Ph-positive metaphases, and complete response: 0% Ph-positive metaphases. Complete and partial responses may be grouped together as 'major responses'.

The MD Anderson group have shown that 70–80% of patients treated with IFN-α achieve a complete haematological response and 26–38% achieve a major cytogenetic response (<35% Ph-positive) [16, 17]. Similar response rates have been seen by some other groups although in general responses have been lower in studies of patients with more advanced disease and in most multicentre studies.

The clinical advantage of obtaining a cytogenetic response has varied in different trials. However in general, particularly for patients with early disease, obtaining a partial or complete cytogenetic response correlates with increased survival [17, 18], an observation that demonstrates the usefulness of monitoring IFN-α-treated patients with cytogenetic analysis. Complete cytogenetic responders fare particularly well with a 10-year survival rate (from date of achieving complete remission) of 72% [19]. The use of techniques with increased sensitivity such as FISH or RT-PCR can provide additional information of clinical value.

*Imatinib Mesylate*

Treatment with imatinib mesylate has shown encouraging results in early trials although published trials include only preliminary data. In one study a substantial proportion (31%) of patients resistant to or intolerant of IFN-α showed major cytogenetic responses [20]. Newly diagnosed patients fare particularly well, and as many as 60% may achieve a complete cytogenetic response [unpublished observations, J. Goldman]. Although it is expected that these responses will result in increased survival, this has not yet been formally demonstrated, and the clinical relevance of cytogenetic analysis and other laboratory tests as reliable 'surrogate markers' for survival has still to be determined.

## Molecular Techniques

Though cytogenetic analysis and FISH studies remain extremely valuable in the initial investigation of malignant haematopoietic disorders, their role in monitoring MRD has been reduced with the introduction of molecular techniques described in more detail below. The application of Southern and Western blot analysis will be mentioned only briefly as these have largely been rendered obsolete by the introduction of RT-PCR for monitoring MRD.

### *Methods*

*Southern Blot*

The breakpoints in the chromosome 22 are largely clustered within a 5.8-kb sequence of the *BCR* gene, a region designated the 'major breakpoint cluster region' (M-*BCR*) [21]. Therefore, it is possible to detect any rearrangements involving the *BCR* gene at the genomic DNA level using Southern blot analysis with a *BCR* probe. This technique will pick up *BCR* rearrangement and by implication a *BCR-ABL* fusion gene in >95% of Ph-positive cases and in a significant number of Ph-negative cases [22]. The rearranged fragment corresponding to *BCR-ABL* can then be followed in serial samples obtained during treatment using selected restriction enzymes. The intensity of the rearranged *BCR-ABL* fragment can be measured by densitometry or phosphoimager analysis relative to the germ line fragment, i.e. the unrearranged *BCR* gene. A sensitivity of 1% can be achieved by Southern blot analysis in monitoring MRD [22–24]. However, the expense, the need to use [$\alpha^{32}$P]dCTP and, most importantly, its relative insensitivity mean that it is now rarely used for MRD studies.

*Western Blot*

The *BCR-ABL* fusion gene is transcribed and translated to yield BCR-ABL oncoprotein. The presence of the novel BCR-ABL protein in peripheral blood or bone marrow aspirates can be detected using an anti-ABL antibody following Western blotting. In addition to the normal $p145^{ABL}$, a larger BCR-ABL protein is detectable in samples from Ph-positive individuals. The intensities of bands corresponding to the proteins can be measured and thus serial samples can be analysed in patients undergoing treatment. Sensitivity of the order of 1–10% has been reported. In practice the inordinate complexities and the relative insensitivity of this technique make it unsuitable for routine monitoring of MRD [25, 26].

*RT-PCR*

The application of PCR has revolutionised the monitoring of MRD. PCR enables rapid, specific detection of target DNA with greatly enhanced sensitivity. However, as with a number of chromosomal translocations, the breakpoints in CML are too widely distributed *at the DNA level* to allow direct amplification of the *BCR-ABL* fusion gene by a single PCR analysis. This difficulty has been overcome by exploiting the molecular homogeneity of CML. In the vast majority of cases the t(9;22) translocation juxtaposes exons e13 (previously known as b2) or e14 (previously known as b3) adjacent to exon 2 of the *ABL* gene. The novel fusion genes are transcribed and the introns excised to yield *BCR-ABL* fusion transcripts referred to as e13a2 or e14a2 (previously b2a2 or b3a2), respectively [27]. Thus junctional *BCR* and *ABL* exons that may be widely separated at the DNA level are brought into close proximity such that they can be detected by a single PCR using cDNA as the template following reverse transcription of mRNA. [One should note parenthetically that about one-half of patients with Ph-positive acute lymphoblastic leukaemia have a translocation that juxtaposes the first *BCR* exon (e1) with *ABL* exon 2 (a2), so-called e1a2; this is expressed as a p190 oncoprotein ($p190^{BCR\text{-}ABL}$)]. The RT-PCR has been widely exploited to detect the different *BCR-ABL* transcripts by multiplex PCR. More importantly it has greatly enhanced the level of sensitivity of MRD detection. The level of sensitivity achievable by multiplex PCR is of the order of $10^{-2}$–$10^{-3}$. Having determined the type of *BCR-ABL* transcript expressed, the level of sensitivity can be increased by 2–3 logs by subjecting the cDNA to transcript-specific nested-primer PCR; this technique involves submitting an aliquot of amplicons to a second round of PCR using a new set of primers internal to those used in the first amplification [28].

*Sensitivity of RT-PCR.* A sensitivity of $10^{-5}$–$10^{-6}$ is achievable by nested PCR in a clinical laboratory [28]. Indeed, using modified and extremely sensitive PCR techniques, sensitivity levels of $10^{-8}$ have been achieved. With these conditions *BCR-ABL* mRNA was detected in a high proportion of normal healthy individuals [29, 30]. In CML patients there is little or no difference in *BCR-ABL* expression in bone marrow aspirates and peripheral blood for the purposes of studying MRD in CML [31] (although bone marrow specimens are clearly more informative for monitoring Ph-positive acute leukaemia). These findings eliminate the need for routine bone marrow samples and thereby permit a relatively high frequency of testing based on blood specimens.

However an important caveat to the achievable sensitivity by RT-PCR is the fact that stringent precautions are necessary to avoid contamination which can give false-positive results. Similarly, false-negative data due to lack of mRNA and cDNA integrity, i.e. quality and quantity, need to be excluded by including as a control a ubiquitously expressed housekeeping gene, such as *ABL*, $\beta_2$-microglobulin, glucose-6-phosphate dehydrogenase, porphobilinogen deaminase or β-glucuronidase. There is no general agreement as to the ideal endogenous control for haematological malignancies. It is also important to note that even when *BCR-ABL* transcripts cannot be detected by RT-PCR, which is 4 logs more sensitive than conventional techniques, patients may still harbour up to a million malignant cells in their body (see below).

*Quantitative RT-PCR* Because different PCR techniques may have different reaction dynamics, it is difficult to compare data from separate experiments and comparisons of levels of mRNA transcripts amplified with different primer pairs is an imprecise science. Moreover, fusion genes may be detectable in normal individuals (see above), as illustrated by *BCR-ABL* and re-arranged *BCL2*, commonly seen in association with t(14;18). These observations make the mere finding of residual *BCR-ABL* transcripts in CML patients in complete cytogenetic and haematological remission difficult to interpret and means that *qualitative* MRD analysis provides limited information. Therefore, several groups developed *quantitative* or *semi-quantitative* RT-PCR assays which enable the kinetics of residual disease to be monitored over time. In the early 1990s a number of groups developed *competitive* PCR strategies to measure MRD and quantify an endogenous control gene for each sample [32, 33]. The quantification of expression of a control gene made it possible to control for inter-sample PCR variability attributable to varying PCR dynamics and varying integrity and quantity of mRNA and cDNA.

Competitive PCR involves the co-amplification of an internal control, the competitor, at graded concentrations together with the target gene. The number of *BCR-ABL* transcripts in a given sample is estimated by determining the concentration of the internal control at which equivalence is obtained between the target and the competitor. The endogenous control is quantified likewise. The *BCR-ABL* copy number is normalised to the number of endogenous control gene transcripts and the data are expressed as a percentage ratio. To test for the quality of a given sample when the *BCR-ABL* is undetectable the control gene is quantified. A minimum number of control gene transcripts below which a negative RT-PCR is unreliable needs to be determined. In our laboratory this lower limit is $1 \times 10^4$ copies of *ABL*.

Until recently, the complexity of competitive PCR analysis has limited its use to a few specialised centres, despite its proven clinical utility (see below). The introduction of real-time quantitative PCR (RQ-PCR) [34, 35] approximately 3 years ago made quantitation of MRD accessible to a larger number of centres. Significantly, in contrast to endpoint PCR technology, RQ-PCR permits an increased precision and accuracy by quantifying during the exponential phase of PCR amplification. It offers a dynamic detection range over five orders of magnitude. Moreover the real time detection of accumulating amplicons eliminates post PCR manipulation and so in theory minimises the risk of contamination. The accumulation of amplicons is detected by including a sequence-specific probe labelled with fluorochromes in addition to the primers in PCR (fig. 1). The two commonly used systems are Taq-Man and fluorescence resonance energy transfer (FRET). In the Taq-Man system the emission of the fluorescence arises as a result of hydrolysis of the probe by inherent 5′ to 3′ exonuclease activity of Taq polymerase, resulting in dissociation of the reporter dye from the quencher moiety as the primer is extended. In the FRET system, the amplicons are detected by emission of fluorescence as the two dyes are brought into close proximity by hybridisation to the newly synthesised DNA strand. In

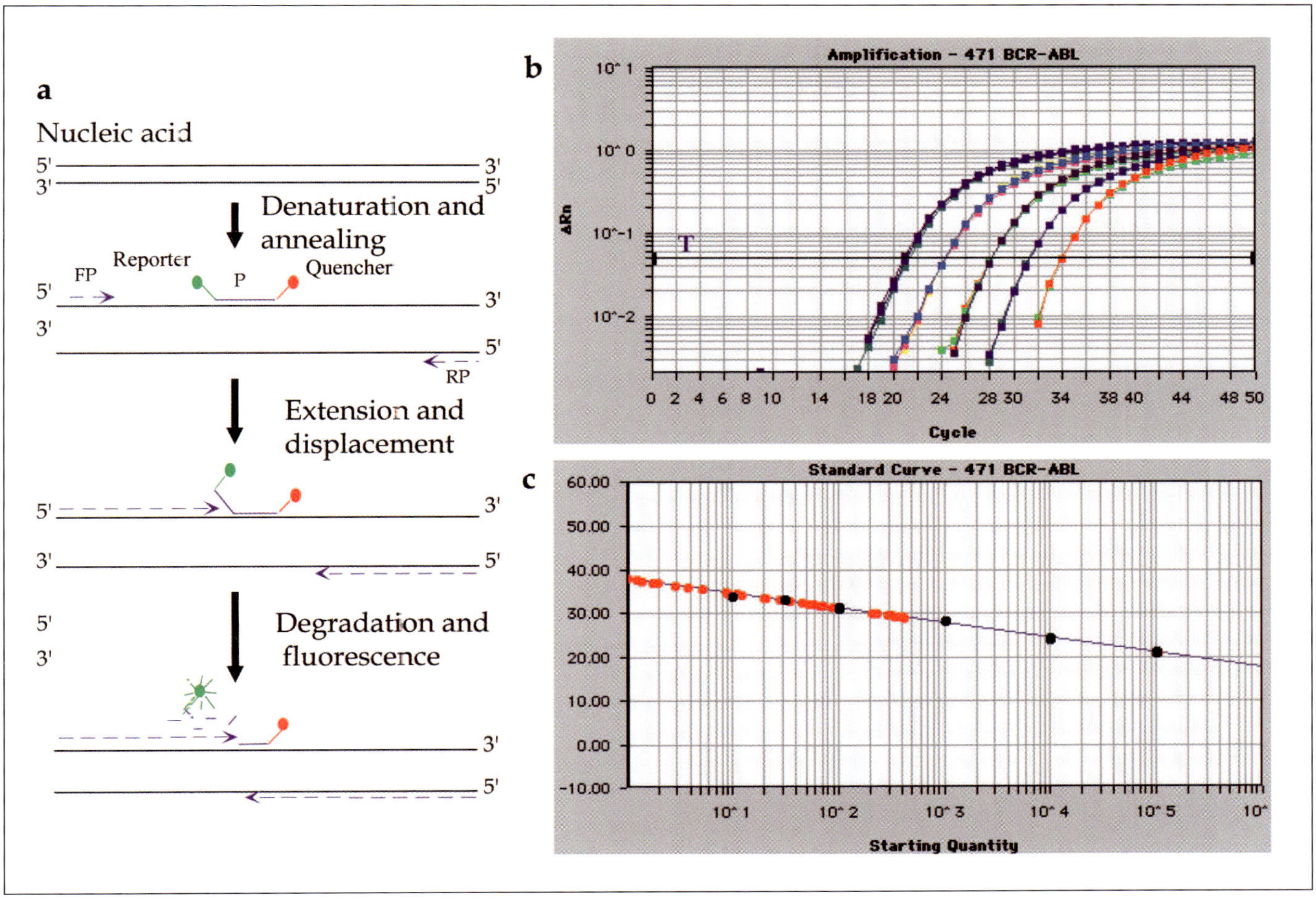


**Fig. 1.** Real-time PCR using the Taq-Man system. **a** Schematic representation of how the increase in amplicons leads to the increase in fluorescence in RQ-PCR. The close proximity of the 'quencher' moiety suppresses the fluorescence from the reporter moiety. On degradation of the annealed probe (P) by the 5′ to 3′ exonuclease activity of Taq DNA polymerase the reporter and the quencher dissociate and so allow the former to fluoresce. As the number of amplicons increases so does the number of probes which can anneal and be degraded with a resulting increase in fluorescence. **b** The real time PCR log plot of serially diluted BCR-ABL standards, in log increments. Each standard is analysed in triplicate. The greater the number of PCR cycles taken for the fluorescence to cross the threshold (T) the fewer the number of *BCR-ABL* copies in the original sample. **c** A standard curve generated by subjecting serial dilutions of target gene to real-time PCR (black symbols). The number of *BCR-ABL* transcripts in unknown samples (red symbols) can then be read off the standard curve. FP = Forward primer; RP = reverse primer.

each case the intensity of the fluorescence is proportional to the number of amplicons, and the efficiency with which they accumulate is related to the quantity of the target, i.e. *BCR-ABL*, in the starting sample. The latter relationship is only valid in the exponential phase of PCR. The *BCR-ABL* copy number in the sample is read off the standard curve which is generated for each assay. The standard curve can be generated using serial dilutions of plasmid-with *BCR-ABL* insertion or cDNA prepared from mRNA derived from a cell line, e.g. K562, as template. There is an ongoing debate as to the ideal starting material for the standard curve and indeed debate as to whether a standard curve is really necessary. In a clinical laboratory a standard curve is strongly recommended, and use of plasmids provides stability over time.

Opinion remains divided also as to whether a *BCR-ABL* transcript type-specific assay is preferable to a single Taq-Man assay to quantify e13a2 and e14a2 transcripts

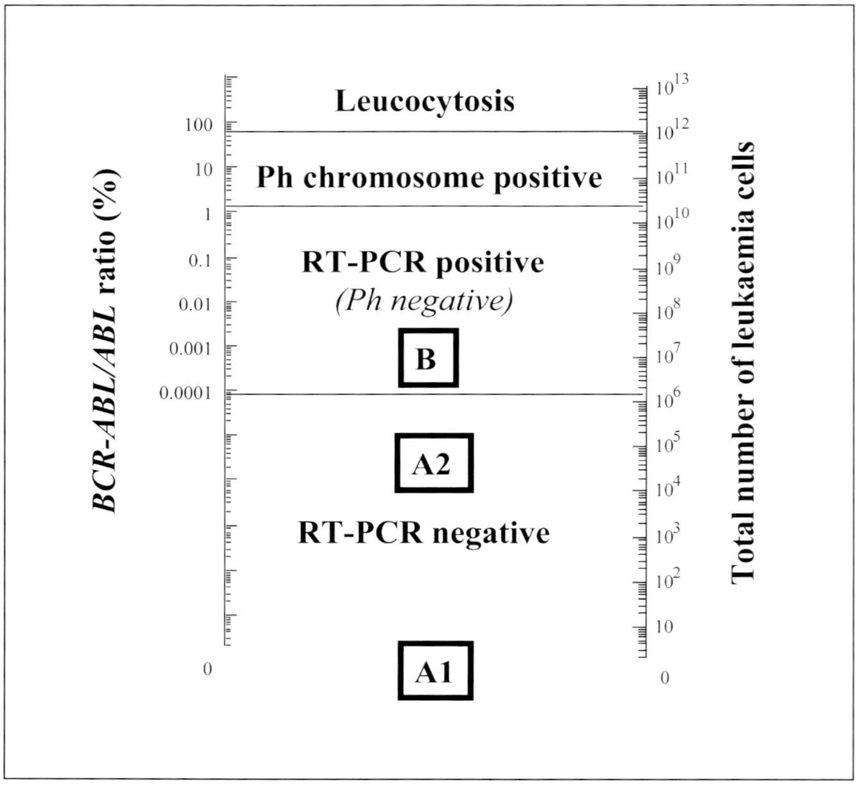


**Fig. 2.** A hypothetical relationship linking the *BCR-ABL/ABL* ratio and the estimated number of leukaemia cells in a patient's body with the haematological, cytogenetic and molecular manifestations of disease. The box A1 reflects theoretical eradication of all residual disease, while A2 represents residual disease below the level of detection by RT-PCR. B represents disease at a low level which may or may not proceed to satisfy criteria for molecular relapse (see text). (Figure modifed from original figure designed by Professor Nick Cross.)

[34, 35]. Whilst there is little or no difference in sensitivity, a single assay is cost effective and probably more accurate as it will quantify e14a2 and e13a2 in patients who express both transcripts. However, because the RQ-PCR assays are *BCR-ABL* transcript specific, i.e. e1a2, e13a2 and/or e14a2, it is essential to perform a multiplex PCR at presentation to assign the transcript type so as to use the appropriate RQ-PCR assay. An RQ-PCR assay designed for e13a2 and/or e14a2 transcripts would give rise to a false-negative result in patients expressing rare transcripts such as e13a3 and b14a3. Therefore a negative result should be interpreted with caution if the *BCR-ABL* transcript type is not documented prior to commencing treatment.

It is arguable whether RQ-PCR is less or more sensitive than nested PCR. Preliminary data support the view that RQ-PCR is at least as sensitive as nested PCR [unpubl. obs.]. Significantly, the simplicity of RQ-PCR analysis affords considerable advantages: it is rapid and cost effective and enhances MRD monitoring by minimising the difficulties associated with more frequent testing during treatment. RQ-PCR makes the inter-laboratory standardisation of protocols easier and so should improve quality control of monitoring MRD. Such advances should make interpretation of data from multicentre studies more feasible. As a consequence, statistically valid data measuring therapeutic effectiveness of novel treatments will be readily obtainable.

### *Clinical Results and Definition of Cure*

RT-PCR assays have been used extensively during the last 10 years to monitor the results of treating patients with CML by IFN-α and allogeneic SCT. This inevitably leads to the question as to whether PCR techniques help to define the patient who has been cured by allografting or possibly by administration of IFN-α. Most recently the question has arisen whether imatinib mesylate alone or in combination with other agents may cure some patients with CML. There are various possible definitions of 'cure' (fig. 2). The purist may require that every single copy of the *BCR-ABL* fusion gene and every *BCR-ABL* transcript should have been eliminated from a patient's body

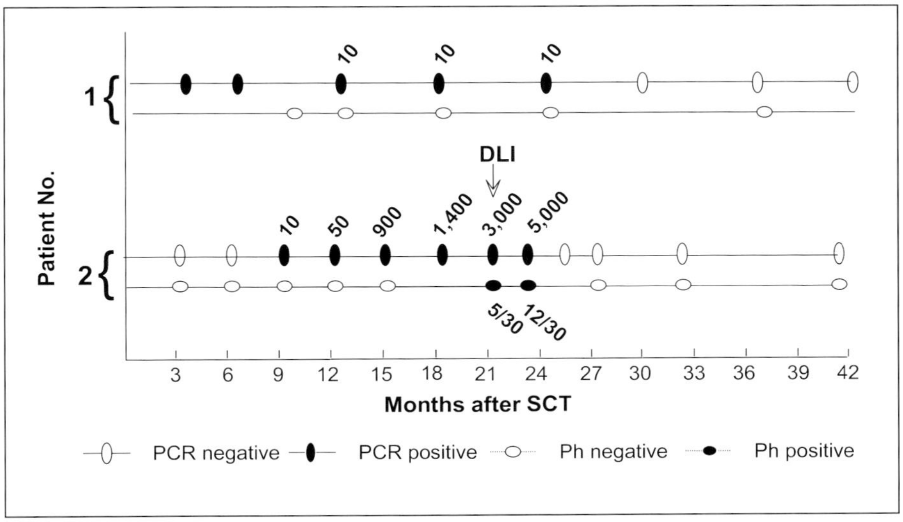


**Fig. 3.** The sequence of PCR and cytogenetic findings in 2 patients shown during the first 42 months after stem cell transplantation (SCT). Patient 1 was initially *BCR-ABL* positive by RT-PCR but became negative and remained so beyond 42 months (full data not shown). In patient 2 rising *BCR-ABL* transcript numbers preceded relapse defined at the cytogenetic level. The patient was treated with a single dose of donor lymphocyte infusion (DLI) and achieved cytogenetic and molecular remission. This patient has remained in molecular remission for 8 years following SCT (full data not shown). (Figure prepared by Professor Nick Cross and reproduced with his permission.)

(equivalent to A1 in fig. 2) but this is probably impossible to ascertain in practice. Thus the failure to detect *BCR-ABL* transcripts could mean either that all transcripts have indeed been eradicated in the patient's body (A1 in fig. 2) or merely that their number is below the limit of detection given the limitation of the sampling or the assay (A2 in fig. 2). It is also consistent with the notion that significant numbers of Ph-positive cells survive in the patient's body but are transcriptionally silent. There is even a fourth possibility – that transcripts are detectable at very low level but that their numbers do not increase significantly during longer periods of follow-up – in some cases even in the absence of further treatment (B in fig. 2). Although one might speculate that such patients could be at risk of relapse to chronic phase or even to blastic transformation at some time in the future, the risk could be exceedingly low and such patients might be regarded as 'operationally cured'. These issues will be addressed further in relation to the specific treatments considered below.

*Allogeneic SCT*

MRD studies assist very greatly in classifying patients in accordance with their risk of subsequent relapse after allogeneic SCT. For example, a given patient may have a low number of transcripts detectable in his or her blood for some months after transplant but can become durably PCR negative thereafter (fig. 3). In contrast, another patient who is initially PCR negative may show rising numbers of transcripts before chromosomal relapse. The observation of increasing transcript numbers may permit early intervention with appropriate treatment (fig. 3). Since the first report in 1990 there are many data supporting the notion that serial quantitative RT-PCR analyses of peripheral blood specimens following SCT can distinguish patients at high and low risk of relapse [28]. The scant contradictory data may simply reflect the difference in RT-PCR sensitivity, the heterogeneous SCT regimens and the phase of the disease at time of SCT. Patients who remain in remission after SCT have persistently undetectable, low or falling levels on sequential analysis. Patients in whom *BCR-ABL* is undetectable 6–12 months after SCT, remain so for up to 5 years and are at low risk of relapse [36, 37]. In such patients, analysis of a bone marrow aspirate is very rarely indicated. Conversely, patients who have occasional weakly positive PCR studies from 1 to 5 years after SCT are at relatively high risk of relapsing thereafter. Patients who have increasing or persistently high levels of *BCR-ABL* transcripts are very likely to progress within months to Ph chromosome positivity.

The early recognition of disease progression by RT-PCR provides a window of opportunity for therapeutic intervention while the tumour load is still low. Once a decision to intervene has been taken, there are various possibilities. For the patient who is still receiving some form of immune suppression for prevention of graft-versus-host disease, such as cyclosporin A or tacrolimus, discontinuing the medication may be followed by restoration of molecular remission [36]. Patients not receiving immunosuppressive drugs may be treated with IFN-α. Currently most groups favour the administration of lymphocytes collected from the original transplant donor [38]. Given in relatively large doses, such donor lymphocyte infusions (DLI) may induce graft-versus-host disease, but escalating dose DLI carries only a very small risk of inducing severe graft-versus-host disese and results in complete response rates of 70–80%. Molecular remissions induced in this way seem to be very durable with a median follow-up of 29 months [39–41]. MRD monitoring is particularly effective if RT-PCR analyses are performed frequently with the rapid turn round time which is now logistically possible with the introduction of RQ-PCR.

A Working Group convened by members of the International Bone Marrow Transplant Registry under the leadership of Sergio Giralt has recently proposed a set of recommendations for monitoring disease response and relapse after allogeneic SCT [42]. They recommend inter alia that RQ-PCR should be carried out on peripheral blood specimens every 3–6 months for the first 2 years after transplantation, every 6–12 months from years 2 to 5 and yearly thereafter. Patients with persistently positive PCR results or increasing levels of *BCR-ABL* transcripts should undergo bone marrow examination for cytogenetic analysis.

*Definition of Molecular Remission and Relapse*

Although the clinical utility of RT-PCR MRD studies is widely accepted, the precise definition of molecular relapse remains controversial, partly because a number of different factors can influence the actual results. These include the frequency of analysis, the choice of endogenous control gene, the technique of RT and the details of the PCR protocol. Some of these issues will undoubtedly be resolved with the inter-laboratory standardisation of PCR following the introduction of RQ-PCR. Interestingly, RQ-PCR has now shifted the focus from PCR to standardisation conditions for the reverse transcription step. This and the choice of technology used to isolate mRNA have a disproportionate effect on the level of sensitivity and therefore on the definition of molecular relapse. In the absence of consensus and with the knowledge that a low level of RT-PCR positivity does not always predict disease progression, we have established a set of criteria for molecular relapse. A *BCR-ABL/ABL* ratio of 0.02% in a patient who was previously negative or previously had a lower level of transcripts raises the possibility but does not yet satisfy criteria for molecular relapse [36]. In such cases the analysis is repeated twice within 2–3 months to establish the course of the disease. However, if the *BCR-ABL/ABL* ratio is ≥0.05% then one more independent confirmatory analysis is sufficient to satisfy criteria for molecular relapse. The close monitoring of such cases takes account of the inherent variability of the technique and minimises the risk of unnecessary clinical intervention in patients with oscillating levels of *BCR-ABL* transcripts.

There are no agreed criteria for molecular remission principally because of the variable levels of sensitivity among centres. Clearly, molecular remission implies RT-PCR negativity with an assay which achieves a sensitivity level of $10^{-5}$ to $10^{-6}$, i.e. nested RT-PCR or RQ-PCR. As with molecular relapse, data should be confirmed by an independent sample.

*IFN-α*

Virtually all patients receiving IFN-α treatment, including those in complete cytogenetic remission (CCR), are persistently RT-PCR positive [43]. These observations necessitate quantitative RT-PCR to monitor patient response to IFN-α therapy. Furthermore, the need for dividing cells and the fact that IFN-α treatment can lead to hypoplastic fibrotic marrows make close monitoring by cytogenetic analysis impractical. Hence, quantitative RT-PCR has been invaluable in assessing a patient's response to therapy. Investigators have shown that MRD levels can vary over as much as 10,000-fold among patients in CCR [43]. Significantly, the level of *BCR-ABL* mRNA among patients in CCR is related to the probability of relapse. Therefore, for the 6–20% of CML patients who achieve CCR, serial quantitative RT-PCR is the best way to identify individuals at risk of relapse [28]. Conversely it has been reported that those with *BCR-ABL/ABL* ratios of <0.045% have a considerably reduced risk of cytogenetic relapse [19, 43]. With time such observations may identify a group of patients for whom IFN-α therapy can be withdrawn without the risk of later relapse. Also, demonstration that patients achieving *BCR-ABL/ABL* ratios <2.0% are invariably in CCR means that routine bone marrow examinations for cytogenetic analysis become redundant, a great benefit for the patients [43]. However, caution is required in attempting to correlate *BCR-ABL* copy numbers or *BCR-ABL/ABL* ratios to the number of Ph-positive cells (despite the attempt in fig. 2 above). Since the number of steps involved in the derivation of *BCR-ABL/ABL* ratio or the *BCR-ABL* copy number are numerous there are many possible variables, e.g. the quality of the specimen, the quality and quantity of RNA, the efficiency of RT and PCR steps, including operator variation.

*Imatinib Mesylate*

RQ-PCR is now being used to monitor and evaluate the treatment of CML with imatinib mesylate in a clinical phase III trial [44]. This ongoing prospective randomised study, comparing imatinib mesylate to IFN-α plus cytarabine, addresses the question whether imatinib mesylate is equal to or more effective than the best available alternative drug combination. Interestingly, *BCR-ABL* transcripts were undetectable 6–9 months after starting treatment with imatinib mesylate therapy in a small number of patients [unpubl. obs.]. These findings await confirmation and it remains to be seen whether the observed RT-PCR negativity is sustainable. With time it may be possible to use the kinetics of *BCR-ABL* transcript numbers to distinguish between responders and non-responders early after starting treatment with imatinib mesylate with a view to increasing the dosage, adding other drugs or resorting instead to allogeneic SCT if necessary.

## Conclusion

The identification of a Ph chromosome remains a crucial step in confirming the diagnosis of CML but it should be backed today by demonstrating also the presence of a *BCR-ABL* fusion gene. The identification by FISH of an interstitial deletion in the derivative 9q+ chromosome promises to have important prognostic value and the finding after diagnosis of evidence of cytogenetic evolution, especially the acquisition of an iso(17q) chromosome, in a patient who is otherwise apparently in chronic phase may also have prognostic significance. However, the lower levels of residual disease that may be present in patients treated by allogeneic SCT or in patients who have responded well to IFN-α or imatinib mesylate must be monitored with more sensitive techniques, notably RQ-PCR.

One may speculate that since RT-PCR measures the changing level of *BCR-ABL* mRNA in response to therapy it may reflect the intrinsic kinetics of the disease and could thus turn out to be a good prognostic indicator. If, for example, the observed clinical heterogeneity of CML is multifactorial and due in part to molecular features intrinsic to the leukaemia and in part to the patient's response to that leukaemia, it is possible that the best single measure of 'aggressiveness' would be the kinetics of response to standard therapy, such as imatinib mesylate, as measured by RQ-PCR. Some evidence already suggests the rapidity of a patient's response to donor lymphocyte infusions correlates with prognosis [45] and the same may apply after treatment with other effective agents. The introduction of RQ-PCR makes it possible to monitor response to treatment with greater precision and this in turn could facilitate earlier recognition of relative resistance and thus the need to modify therapy.

As quantitative RT-PCR to monitor MRD is applied in an increasing number of centres as a consequence of the introduction of semi-automated RQ-PCR the standardisation of PCR protocols will be made easier. This will undoubtedly make a uniform definition of molecular remission and relapse easier. The proven clinical utility of quantitative RT-PCR to monitor MRD for CML over the past decade has led to an increased interest in the introduction of RQ-PCR analysis.

## References


1 Chase A, Grand F, Zhang JG, Blackett N, Goldman J, Gordon M: Factors influencing the false positive and negative rates of *BCR-ABL* fluorescence in situ hybridization. Genes Chromosomes Cancer 1997;4:246–253.

2 Dewald GW, Wyatt WA, Juneau AL, Carlson RO, Zinsmeister AR, Jalal SM, Spurbeck JL, Silver RT: Highly sensitive fluorescence in situ hybridization method to detect double *BCR-ABL* fusion and monitor response to therapy in chronic myeloid leukemia. Blood 1998;91:3357–3365.

3 Grand FH, Chase A, Iqbal S, Nguyen DX, Lewis JL, Marley SB, Davidson RJ, Goldman JM, Gordon MY: A two-color *BCR-ABL* probe that greatly reduces the false positive and false negative rates for fluorescence in situ hybridization in chronic myeloid leukemia. Genes Chromosomes Cancer 1998;23:109–115.

4 Sinclair PB, Nacheva EP, Leversha M et al: Large deletions at the t(9;22) breakpoint are common and may identify a poor-prognosis subgroup of patients with chronic myeloid leukemia. Blood 2000;95:738–743.

5 Huntly BJ, Reid AG, Bench AJ, Campbell LJ, Telford N, Shepherd P, Szer J, Prince HM, Turner P, Grace C, Nacheva EP, Green AR: Deletions of the derivative chromosome 9 occur at the time of the Ph translocation and provide a powerful and independent prognostic indicator in chronic myeloid leukemia. Blood 2001;98:1732–1738.

6 Kolomietz E, Al-Maghrabi J, Brennan S, Karaskova J, Minkin S, Lipton J, Squire JA: Primary chromosomal rearrangements of leukemia are frequently accompanied by extensive submicroscopic deletions and may lead to altered prognosis. Blood 2001;97:3581–3588.

7 Dewald GW, Wyatt WA, Silver RT: Atypical *BCR* and *ABL* D-FISH patterns in chronic myeloid leukemia and their possible role in therapy. Leuk Lymphoma 1999;34:481–491.

8 Tchirkov A, Giollant M, Tavernier F, Briancon G, Tournilhac O, Kwiatkowski F, Philippe P, Choufi B, Demeocq F, Travade P, Malet P: Interphase cytogenetic and competitive RT-PCR for residual disease monitoring in patients with chronic myeloid leukaemia during interferon-alpha therapy. Br J Haematol 1998;101:552–557.

9 Sinclair PB, Green AR, Grace C, Nacheva EP: Improved sensitivity of *BCR-ABL* detection: A triple-probe three-color fluorescence in situ hybridization system. Blood 1997;90:1395–1402.
10 Seong D, Thall P, Kantarjian HM, Talpaz M, Swantkowski J, Xu J, Shen Y, Glassman A, Ramagli L, Siciliano MJ: Philadelphia chromosome-positive myeloid cells in the peripheral blood of chronic myelogenous leukemia patients: Comparison with the frequency detected in cycling cells of the bone marrow. Clin Cancer Res 1998;4:861–867.
11 Sick C, Schultheis B, Pasternak G, Kottke I, Horner S, Heissig B, Hehlmann R: Predominantly *BCR-ABL* negative myeloid precursors in interferon-alpha treated chronic myeloid leukemia: A follow-up study of peripheral blood colony-forming cells with fluorescence in situ hybridization. Ann Hematol. 2001;80:9–16.
12 Itoh T, Tamura S, Takemoto Y, Kakishita E: A cytogenetic and fluorescence in situ hybridization evaluation of interferon-alpha in the treatment of chronic myeloid leukemia. Int J Mol Med 1999;4:659–663.
13 Garicochea B, Chase A, Lazaridou A, Goldman JM: T lymphocytes in chronic myelogenous leukaemia (CML): No evidence of the *BCR/ABL* fusion gene detected by fluorescence in situ hybridization in 14 patients. Leukemia 1994;8:1197–1201.
14 Dewald G, Stallard R, Alsaadi A, et al: A multicenter investigation with D-FISH *BCR/ABL*1 probes. Cancer Genet Cytogenet. 2000;116: 97–104.
15 Seong DC, Kantarjian HM, Ro JY, Talpaz M, Xu J, Robinson JR, Deisseroth AB, Champlin RE, Siciliano MJ: Hypermetaphase fluorescence in situ hybridization for quantitative monitoring of Philadelphia chromosome-positive cells in patients with chronic myelogenous leukemia during treatment. Blood 1995;86: 2343–2349.
16 Talpaz M, Kantarjian H, Kurzrock R, Trujillo JM, Gutterman JU: Interferon-alpha produces sustained cytogenetic responses in chronic myelogenous leukemia. Ann Intern Med 1991; 114:532–538.
17 Kantarjian HM, Smith TL, O'Brien SO, Beran M, Pierce S, Talpaz M: Prolonged survival in chronic myelogenous leukemia after cytogenetic response to interferon-alphatherapy. Ann Intern Med 1995;122:254–261.
18 Mahon F, Monstastruc M, Faberes C, Reiffers J: Predicting complete cytogenetic response in chronic myelogenous leukemia patients treated with recombinant interferon-α. Blood 1994;84: 3592–3594.
19 Bonifazi F, de Vivo A, Rosti G, Guilhot F, Guilhot J, Trabacchi E, Hehlmann R, Hochhaus A, Shepherd PCA., Steegmann JL, Kluin-Nelemans HC, Thaler J, Simonsson B, Louwagie A, Reiffers J, Mahon F-X, Montefusco E, Alimena G, Hasford J, Richards S, Saglio G, Testoni N, Martinelli G, Tura S, Baccarani M for the European Study of Interferon in Chronic Myeloid Leukemia: Chronic myeloid leukemia and interferon-α: A study of complete cytogenetic responders. Blood 2001;98:3074–3081.
20 Druker BJ, Talpaz M, Resta DJ, Peng B, Buchdunger E, Ford JM, Lydon NA, Kantarjian H, Capdeville R, Ohno-Jones S, Sawyers CL: Efficacy and safety of a specific inhibitor of the *BCR-ABL* tyrosine kinase in chronic myeloid leukemia. N Engl J Med 2001;344:1031–1037.
21 Groffen J, Stephenson JR, Heisterkamp N, de Klein A, Bartram CR, Growsveld G: Philadelphia chromosome breakpoints are clustered within a limited region, bcr, on chromosome 22. Cell 1984;36:93–99.
22 Reiter A, Skladny H, Hochhaus A, Seifarth, W, Heimpel H, Bartram CR, Cross NCP, Hehlmann R and the German CML Study Group. III. Molecular response of CML patients treated with interferon-α monitored by quantitative Southern blot analysis. Br J Haematol 1997;99:945–950.
23 Vershragen CF, Talpaz M., Hirsch-Ginsberg CF, Pherwani R, Rios MB, Stass SA, Kantarjian HM: Quantification of the breakpoint cluster region rearrangment for clinical monitoring in Philadelphia chromosome-positive chronic myeloid leukemia. Blood 1995;85: 2705–2710.
24 Stock W, Westbrook CA, Peterson B, Arthur DC, Szatrowski TP, Silver RT, Sher DA, Wu D, Le Beau MM, Schiffer CA, Bloomfield CD: Value of molecular monitoring during treatment of chronic myeloid leukemia: A cancer and leukemia group B study. J Clin Oncol 1997;15:26–36.
25 Guo JQ, Lian J, Glassman A: Comparison of *BCR-ABL* protein expression and Philadelphia chromosome in patients with chronic myeloid leukemia patients. Am J Clin Path 1996;106: 442–448.
26 Hochhaus A, Lin F, Reiter A, Skladny H, Mason P, van Rhee F, Shepherd PCA, Allan NC, Hehlmann R, Goldman JM, Cross NCP: Quantification of residual disease in chronic myelogenous leukemia on interferon-α therapy by competitive polymerase chain reaction. Blood 1996;87:1549–1555.
27 Melo JV: *BCR-ABL* gene variants. Baillière's Clin Haematol 1997;10:203–222.
28 Cross NCP: Assessing residual leukaemia. Baillière's Clin Haematol 1997;10:389–402.
29 Biernaux C, Loos M, Sels A, Stryckmans P: Detection of major *BCR-ABL* gene expression at a very low level in blood cells of healthy individuals. Blood 1995;88:3118–3122.
30 Bose S, Deininger M, Gora-Tybor J, Goldman JM, Melo JV: The presence of typical and atypical *BCR-ABL* fusion genes in leukocytes of normal individuals: Biological significance and implications for the assessment of minimal residual disease. Blood 1998;92:3362–3367.
31 Lin F, Goldman JM, Cross NCP: A comparison of the sensitivity of blood and bone marrow for the detection of minimal residual disease in chronic myeloid leukaemia. Br J Haematol 1994;86:683–685.
32 Cross NCP, Lin F, Chase A, Bungey J, Hughes TP, Goldman, JM: Competitive polymerase chain reaction to estimate the number of *BCR-ABL* transcripts in chronic myeloid leukaemia after bone marrow transplantation. Blood 1993;82:1929–1936.
33 Lion T, Henn T, Gaiger A, Kalhs P, Gadner H: Early detection of relapse after bone marrow transplantation in patients with chronic myelogenous leukaemia. Lancet 1993;341:275–276.
34 Menisk E, van de Locht A, Schattenberg A: Quantitation of minimal residual disease in Philadelphia chromosome positive chronic myeloid leukaemia patients using real-time quantitative RT-PCR. Br J Haematol 1998; 102:768–774.
35 Branford S, Hughes T, Rudzki Z: Monitoring chronic myeloid leukaemia therapy by real time quantitative PCR in blood is a reliable alternative to bone marrow cytogenetics. Br J Haematol 1999;107:587–599.
36 Olavarria E, Kanfer E, Szydlo R, Kaeda J, Rezvani K, Cwynarski K, Pocock C, Dazzi F, Craddock C, Apperley J, Cross NCP, Goldman JM: Early detection of *BCR-ABL* transcripts by quantitative reverse transcriptase-polymerase chain reaction predicts outcome after allogeneic stem cell transplantation for chronic myeloid leukemia. Blood 2001;97:1560–1565.
37 Mughal TI, Yong A, Syzdlo RM, Dazzi F, Olavarria E, van Rhee F, Kaeda J, Cross NCP, Craddock C, Kanfer E, Apperley J, Goldman J: Molecular studies in patients with chronic myeloid leukaemia in remission 5 years after allogeneic stem cell transplant define the risk of subsequent relapse. Br J Haematol 2001;115: 569–574.
38 Kolb H, Schattenberg A, Goldman JM, Hertenstein B, Jacobsen N, Arcese W, Ljungman P, Ferrant A, Verdonck L, Niederwieser D: Graft-versus-leukemia effect of donor lymphocyte transfusions in marrow grafted patients. Blood 1995;86:2041–2050.
39 Dazzi F, Szydlo RM, Craddock C, Cross NCP, Kaeda J, Chase A, Olavarria E, van Rhee F, Kanfer E, Apperley JF, Goldman, JM: Comparison of single-dose and escalating-dose regimens of donor lymphocyte infusion for relapse after allografting for chronic myleoid leukemia. Blood 2000;95:67–71.
40 Dazzi F, Szydlo RM, Cross NC, Craddock C, Kaeda J, Kanfer E, Cwynarski K, Olavarria E, Yong A, Apperley JF, Goldman JM: Durability of responses following donor lymphocyte infusions for patients who relapse after allogeneic stem cell transplantation for chronic myeloid leukemia. Blood 2000;96:2712–2716.
41 Craddock C, Szydlo RM, Klein JP, Dazzi F, Olavarria E, van Rhee F, Pocock C, Cwynarski K, Apperley JF, Goldman JM: Estimating leukemia-free survival after allografting for chronic myeloid leukemia: A new method that takes into account patients who relapse and are restored to complete remission. Blood 2000;96: 86–90.

42 Giralt SA, Goldman J, Anasetti C, Cervantes F, Champlin R, Cross N, Hochhaus A, Klein JP, Papadopoulos E, Sobocinski K, Weisdorf D, Horowitz M, on behalf of the Chronic Leukemia Working Committee of the IBMTR: Recommendations for assessment and definitions of response and relapse in CML: A report from the Chronic Leukemia Working Committee of the IBMTR; in Carella AM, Daley GQ, Eaves CJ, Goldman JM, Hehlmann R (eds): Chronic Myeloid Leukemia: Biology and Treatment. London, Dunitz, 2001, pp 357–368.

43 Hochhaus A, Lin F, Reiter A, Saussele S, Reichert A, Emig M, Kaeda J, Schultheis B, Berger U, Shepherd PC, Allan NC, Hehlmann R, Goldman JM, Cross NCP for the German CML Study Group and the UK MRC CML Study Group: Molecular heterogeneity in complete cytogenetic responders after interferon-α therapy for chronic myeloid leukemia: Low levels of minimal residual disease are associated with continuing remission. Blood 2000;95:62–66.

44 Goldman JM, Druker BJ: Chronic myeloid leukemia: Current treatment options. Blood 2001; 98:2039–2042.

45 Raanani P, Dazzi F, Sohal J, Szydlo R M, van Rhee F, Reiter A, Lin F, Goldman JM, Cross NCP: The rate and kinetics of molecular response to donor leucocyte transfusions in chronic myeloid leukaemia patients treated for relapse after allogeneic bone marrow transplantation. Br J Haematol 1997;97:86–93.

Prof. John M. Goldman
Department of Haematology
Imperial College at Hammersmith Hospital
Du Cane Road, London W12 0NN (UK)
E-Mail j.goldman@ic.ac.uk

Bain BJ (ed): Chronic Myeloproliferative Disorders.
Basel, Karger, 2003, pp 44–61

# Cytogenetic and Molecular Genetic Evolution of Philadelphia-Chromosome-Positive Chronic Myeloid Leukaemia

Bertil Johansson Thoas Fioretos Felix Mitelman

Department of Clinical Genetics, Lund University Hospital, Lund, Sweden

**Key Words**
Chronic myeloid leukaemia · Progression · Chromosomes · Genes

**Abstract**
Chronic myeloid leukaemia (CML) is genetically characterised by the presence of the reciprocal translocation t(9;22)(q34;q11), resulting in a *BCR/ABL* gene fusion on the derivative chromosome 22 called the Philadelphia (Ph) chromosome. In 2–10% of the cases, this chimeric gene is generated by variant rearrangements, involving 9q34, 22q11, and one or several other genomic regions. All chromosomes have been described as participating in these variants, but there is a marked breakpoint clustering to chromosome bands 1p36, 3p21, 5q13, 6p21, 9q22, 10q22, 11q13, 12p13, 17p13, 17q21, 17q25, 19q13, 21q22, 22q12, and 22q13. Despite their genetically complex nature, available data indicate that variant rearrangements do not confer any specific phenotypic or prognostic impact as compared to CML with a standard Ph chromosome. In most instances, the t(9;22), or a variant thereof, is the sole chromosomal anomaly during the chronic phase (CP) of the disease, whereas additional genetic changes are demonstrable in 60–80% of cases in blast crisis (BC). The secondary chromosomal aberrations are clearly non-random, with the most common chromosomal abnormalities being +8 (33% of cases with additional changes), + Ph (30%), i(17q) (20%), +19 (12%), –Y (8% of males), +21 (7%), and –7 (5%). We suggest that all these aberrations, occurring in ≥5% of CML with secondary changes, should be denoted major route abnormalities. Chromosome segments often involved in structural rearrangements include 1p36, 1q12–32, 3q21, 3q26, 7p15, 11q23, 12p13, 12q24, 13q14, 14q32, 17p11–13, 17q10–11, 21q22, and 22q10. No major differences as regards type and prevalence of additional aberrations seem to exist between CML with standard t(9;22) and CML with variants. The temporal order of the secondary changes varies, but the preferred pathway appears to start with i(17q), followed by +8 and +Ph, and then +19. Molecular genetic abnormalities preceding, or occurring during, BC include overexpression of the *BCR/ABL* transcript, up-regulation of the *EVI1* gene, increased telomerase activity, and mutations of the tumour suppressor genes *RB1, TP53*, and *CDKN2A*. The cytogenetic evolution patterns vary significantly in relation to treatment given during CP. For example, +8 is more common after busulphan than hydroxyurea therapy, and the secondary changes seen after interferon-α treatment or bone marrow transplantation are often unusual, seemingly random, and occasionally transient. Apart from the strong phenotypic impact of addition of acute myeloid leukaemia/myelodysplasia-associated translocations and inversions, such as inv(3)(q21q26), t(3;21)(q26;q22), and t(15;17)(q22;q12–21), in CML BC, only a few significant differences between myeloid and lymphoid BC are discerned, with +8, i(17q) and *TP53* mutations being more common in myeloid BC and monosomy 7,

This article is based on a review first published in *Acta Haematologica* 2002;107:76–94.

hypodiploidy, and *CDKN2A* deletions being more frequent in lymphoid BC. The prognostic significance of the secondary genetic changes is not uniform, although abnormalities involving chromosome 17, e.g., i(17q), have repeatedly been shown to be ominous. However, the clinical impact of additional cytogenetic and molecular genetic aberrations is most likely modified by the treatment modalities used.

## Introduction

Ever since 1960, when Nowell and Hungerford [1, 2] first described the Philadelphia (Ph) chromosome, later shown to be a deleted chromosome 22 [3] and then to be the derivative chromosome 22 of a t(9;22)(q34;q11) [4], in chronic myeloid leukaemia (CML), this translocation and its genetic, biologic and clinical implications have remained a paradigmatic example of the importance of cytogenetic changes in neoplastic disorders. The t(9;22) was the first specific, almost pathognomonic, chromosomal abnormality identified in neoplasia, it was the first aberration to be associated with a characteristic cytogenetic evolution pattern that correlated with the clinical behaviour of a disease, it was the first translocation to be cloned and characterised at the molecular level, and, very recently, it has become the first example of how our molecular understanding of the functional outcome of a chromosomal abnormality can result in a treatment that specifically targets the cells harbouring a genetic change [5–9]. Thus, the 'Philadelphia chromosome story' is truly the first success story of cytogenetic and molecular genetic studies of human neoplasia.

The aforementioned notwithstanding, numerous questions on the genetics of Ph-positive CML still remain to be answered. Why does the t(9;22) arise? Is it a totally stochastic event occurring in all cell types at all times, with only selection being the name of the game, or are the *BCR* and *ABL* genes particularly prone to fuse in certain cells at certain stages of differentiation? Is the formation of this chimeric gene facilitated by internal or external factors? How do the additional chromosomal and molecular genetic changes originate? Is the clonal evolution merely a Darwinian process of random mutations followed by a selection of the fitter clones or are the secondary aberrations specifically induced? Are all additional abnormalities pathogenetically important or may some be neutral and innocent bystanders? Do the types and frequencies of secondary anomalies differ between CML with the standard t(9;22) and CML with variant translocations? Does previous treatment influence the cytogenetic evolution pattern? Are myeloid and lymphoid blast crises (BCs), or other morphologic subtypes, of CML associated with specific additional changes? What are the molecular genetic consequences of the secondary chromosomal aberrations? Are there other, cytogenetically cryptic, molecular genetic abnormalities that correlate with disease progression? In this review, we will try to answer some of these questions, focussing on the cytogenetic and the molecular genetic evolution of CML.

## The Ph Chromosome and Cytogenetic Evolution in the Pre-Banding Era

Soon after the discovery of the Ph chromosome in the malignant cells from a total of seven CML patients in chronic phase (CP) or in terminal BC [1, 2], its presence in the great majority of cases was rapidly confirmed [10–13]. These early studies also provided evidence, albeit circumstantial, that the abnormality was present not only in granulocyte precursor cells but also in erythropoietic cells, macrophages, and megakaryocytes, suggesting that these cell types are derived from a common ancestral cell in which the Ph chromosome arises [11, 12, 14–18]. In line with this, the Ph chromosome was shown to be of clonal origin, based on X chromosome inactivation studies as well as on cytogenetic analyses of CML patients mosaic for constitutional sex chromosome aberrations [16, 19–21].

The karyotypic investigations also revealed that the Ph chromosome most often was the only identifiable chromosomal change during CP, whereas additional changes frequently accompanied, or even preceded, the acute transformation, in some instances disappearing after treatment, indicating a close relationship between karyotypic evolution and disease progression [22–32]. Exceptions were noted, however. In several patients, loss of the Y chromosome and gain of a C group chromosome, for example, could be seen in CP without heralding BC, and it was suggested that initial aneuploidy does not necessarily represent an emerging transformation but that abnormalities developing at a later stage do [33]. Although no detailed cytogenetic characterisation could be performed in the pre-banding era, the secondary aberrations seemed, at least to some investigators, non-random. Among the anomalies that could be identified with some certainty were duplication of the Ph chromosome (+Ph), gains of C and F group chromosomes, loss of an E chromosome, and the presence of a marker chromosome the size of a C chromosome thought to represent an isochromosome – abnormalities that we today would recognise as +der(22)t(9;22),

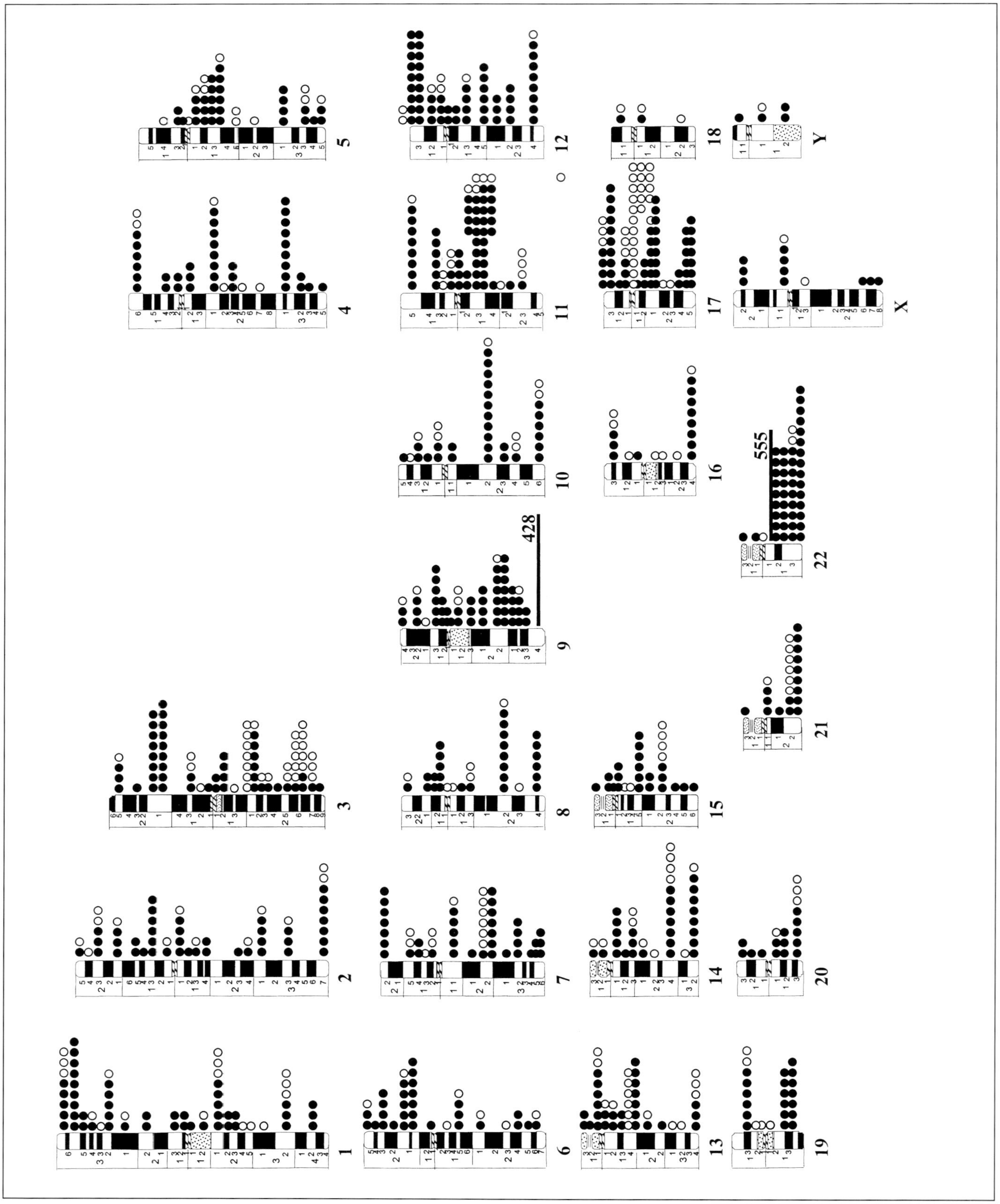
1
2
3
4
5
6
7
8
9
428
10
11
12
13
14
15
16
17
18
19
20
21
22
555
X
Y

+8, +19, monosomy 17, and i(17q), respectively. The secondary chromosomal changes observed in CML during disease progression thus agreed well with the stemline concept proposed by Winge in 1930 and then further elaborated by, e.g., Hauschka and Levan, in the 1950s [reviewed in ref. 34, 35].

However, despite all the above-mentioned achievements during the 1960s, it soon became apparent that renewed progress would have to await the development of new methods of chromosome study.

### Chromosome Banding Revealed Variant Translocations in Chronic Myeloid Leukaemia

In the early 1970s, when chromosome-banding techniques were introduced and the Ph chromosome identified, and confirmed, as being the result of a t(9;22)(q34;q11) [4, 36–39], it was realised that some Ph chromosomes were not generated by this abnormality. Instead, either simple aberrations involving chromosome band 22q11 or complex changes affecting 9q34 and 22q11, and a third, or even a fourth and a fifth, chromosome were found [40–47]. It was initially debated whether these variant rearrangements, found in 2–10% of larger series of Ph-positive CML [5, 48–61], had any specific phenotypic impact. Potter et al. [55] reported that the CP duration was significantly shorter for patients with CML with non-standard Ph chromosomes than for those with t(9;22), whereas Sandberg [62] in a review of his own and published variants concluded that the survival of CML cases with variant rearrangements was not significantly different from those with standard Ph translocation. The latter view quickly gained the upper hand, and it is generally accepted today that the clinical, prognostic, and haematological features of CML with variant translocations are not distinct from those with the typical t(9;22) [58, 63–65]. It should be stressed, however, that the reported studies on the phenotypic impact of variants were either based on very few patients or on literature reviews. Having tried in vain to find any larger, well-conducted clinical studies on variant Ph chromosomes one cannot help but conclude that the 'ruling' that they are of no prognostic significance was made on rather flimsy grounds.

A priori, one would perhaps have expected variant translocations to be associated with particular phenotypic features considering the fact that more genetic changes are required in order to generate these abnormalities compared to the standard t(9;22). That more than two breaks are involved holds true both for simple and complex variant rearrangements because the former aberrations are also now known to involve both 9q34 and 22q11, i.e., all simple variants are undetectable complex ones involving also 9q34 [66–68]. Thus, the origin of variant Ph chromosomes may be considered to be part of a cytogenetic evolution. In fact, coexistence of standard and variant Ph translocations in the same patient has been reported in several instances [68, 69], and many variants characterised at the molecular level have been shown to be the result of a t(9;22) undergoing additional genetic changes, i.e., arising from two consecutive translocations [70, 71]. However, it has also been proposed, and demonstrated, that complex translocations may be caused by a single event [65, 72, 73]. Hence, no single mechanism is involved – some variants are formed by multiple simultaneous breaks, some arise as a result of two, or even more, genetic events in close succession [74].

If the additional breakpoints that occur in variants provide no selective edge, one would have thought that all chromosome regions would be equally involved. This, however, is not the case. To date, more than 600 CML cases with variant rearrangements have been reported [47]. Although all chromosomes, including the Y chromosome that for many years was the only exception, have been described as participating in variants, the distribution of the breakpoints clearly exhibits a non-random pattern, with marked clustering to some chromosome bands, such as 1p36, 3p21, 5q13, 6p21, 9q22, 10q22, 11q13, 12p13, 17p13, 17q21, 17q25, 19q13, 21q22, 22q12, and 22q13 (fig. 1), suggesting that these regions may be particularly prone to breakage. In addition, some specific variants are also known to be recurrent, with the two translocations t(3;9;22)(p21;q34;q11) and t(17;22)(q25;q11) having both been reported in more than 10 cases [47].

Considering the fact that the genetic changes present in addition to breaks in 9q34 and 22q11 in variants often are secondary events, it would seem reasonable to expect that the additional aberrations occurring during the clonal evolution would be different from the ones observed in standard Ph-positive CML. This issue will be reviewed below on the basis of literature data presently available.

**Fig. 1.** Breakpoint map of structural abnormalities in CML with variant rearrangements. Closed circles represent breakpoints involved in variant rearrangements, open circles breakpoints involved in secondary aberrations. The thick lines and numbers indicate breakpoints in 9q34 and 22q11. The map is based on all 604 CML cases with variant rearrangements, 182 of which with secondary changes, reported in the literature [47].

**Table 1.** Secondary aberrations in CML with t(9;22) or variant Ph rearrangements[a]

| Aberration | t(9;22) n = 1,674 | Variant Ph n = 182 | Aberration | t(9;22) n = 1,674 | Variant Ph n = 182 |
|---|---|---|---|---|---|
| +8 | 574 (34.3)** | 42 (23.1)** | +1 | 26 (1.6) | 1 (0.5) |
| +Ph[b] | 515 (30.8)*** | 34 (18.7)*** | −5 | 25 (1.5) | 2 (1.1) |
| i(17q)[c] | 340 (20.3)* | 23 (12.6)* | 7p−[e] | 24 (1.4) | 1 (0.5) |
| +19 | 219 (13.1)* | 12 (6.6)* | −13 | 24 (1.4) | 1 (0.5) |
| −Y | 85 (8.6)[d] | 6 (5.2)[d] | 13q−[e] | 24 (1.4)*** | 15 (8.2)*** |
| +21 | 114 (6.8) | 8 (4.4) | inv(3)(q21q26)[f] | 23 (1.4) | 4 (2.2) |
| −7 | 82 (4.9) | 5 (2.7) | 5q−[e] | 23 (1.4) | 3 (1.6) |
| +17 | 78 (4.7) | 3 (1.6) | 6q−[e] | 23 (1.4) | 4 (2.2) |
| −17 | 65 (3.9)* | 1 (0.5)* | −8 | 23 (1.4) | 1 (0.5) |
| +10 | 58 (3.5) | 4 (2.2) | −15 | 23 (1.4) | 4 (2.2) |
| +6 | 57 (3.4) | 2 (1.1) | 1p−[e] | 22 (1.3) | 3 (1.6) |
| +14 | 52 (3.1)* | 0* | +3 | 22 (1.3) | 1 (0.5) |
| +12 | 46 (2.7) | 3 (1.6) | +5 | 22 (1.3) | 1 (0.5) |
| −18 | 42 (2.5) | 6 (3.3) | 3q−[e] | 21 (1.3) | 4 (2.2) |
| 17p−[e] | 40 (2.4) | 3 (1.6) | −4 | 21 (1.3) | 3 (1.6) |
| +Y | 22 (2.2)[d] | 3 (2.6)[d] | −19 | 20 (1.2) | 2 (1.1) |
| +20 | 37 (2.2) | 1 (0.5) | −22 | 20 (1.2) | 3 (1.6) |
| +11 | 36 (2.2) | 3 (1.6) | −X | 19 (1.1) | 2 (1.1) |
| +22 | 36 (2.2) | 5 (2.7) | 2q−[e] | 19 (1.1) | 2 (1.1) |
| −9 | 34 (2.0) | 4 (2.2) | +7 | 19 (1.1) | 1 (0.5) |
| +15 | 33 (2.0) | 2 (1.1) | −3 | 18 (1.1) | 1 (0.5) |
| +9 | 32 (1.9) | 0 | −11 | 18 (1.1) | 0 |
| +18 | 32 (1.9) | 1 (0.5) | −12 | 18 (1.1) | 3 (1.6) |
| +X | 31 (1.9) | 3 (1.6) | +16 | 18 (1.1) | 0 |
| −16 | 30 (1.8) | 1 (0.5) | −21 | 17 (1.0) | 0 |
| −20 | 30 (1.8) | 3 (1.6) | +4 | 16 (1.0) | 1 (0.5) |
| +13 | 29 (1.7) | 4 (2.2) | −14 | 13 (0.8) | 2 (1.1) |
| 1q−[e] | 28 (1.7) | 3 (1.6) | 2p−[e] | 7 (0.4) | 2 (1.1) |
| 7q−[e] | 28 (1.7) | 7 (3.8) | 12p−[e] | 6 (0.4) | 2 (1.1) |
| t(3;21)(q26;q22) | 27 (1.6) | 2 (1.1) | 12q−[e] | 6 (0.4) | 2 (1.1) |

Figures in parentheses are percentages. Significant frequency differences ($\chi^2$ with Yates' correction): * $p < 0.05$, ** $p < 0.01$, *** $p < 0.001$.

[a] Only aberrations reported in more than 1% of the t(9;22)- or variant-positive CML are included. The table is based on all reported cases in the literature [47].

[b] Includes also ider(22)(q10)t(9;22)(q34;q11).

[c] Includes also idic(17)(p11).

[d] Percentage of males only.

[e] Includes deletions as well as abnormalities denoted as 'additions' involving this chromosome arm.

[f] Includes also t(3;3)(q21;q26).

### Cytogenetic Evolution in the Banding Era

Based on a series of 10 patients as well as on a literature review of 57 cases with additional aberrations, all investigated using banding techniques, Mitelman et al. [75] identified +Ph, +8, and i(17q) as the most common secondary changes, occurring in close to 90% of the cases with additional abnormalities. These three abnormalities were classified as following the 'major route' of clonal evolution, whereas other changes evolving more rarely were classified as following the 'minor route' [75]. It should thus be stressed that the terms 'major' and 'minor' refer to frequencies and not to the aberrations' presumed pathogenetic impact. That the three major route changes, followed by +19, were the most common ones was corroborated by several studies during the late 1970s and early

**Table 2.** Secondary abnormality patterns in CML with t(9;22) or variant Ph rearrangements[a]

| Abnormality pattern | t(9;22) n = 1,674 | Variant Ph n = 182 |
|---|---|---|
| +8, +Ph[b] | 250 (14.9)** | 13 (7.1)** |
| +8, i(17q)[c] | 168 (10.0) | 11 (6.0) |
| +8, +19 | 148 (8.8)** | 4 (2.2)** |
| +19, +Ph[b] | 139 (8.3)* | 6 (3.3)* |
| i(17q)[c], +Ph[b] | 68 (4.1) | 7 (3.8) |
| i(17q)[c], +19 | 39 (2.3) | 2 (1.1) |
| +8, +19, +Ph[b] | 94 (5.6)** | 1 (0.5)** |
| +8, i(17q)[c], +Ph[b] | 56 (3.3) | 5 (2.7) |
| +8, i(17q)[c], +19 | 37 (2.2) | 1 (0.5) |
| i(17q)[c], +19, +Ph[b] | 20 (1.2) | 1 (0.5) |
| +8, i(17q)[c], +19, +Ph[b] | 19 (1.1) | 0 |
| ≥81 chromosomes[d] | 2 (0.1) | 1 (0.5) |
| 58–80 chromosomes[d] | 21 (1.3) | 2 (1.1) |
| 51–57 chromosomes[d] | 113 (6.8) | 5 (2.7) |
| 47–50 chromosomes[d] | 819 (48.9) | 72 (39.6) |
| 46 chromosomes[d] | 736 (44.0) | 76 (41.8) |
| 41–45 chromosomes[d] | 235 (14.0) | 22 (12.1) |
| ≤40 chromosomes[d] | 12 (0.7) | 0 |

Figures in parentheses are percentages. Significant frequency differences ($\chi^2$ with Yates' correction): * $p < 0.05$, ** $p < 0.01$.

[a] Based on all reported cases in the literature [47].

[b] Includes also ider(22)(q10)t(9;22)(q34;q11).

[c] Includes also idic(17)(p11).

[d] Includes all cases with at least one subclone with that chromosome number. A case with subclones of different modal numbers may hence be included in more than one chromosome number group.

1980s [5, 51–53, 57, 76–82]. Among the close to 1,700 CML cases with standard Ph and secondary changes published to date [47], +8, +Ph, i(17q), and +19 have been described in 34, 31, 20 and 13% of the cases, respectively (table 1). All other additional chromosomal abnormalities occur in less than 10% of the cases, the most frequent being –Y, +21, –7, +17, and –17. These aberrations are also the most common ones in the roughly 180 CML cases with variant translocations and additional abnormalities reported in the literature [47] (table 1). Apart from slightly lower prevalences of +8, +Ph, i(17q), +19, –17, and +14, and a higher frequency of 13q– in CML with variant translocations, no apparent differences in secondary changes are discerned between CML cases with standard Ph and CML cases with variants (table 1). Combining these two CML groups, the most common additional chromosomal changes are +8 (33%), +Ph (30%), i(17q) (20%), +19 (12%), –Y (8% of males), +21 (7%), and –7 (5%). In view of our present knowledge, it may be reasonable to expand the major evolutionary route to include all these aberrations, using 5% as a reasonable cut-off value.

That there may be no significant differences in the pattern and distribution of various types of clonal change between the standard and the variant groups has been noted in one previous study [58]. Also the chromosome bands frequently involved in secondary structural changes seem similar between the two groups, albeit few such changes have been reported in cases with variant translocations (fig. 1, 2). Looking only at the breakpoint distribution in CML with standard Ph, it is apparent that some chromosome segments are particularly 'breakprone', i.e., 1p36, 1q12–32, 3q21, 3q26, 7p15, 11q23, 12p13, 12q24, 13q14, 14q32, 17p11–13, 17q10–11, 21q22, and 22q10 (fig. 2).

The secondary aberrations are found in various combinations (table 2), and attempts have been made to infer the order of appearance of the changes based on the combinations found. For example, it has been suggested that i(17q) followed by +8 are rather early changes, whereas trisomy 19 is a relatively late one [8, 78]. Although this temporal order of secondary changes appears to be common, exceptions clearly exist. There are several examples of Ph-positive CML cases with +8 in all metaphases but i(17q) only in subclones, and trisomy 19 may well occur, albeit rarely, before +8 and i(17q) [47]. Also, some combinations are more frequent than others. Whereas +8, +Ph, and i(17q) often occur together, with slightly lower incidences in the variant group (table 2), the combination i(17q),+19 (without +8/+Ph) has only been reported in two cases, +8,i(17q),+19 (without +Ph) in 16 cases, and i(17q),+19,+Ph (without +8) in two cases [47]. Hashimoto et al. [83], using hierarchical log-linear models, showed that some combinations (+8,i(17q); +8,+19; +19,+Ph) are positively associated, while i(17q),+19 and i(17q),+Ph are negatively associated. Thus, the selective advantage of a secondary change seems to vary, depending on other aberrations present in the combination. Furthermore, available data indicate that the most common secondary abnormalities occur in a step-wise, well-ordered manner, with a putative frequent pathway starting with i(17q), followed by +8 and +Ph, and then +19.

Because three of the four most frequent additional changes involve gains, i.e., +8, +19, and +Ph (table 1), it is not surprising that the modal number is most often greater than 46 in CML with secondary aberrations [5, 81]; close to 50% have 47–50 chromosomes (table 2). Pseudodiploidy is seen in approximately 40% of the cases, whereas other modal numbers are quite rare (table 2). It is

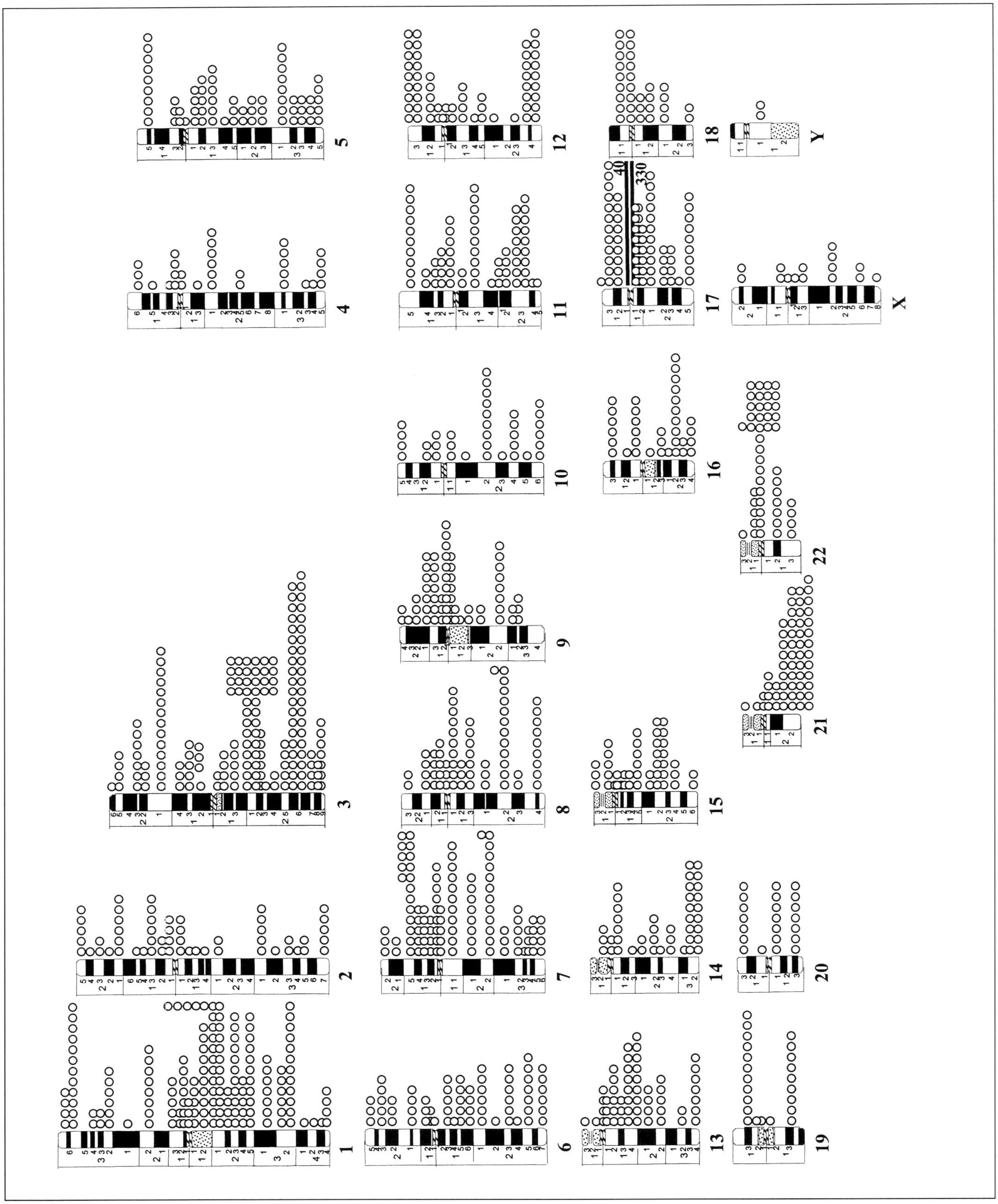
1
2
3
4
5
6
7
8
9
10
11
12
13
14
15
16
17
18
19
20
21
22
X
Y

noteworthy that the distributions of ploidy levels do not differ significantly between CML cases with standard Ph and variant translocations, again strongly indicating that there are no major differences between these two groups as regards cytogenetic evolution patterns.

The vast majority of the secondary changes in CML are genomically unbalanced, i.e., trisomies, monosomies, and deletions (table 1). In fact, this is a common theme in secondary cytogenetic changes in both haematological malignancies and solid tumours, and it has been proposed that primary changes are balanced, secondary unbalanced [84]. However, there are some notable exceptions in CML. In 1983, Berger et al. [85] described the presence of t(15;17)(q22;q12–21), almost pathognomonic for acute promyelocytic leukaemia, in addition to t(9;22) in a case of CML with promyelocytic BC, and since then, close to ten such cases have been reported [47]. Other balanced rearrangements characteristically found in acute myeloid leukaemia (AML) or myelodysplastic syndromes (MDS) but occasionally also found in CML BC include inv(3)(q21q26)/t(3;3)(q21;q26), t(3;21)(q26;q22), t(7;11)(p15;p15), t(8;21)(q22;q22), and inv(16)(p13q22) [47, 86–90]. Although secondary by definition, these additional changes do not behave as ordinary secondary changes. First, they are strongly related to quite specific phenotypic features, which other secondary aberrations are not, as will be discussed below. Second, generally they do not occur together with other common additional changes, such as +8, i(17q), and +Ph, strongly indicating that they follow a distinct evolutionary pathway. Instead of being secondary changes, they may rather be seen as second primary changes. Quite surprisingly, and noteworthy, is the fact that balanced aberrations characteristic of acute lymphoblastic leukaemia (ALL), such as t(4;11)(q21;q23) and t(8;14)(q24;q32), have, as yet, never been reported in CML BC [47]. It is tempting to speculate that ALL-associated translocations cannot originate during the particular differentiation stage that characterises the Ph-positive stem cell population. Alternatively, and perhaps equally likely, ALL-associated translocations may not provide any selective advantage for cells expressing the BCR/ABL chimeric protein

**Fig. 2.** Breakpoint map of secondary structural rearrangements in CML cases with t(9;22)(q34;q11). The thick lines and numbers indicate breakpoints in 17p11 and 17q10. The map is based on all 1,674 t(9;22)-positive CML cases with additional abnormalities reported in the literature [47].

Our corresponding knowledge of additional molecular genetic aberrations occurring in CML during progression is presently not on par with the wealth of information on the cytogenetic evolution patterns. However, a substantial amount of data has been generated during the last decade, revealing quite a complex array of molecular genetic abnormalities associated with CML transformation.

## Molecular Genetic Changes Involved in Chronic Myeloid Leukaemia Progression

In principle, secondary molecular genetic abnormalities in CML may be dichotomised into those associated with identifiable chromosomal changes, i.e., being the molecular genetic consequences of the abnormalities listed in table 1, and those that are cytogenetically cryptic.

Next to nothing is known about the molecular genetic correlates of the unbalanced secondary chromosomal changes. Needless to say, trisomies are not easily reduced to particular molecular genetic events [91]. Obviously, a gain of a chromosome results in duplication of the genes located on that chromosome, but few would argue that gain of all genes on a chromosome is the molecularly pathogenetically important outcome of a trisomy. However, it seems equally simplistic to argue that duplication of one specific gene is the functionally essential consequence of a trisomy. This notwithstanding, attempts have been made to reduce the effect of trisomy 8 into gain, and overexpression, of the *MYC* gene located at 8q24. Although it has been shown that this gene occasionally is amplified and up-regulated in CML during BC, and that enhanced expression of *MYC* may be associated with trisomy 8, there is no clear-cut correlation between trisomy 8 and *MYC* amplification/overexpression [92–95]. Furthermore, a recent microarray analysis of AML with trisomy 8 as the sole chromosomal change revealed a complex expression pattern, with not only up-regulation of several genes on chromosome 8, as well as on other chromosomes, but also down-regulation of *MYC* [96]. Hence, it seems reasonable to conclude that trisomy 8 in CML is reducible neither to overexpression of all genes on chromosome 8 nor to overexpression of only one gene. Hopefully, microarray analyses in CML [97] will provide important information on this issue, although interpretation of the complex set of data generated by such investigations will be an arduous and daunting task.

Also the pathogenetically essential molecular genetic consequences of monosomies and deletions remain elusive. Although loss of genetic material, as detected by cytogenetic analyses or loss of heterozygosity (LOH) studies, often is claimed to represent deletion of a tumour suppressor gene (TSG), few TSGs have been shown to be inactivated in CML. The frequent loss of 17p, due to i(17q), –17, or other changes resulting in 17p– such as the recently described der(17;18)(q10;q10) [98] (table 1), has made the *TP53* gene an obvious candidate. *TP53* mutations are found in approximately 10% (0–30%) of CML BC, and it has hence been suggested that abrogation of TP53 function plays an important role in CML progression [99–108]. However, it should be noted that a close correlation between loss of 17p and *TP53* mutations has been questioned [109–111]. For example, Fioretos et al. [111] detected no coding *TP53* mutations in CML cases with i(17q). Their detailed fluorescence in situ hybridisation (FISH) analysis also revealed that the majority of the i(17q) chromosomes were dicentric, with breakpoint clustering in 17p11; hence, the i(17q) should in fact be designated idic(17)(p11). In conclusion, *TP53* mutations are undoubtedly involved in the progression of some CML cases but other, as yet unidentified, genes on 17p are most likely of greater pathogenetic importance in i(17q)-positive CML.

The only other TSGs that have been investigated in CML in some detail are the *CDKN2A* ('p16') and *RB1* genes, located at 9p21 and 13q14, respectively. Sill et al. [112] reported homozygous deletions of *CDKN2A* in a large proportion (5/10) of CML cases with lymphoid, but not myeloid, BC. Frequent homozygous loss (20–40%) of this gene in lymphoid BC has since been confirmed [113, 114], but otherwise involvement of *CDKN2A* is quite rare in CML [115, 116]. As regards *RB1* abnormalities, the available data are somewhat conflicting. On the one hand, Towatari et al. [117] reported that the RB1 protein is absent in cases with megakaryoblastic BC but not in CML with other BC phenotypes. On the other hand, Gaidano et al. [118] did not detect any *RB1* mutations in a series of 26 cases of CML BC, including one with a megakaryoblastic crisis. Other studies indicate that aberrations of *RB1* are very rare during CP, but that losses or mutations of this gene may be found in up to 20% of cases in BC [119–121]. Taken together, functional abrogation of CDKN2A and RB1 may play a role in the development of CML BC in some patients. It should be stressed that cytogenetically identifiable losses of 9p and 13q are quite rare in CML [47], indicating that most *CDKN2A* and *RB1* deletions are submicroscopic. Recently, mutations and decreased activity of the TSG *IKAROS*, encoding a family of zinc-finger DNA-binding proteins, and mutations of *TP63* ('p51/p63'), a member of the *TP53* gene family, were found to be associated with disease progression [108, 122], but it remains to be seen whether these associations can be verified in larger patient series.

Attempts have been made to identify frequently deleted regions in CML BC using LOH, or rather allelic imbalance, studies. Such analyses have revealed frequent allelic loss at 1p (particularly 1p36), 7p, 19p, and 20q [123, 124], but so far no TSGs at these regions have been identified. Without being too pessimistic about the prospect of identifying a TSG through LOH analyses, it should be noted that the success rate in cloning TSGs, in non-inherited malignant disorders, based on LOH mapping is quite close to zero [125]. It has also been reported that microsatellite instability (MSI) is associated with CML evolution [126, 127], although this has not been confirmed by other groups [128, 129]. Thus, LOH and MSI studies have not, as yet, proved particularly fruitful in the elucidation of molecular genetic events occurring during CML progression.

Signalling via RAS may be a central pathway in CML pathogenesis [130], but mutations involving the *RAS* gene family are very rare in CML [103, 106, 131–134], although a possible association between the presence of *RAS* mutations and development of granulocytic sarcomas during BC has been suggested [135]. Recently, another 'dominantly acting' gene has received much attention in hematologic malignancies, namely *CBFA2 (AML1)*. This gene is known to be mutated in some familial as well as in some sporadic AML cases, but mutations of this transcription factor appear to be infrequent in CML [136].

Apart from the detailed mutation analyses of particular genes in CML BC referred to above, much information of biologic and clinical importance has been obtained from investigations focussing on identification, and expression analyses, of fusion genes. Soon after the discovery of the *BCR/ABL* fusion, an increased expression of the chimeric *BCR/ABL* transcript during, or even preceding, BC was noted. However, there does not seem to be a clear-cut association between the number of Ph chromosomes and the *BCR/ABL* expression levels [137–139]. Thus, the pathogenetically important outcome of +Ph in CML BC may not necessarily be amplification with subsequent overexpression of the *BCR/ABL* fusion gene. Perhaps the genomic imbalance as such is the important outcome [140]? As regards other fusion genes found in addition to the *BCR/ABL* in CML BC, most studies have focussed on the *CBFA2/EVI1* fusion generated by the t(3;21)(q26;q22), found in approximately 2% of cases with myeloid BC (table 1) [141, 142]. Occasionally, the t(3;21) results in *CBFA2/EAP* and *CBFA2/MDS1* chimeras, but most often the *EVI1* gene is involved. This gene is also up-regulated by inv(3)(q21q26) and by other abnormalities involving 3q26 [143, 144]. However, it should be stressed that overexpression of this gene occurs in most CML BC cases, i.e., also in cases without cytogenetically detectable involvement of 3q26 [145]. To date, only four other AML-associated fusion genes have been described in CML, namely *CBFA2/ETO* as a result of t(8;21)(q22;q22), *CBFB/MYH11* generated by inv(16)(p13q22), and *NUP98/HOXA9* and *NUP98/HOXA11*, both corresponding to t(7;11)(p15;p15) [47, 146–150]. In addition, single cases with rearrangements of the *ETV6*, *MLL*, and *NUP98* genes have been reported [151–155]. Although the fusion genes most likely are of great pathogenetic importance, only a minority of CML BC cases display such chimeric genes. Obviously, balanced chromosomal changes resulting in fusion genes may be more common than chromosome-banding analyses indicate. In fact, a recent multicolour FISH analysis of CML BC cases revealed two novel cryptic balanced translocations – t(7;17)(p15;q23) and t(7;17)(q32–34;q23) – which most likely result in fusion genes [155].

During the last few years, much attention has been focussed on epigenetic events occurring at, or prior to, BC. For example, data have been forthcoming, indicating that loss of imprinting of the *IGF2* gene [156], hypermethylation of the calcitonin gene [157, 158], methylation of the proximal promoter of the *ABL* gene [159, 160], overexpression of the *SOCS2* gene [161], and increased telomerase activity [162, 163] are involved in CML progression. Interestingly, there seems to be an association between telomerase activity and secondary chromosomal changes, with additional aberrations being more common in cases with telomerase up-regulation [163].

As should now be evident, transformation of CML is strongly associated with numerous, and occasionally quite complex, additional cytogenetic and molecular genetic changes. Although we still do not know why and how these changes originate, exogenous factors appear to play a role, as suggested by the fact that the frequencies of the secondary abnormalities +8, i(17q), +19, and +Ph display striking geographic differences [164] and by studies showing that the type and incidence of additional aberrations vary in relation to the treatment given during CP.

### The Impact of Previous Treatment on the Cytogenetic Evolution Pattern

Questions as to whether the cytogenetic evolution patterns in CML correlate with type of therapy given during CP have been ongoing ever since secondary chromosomal changes were first identified in this disease. Alimena et al. [165] reported that atypical additional abnormalities, in particular structural changes of chromosome 1, are found mainly in patients treated with intensive chemotherapy. However, such an association has not been confirmed by later studies [53, 57]. Swolin et al. [166] described a higher incidence of trisomy 8 in busulphan-treated CML than in patients receiving hydroxyurea, and Nanjangud et al. [167] found a generally higher frequency of secondary changes in patients treated with busulphan. That +8 is more common after busulphan therapy seems to hold true; this anomaly is found in 44% after treatment with busulphan, but only in 12% after hydroxyurea therapy (table 3).

Several studies have shown that the cytogenetic evolution pattern is quite distinct after allogeneic bone marrow transplantation (BMT), being characterised by seemingly random, sometimes transient, chromosomal changes [168–175]. As regards breakprone chromosome bands in Ph-positive cells after BMT, the available data are rather conflicting. Offit et al. [172] reported a non-random involvement of, especially, 1p22–36, 1q21–25, 9q22, 10q22, and 11q13, whereas other groups subsequently described clustering of breakpoints on chromosomes 1, 4, 7, and 12 [171], 1q32, 6p21–22, 7p21–22, and 10q21 [175], and on chromosome 13, in particular the 13q12–14 [176]. As seen in table 3, secondary abnormalities otherwise common in CML are notably rare post-BMT, whereas the presence of additional balanced abnormalities is frequent. Also, divergent clones, i.e., cell populations with unrelated cytogenetic abnormalities in addition to t(9;22), and pseudidiploidy are much more common after BMT than after treatment with busulphan and/or hydroxyurea (table 3). It has been suggested that the unusual cytogenetic changes observed post-BMT arise as a result of the pre-BMT conditioning regimen, i.e., the clastogenic effect of radiotherapy and cyclophosphamide [171–176]. However, one might also postulate that the emergence of unconventional secondary chromosomal abnormalities is facilitated, or even caused, by the altered environment post-BMT brought about by the donor cells and the concomitant immunosuppression.

Treatment with interferon-α (IFN) also seems to be associated with an aberrant karyotypic evolution pattern.

**Table 3.** Secondary abnormalities/abnormality patterns in relation to previous treatment[a]

| Abnormalities/ abnormality patterns | Bu n = 380 | Hy n = 67 | Bu + Hy n = 82 | IFN n = 66[b] | BMT n = 107[c] | p value[d] |
|---|---|---|---|---|---|---|
| +8 | 168 (44.2) | 8 (11.9) | 26 (31.7) | 12 (18.2) | 10 (9.3) | $p < 0.001$ |
| +Ph[e] | 136 (35.8) | 19 (28.4) | 33 (40.2) | 10 (15.2) | 6 (5.6) | $p < 0.001$ |
| i(17q)[f] | 105 (27.6) | 8 (11.9) | 11 (13.4) | 6 (9.1) | 6 (5.6) | $p < 0.001$ |
| +19 | 61 (16.1) | 1 (1.5) | 11 (13.4) | 2 (3.0) | 5 (4.7) | $p < 0.001$ |
| +21 | 29 (7.6) | 3 (4.5) | 9 (11.0) | 1 (1.5) | 1 (0.9) | – |
| –7 | 15 (3.9) | 6 (9.0) | 6 (7.3) | 4 (6.1) | 3 (2.8) | – |
| +17 | 25 (6.6) | 0 | 4 (4.9) | 1 (1.5) | 1 (0.9) | – |
| –17 | 20 (5.3) | 2 (3.0) | 4 (4.9) | 2 (3.0) | 1 (0.9) | – |
| 51–57 chromosomes[g] | 23 (6.1) | 2 (3.0) | 7 (8.5) | 2 (3.0) | 0 | – |
| 47–50 chromosomes[g] | 215 (56.6) | 23 (34.3) | 41 (50.0) | 18 (27.3) | 20 (18.7) | $p < 0.001$ |
| 46 chromosomes[g] | 128 (33.7) | 35 (52.2) | 36 (43.9) | 39 (59.1) | 86 (80.4) | $p < 0.001$ |
| 41–45 chromosomes[g] | 41 (10.8) | 15 (22.4) | 8 (9.8) | 13 (19.7) | 15 (14.0) | $p < 0.05$ |
| Balanced aberrations | 58 (15.3) | 19 (28.4) | 23 (28.0) | 18 (27.3) | 70 (65.4) | $p < 0.001$ |
| Divergent clones[h] | 18 (4.7) | 6 (9.0) | 6 (7.3) | 12 (18.2) | 29 (27.1) | $p < 0.001$ |

Figures in parentheses are percentages. Bu = Busulphan; Hy = hydroxyurea.

a Based on all reported cases in the literature [47].

b Includes patients treated also with Bu, Hy, or Bu + Hy.

c Includes patients treated also with IFN.

d $\chi^2$. The '–' denotes that statistical analysis was not performed because of too few cases.

e Includes also ider(22)(q10)t(9;22)(q34;q11).

f Includes also idic(17)(p11).

g Includes all cases with at least one subclone with that chromosome number. A case with subclones of different modal numbers may hence be included in more than one chromosome number group.

h Presence of cell populations with unrelated cytogenetic abnormalities in addition to the t(9;22)/variant Ph.

We found increased frequencies of unusual secondary changes and divergent clones, frequently occurring already during CP, in IFN-treated CML patients, i.e., a pattern quite similar to the one observed post-BMT [177]. Although this association has been questioned [178], similar findings have been reported by other groups [179, 180]. Summarising the secondary abnormalities in IFN-treated CML cases (table 3), it becomes clear that the incidences of +8, +Ph, i(17q), +19, +21, and +17 are lower in the IFN-treated group compared to patients treated with busulphan and/or hydroxyurea. Furthermore, and similar to the post-BMT group, the frequencies of divergent clones, and pseudodiploidy are high. It remains to be elucidated why treatment with IFN is associated with such unusual cytogenetic abnormality patterns.

Considering the high frequency of the common secondary changes +8, +Ph, i(17q), and +19 in patients treated with busulphan (table 3) and the decreased use of this drug in the treatment of CML, it is tempting to predict that the incidences of these aberrations will diminish with time. Whether treatment with imatinib mesylate (STI-571) is associated with particular secondary changes is presently unknown, but because clinical resistance to this drug was recently shown to be caused by mutation or amplification of the *BCR/ABL* fusion gene [181] it is not too farfetched to anticipate secondary changes in the form of amplification of this chimera, perhaps cytogenetically identifiable as double minute chromosomes or homogeneously staining regions.

Although the aetiology of secondary changes is of great biologic interest, the phenotypic effects of the additional aberrations may be of greater clinical importance, i.e., do the secondary changes have a morphologic and/or prognostic impact?

### Secondary Aberrations in Relation to Morphology

Studies on the phenotypic impact of various secondary chromosomal changes have generated rather conflicting data. It has been reported that additional abnormalities as such are more common in myeloid than in lymphoid BC

**Table 4.** Secondary abnormalities/abnormality patterns in myeloid and lymphoid BCs[a]

| Abnormalities/ abnormality patterns | Myeloid BCs n = 580 | Lymphoid BCs n = 124 | p value[b] |
|---|---|---|---|
| +8 | 240 (41.4) | 38 (30.6) | $p < 0.05$ |
| +Ph[c] | 221 (38.1) | 45 (36.3) | NS |
| i(17q)[d] | 131 (22.6) | 12 (9.7) | $p < 0.01$ |
| +19 | 92 (15.9) | 19 (15.3) | NS |
| +21 | 44 (7.6) | 15 (12.1) | NS |
| –Y | 17 (2.9) | 17 (13.7) | $p < 0.001$ |
| +17 | 34 (5.9) | 4 (3.2) | NS |
| –7 | 19 (3.3) | 9 (7.3) | NS |
| 51–57 chromosomes[e] | 46 (7.9) | 9 (7.3) | NS |
| 47–50 chromosomes[e] | 322 (55.5) | 59 (47.6) | NS |
| 46 chromosomes[e] | 248 (42.3) | 48 (38.7) | NS |
| 41–45 chromosomes[e] | 54 (9.3) | 32 (25.8) | $p < 0.001$ |
| Balanced aberrations | 127 (21.9) | 25 (20.2) | NS |
| Divergent clones[f] | 39 (6.7) | 9 (7.3) | NS |

Figures in parentheses are percentages. NS = Not significant.
[a] Based on all reported cases in the literature [47].
[b] $\chi^2$ (with Yates' correction).
[c] Includes also ider(22)(q10)t(9;22)(q34;q11).
[d] Includes also idic(17)(p11).
[e] Includes all cases with at least one subclone with that chromosome number. A case with subclones of different modal numbers may hence be included in more than one chromosome number group.
[f] Presence of cell populations with unrelated cytogenetic abnormalities in addition to the t(9;22)/variant Ph.

and that +8, +19, and, in particular, i(17q) characterise myeloid BC, whereas monosomy 7 and various, atypical additional changes are typically found in lymphoid BC [53, 69, 182–189]. However, some studies have not corroborated these associations [51, 52], and reviewing available literature data (table 4), the only significant differences in cytogenetic evolution patterns between myeloid and lymphoid BC are a higher prevalence of +8 and i(17q) in myeloid BC and a higher frequency of monosomy 7 and hypodiploidy in lymphoid BC.

There are no significant frequency differences of balanced abnormalities in addition to t(9;22) between myeloid and lymphoid BC (table 4). However, it should be noted that most balanced aberrations in myeloid BC are the well-known AML/MDS-associated translocations and inversions mentioned previously, such as inv(3), t(3;3), t(3;21), inv(16), and t(15;17) [47, 190], whereas the balanced rearrangements in lymphoid BC are non-recurrent and not characteristic ALL-associated translocations.

In general, the secondary chromosomal changes, with the exception of the AML/MDS-related translocations and inversions, do not seem to confer any characteristic morphologic features, although i(17q) has been associated with marked basophilia and presence of pseudo Pelger-Huët neutrophils [69, 82, 184, 191–193]. That the cytogenetic evolution generally does not determine the phenotypic differentiation during BC has also been shown by serial karyotypic investigations of a case of CML BC revealing cytogenetically distinct clones in phenotypically and immunogenetically identical lymphoblastic transformations [194]. In addition, Anastasi et al. [195] have reported that secondary changes, such as +8, may arise after the acute transformation, in some instances in a different subclone from the acute phase blasts.

As regards molecular genetic changes, only a few studies have addressed possible frequency differences between myeloid and lymphoid BC. The salient features of these investigations seem to be that *TP53* mutations are associated with myeloid BC and homozygous deletions of *CDKN2A* with lymphoid BC [103, 112–114, 118], LOH at 4p, 7p, 9p, 14q32, 18q, and 19p are more common in lymphoid BC [123, 196], and the methylation status of the major breakpoint cluster region of the *BCR* gene may differ between myeloid and lymphoid BC [197].

Taken together, although some genetic differences may be discerned between myeloid and lymphoid BC, cytogenetic and molecular genetic findings in CML BC should be interpreted with caution when trying to predict lineage involvement of the acute transformation.

## Prognostic Impact of Secondary Genetic Changes

Considering the fact that additional abnormalities are found in 10–20% of patients during CP but in 60–80% preceding or accompanying BC [5, 51, 57, 69, 79–81, 166, 198–202], a strong association between secondary changes and transformation undoubtedly exists. In addition, strong support for the clinical importance of additional changes came from early studies of splenic transformations of CML, showing that the secondary aberrations appeared in the spleen before the bone marrow [203–205]. However, the prognostic impact of the secondary abnormalities has proved difficult to assess. In fact, there is an abundance of conflicting results concerning this issue in the literature.

First, although most studies have reported that lack of additional abnormalities during BC confers a better prognosis [53, 82, 191, 201, 202, 206, 207], some investigators have not found any prognostic differences between CML BC cases with and without secondary changes [51, 199].

Second, it has been debated whether additional changes found during CP constitute a dismal prognostic sign, heralding transformation. On the one hand, it has been stressed repeatedly that emergence of new clones does not necessarily imply disease progression and that some clones appear only transiently [51, 52, 208]. On the other hand, it has also been shown that appearance of new chromosome abnormalities or increase of the clone with additional abnormalities in general is associated with clinical deterioration [53, 77, 166]. Furthermore, Sokal et al. [209] reported that the survival of patients with secondary changes already present at the time of diagnosis was shorter, but with survival curves for those with and without additional changes not diverging until after the 2-year point.

Third, the number and types of aberrations present during CP and BC may also play a role. Most often, the karyotypic changes seen in CP are single events, in particular loss of the Y chromosome, +8, +Ph, and i(17q) [47], whereas they are multiple and complex in BC [52, 53, 59, 81, 200]. Among the aberrations, i(17q), or other changes resulting in loss of 17p, seem particularly ominous [48, 51–53, 69, 81, 178, 210]. In addition, trisomy 8 and +Ph occurring in BC have been associated with poor prognosis [81, 211–213].

Thus, the prognostic impact of secondary abnormalities in CML is complex, heterogeneous, and most likely related to several parameters, such as time of appearance, specific abnormalities, and treatment modalities.

### Acknowledgements

The original research on which this article is based was supported by the Swedish Cancer Society and the Swedish Child Cancer Fund.

### References

1 Nowell PC, Hungerford DA: A minute chromosome in human chronic granulocytic leukemia. Science 1960;132:1497.
2 Nowell PC, Hungerford DA: Chromosome studies on normal and leukemic human leukocytes. J Natl Cancer Inst 1960;25:85–109.
3 Caspersson T, Gahrton G, Lindsten J, Zech L: Identification of the Philadelphia chromosome as a number 22 by quinacrine mustard fluorescence analysis. Exp Cell Res 1970;63:238–240.
4 Rowley JD: A new consistent chromosomal abnormality in chronic myelogenous leukaemia identified by quinacrine fluorescence and Giemsa staining. Nature 1973;243:290–293.
5 First International Workshop on Chromosomes in Leukaemia: Chromosomes in Ph[1]-positive chronic granulocytic leukaemia. Br J Haematol 1978;39:305–309.
6 Heisterkamp N, Stam K, Groffen J, de Klein A, Grosveld G: Structural organization of the *bcr* gene and its role in the Ph[1] translocation. Nature 1985;315:758–761.
7 Shtivelman E, Lifshitz B, Gale RP, Canaani E: Fused transcript of abl and bcr genes in chronic myelogenous leukaemia. Nature 1985;315: 550–554.
8 Heim S, Mitelman F: Cancer Cytogenetics, ed 2. New York, Wiley-Liss, 1995.
9 O'Dwyer ME, Mauro MJ, Druker BJ: Recent advancements in the treatment of chronic myelogenous leukemia. Annu Rev Med 2002;53: 369–381.
10 Baikie AG, Court Brown WM, Buckton KE, Harnden DG, Jacobs PA, Tough IM: A possible specific chromosome abnormality in human chronic myeloid leukaemia. Nature 1960; 188:1165–1166.
11 Tough IM, Jacobs PA, Court Brown WM, Baikie AG, Williamson ERD: Cytogenetic studies on bone-marrow in chronic myeloid leukaemia. Lancet 1963;i:844–846.
12 Wang J, Frei III E, Tjio JH, Carbone PP, Brecher G: The distribution of the Philadelphia chromosome in patients with chronic myelogenous leukemia. Blood 1963;22:664–673.
13 Krauss S, Sokal JE, Sandberg AA: Comparison of Philadelphia chromosome-positive and -negative patients with chronic myelocytic leukemia. Ann Intern Med 1964;61:625–635.
14 Trujillo JM, Ohno S: Chromosomal alteration of erythropoietic cells in chronic myeloid leukemia. Acta Haematol 1963;29:311–316.
15 Clein GP, Flemans RJ: Involvement of the erythroid series in blastic crisis of chronic myeloid leukaemia. Br J Haematol 1966;12:754–758.
16 Fialkow PJ, Gartler SM, Yoshida A: Clonal origin of chronic myelocytic leukemia in man. Proc Natl Acad Sci USA 1967;58:1468–1471.
17 Rastrick JM, Fitzgerald PH, Gunz FW: Direct evidence for presence of Ph[1] chromosome in erythroid cells. Br Med J 1968;i:96–98.
18 Golde DW, Burgaleta C, Sparkes RS, Cline MJ: The Philadelphia chromosome in human macrophages. Blood 1977;49:367–370.
19 Fitzgerald PH, Pickering AF, Eiby JR: Clonal origin of the Philadelphia chromosome and chronic myeloid leukaemia: Evidence from a sex chromosome mosaic. Br J Haematol 1971; 21:473–480.
20 Barr RD, Fialkow PJ: Clonal origin of chronic myelocytic leukemia. N Engl J Med 1973;289: 307–309.
21 Moore MAS, Ekert H, Fitzgerald MG, Carmichael A: Evidence for the clonal origin of chronic myeloid leukemia from a sex chromosome mosaic: Clinical, cytogenetic, and marrow culture studies. Blood 1974;43:15–22.
22 Gunz FW, Fitzgerald PH: Chromosomes and leukemia. Blood 1964;23:394–400.
23 Nowell PC: Chromosome changes in primary tumors. Progr Exp Tumor Res 1965;7:83–103.
24 Baikie AG: Chromosomes and leukaemia. Acta Haematol 1966;36:157–173.
25 Lawler SD, Galton DAG: Chromosome changes in the terminal stages of chronic granulocytic leukaemia. Acta Med Scand Suppl 1966;445:312–318.
26 Levan A: Some current problems of cancer cytogenetics. Hereditas 1967;57:343–355.
27 Spiers ASD, Baikie AG: Cytogenetic evolution and clonal proliferation in acute transformation of chronic granulocytic leukaemia. Br J Cancer 1968;22:192–204.
28 De Grouchy J, De Nava C: A chromosomal theory of carcinogenesis. Ann Intern Med 1968;69:381–391.
29 Sakurai M: Chromosome studies in hematological disorders. I. Chromosome findings in chronic myelogenous leukemia with special reference to those after blastic transformation. Acta Haematol Jpn 1970;33:103–155.
30 Sandberg AA, Hossfeld DK: Chromosomal abnormalities in human neoplasia. Annu Rev Med 1970;21:379–408.
31 Pedersen B: The karyotype evolution in chronic granulocytic leukaemia. II. The chromosome and karyotype pattern of advanced evolution. Eur J Cancer 1973;9:509–513.

32 Garson OM, De Gruchy GC: Chromosome changes in the blastic transformation stage of chronic granulocytic leukaemia. Haematologia 1974;8:21–27.

33 Whang-Peng J, Canellos GP, Carbone PP, Tjio JH: Clinical implications of cytogenetic variants in chronic myelocytic leukemia (CML). Blood 1968;32:755–766.

34 Hauschka TS: The chromosomes in ontogeny and oncogeny. Cancer Res 1961;21:957–974.

35 Hsu TC: Chromosomal evolution in cell populations. Int Rev Cytol 1961;12:69–161.

36 Dinauer MC, Pierre RV: Philadelphia chromosome by translocation. Lancet 1973;ii:971.

37 Petit P, Cauchie C: Philadelphia chromosome by translocation. Lancet 1973;ii;94.

38 Van Den Berghe H: The Ph[1]-chromosome: Translocation to chromosome 9. Lancet 1973; ii:1030–1031.

39 Whang-Peng J, Lee EC, Knutsen TA: Genesis of the Ph[1] chromosome. J Natl Cancer Inst 1974;52:1035–1036.

40 Hayata I, Kakati S, Sandberg AA: A new translocation related to the Philadelphia chromosome. Lancet 1973;ii:1385.

41 Bottura C, Couthinho V: G/G translocation and chronic myelocytic leukaemia. Blut 1974; 29:216–218.

42 Foerster W, Medau HJ, Löffler H: Chronische myeloische Leukämie mit Philadelphia-Chromosom und Tandem-Translokation am 2. Chromosom Nr. 22; 46, XX, tan(22q+;22q–). Klin Wochenschr 1974;52:123–126.

43 Gahrton G, Lindsten J, Zech L: Involvement of chromosomes 8, 9, 19 and 22 in Ph[1] positive and Ph[1] negative chronic myelocytic leukemia in the chronic or blastic stage. Acta Med Scand 1974;196:355–360.

44 Mitelman F: Heterogeneity of Ph[1] in chronic myeloid leukaemia. Hereditas 1974;76:315–316.

45 Hayata I, Sakurai M, Kakati S, Sandberg AA: Chromosomes and causation of human cancer and leukemia. XVI. Banding studies of chronic myelocytic leukemia, including five unusual Ph[1] translocations. Cancer 1975;36:1177–1191.

46 Potter AM, Sharp JC, Brown MJ, Sokol RJ: Structural rearrangements associated with the Ph[1] chromosome in chronic granulocytic leukaemia. Humangenetik 1975;29:223–228.

47 Mitelman F, Johansson B, Mertens F. Mitelman Database of Chromosome Aberrations in Cancer (2002). http://cgap.nci.nih.gov/Chromosomes/Mitelman.

48 Lawler SD, O'Malley F, Lobb DS: Chromosome banding studies in Philadelphia chromosome positive myeloid leukaemia. Scand J Haematol 1976;17:17–28.

49 Sonta S, Sandberg AA: Chromosomes and causation of human cancer and leukemia. XXIV. Unusual and complex Ph[1] translocations and their clinical significance. Blood 1977;50:691–697.

50 Pasquali F, Casalone R, Francesconi D, Peretti D, Fraccaro M, Bernasconi C, Lazzarino M: Transposition of 9q34 and 22 (q11-qter) regions has a specific role in chronic myelocytic leukemia. Hum Genet 1979;52:55–67.

51 Bernstein R, Morcom G, Pinto MR, Mendelow B, Dukes I, Penfold G, Bezwoda W: Cytogenetic findings in chronic myeloid leukemia (CML); evaluation of karyotype, blast morphology, and survival in the acute phase. Cancer Genet Cytogenet 1980;2:23–37.

52 Hagemeijer A, Stenfert Kroeze WF, Abels J: Cytogenetic follow-up of patients with nonlymphocytic leukemia. I. Philadelphia chromosome-positive chronic myeloid leukemia. Cancer Genet Cytogenet 1980;2:317–326.

53 Fleischman EW, Prigogina EL, Volkova MA, Frenkel MA, Zakhartchenko NA, Konstantinova LN, Puchkova GP, Balakirev SA: Correlations between the clinical course, characteristics of blast cells, and karyotype patterns in chronic myeloid leukemia. Hum Genet 1981; 58:285–293.

54 Borgström GH: New types of unusual and complex Philadelphia chromosome (Ph[1]) translocations in chronic myeloid leukemia. Cancer Genet Cytogenet 1981;3:19–31.

55 Potter AM, Watmore AE, Cooke P, Lilleyman JS, Sokol RJ: Significance of non-standard Philadelphia chromosomes in chronic granulocytic leukaemia. Br J Cancer 1981;44:51–54.

56 Oshimura M, Ohyashiki K, Terada H, Takaku F, Tonomura A: Variant Ph[1] translocations in CML and their incidence, including two cases with sequential lymphoid and myeloid crises. Cancer Genet Cytogenet 1982;5:187–201.

57 Tomiyasu T, Sasaki M, Kondo K, Okada M: Chromosome banding studies in 106 cases of chronic myelogenous leukemia. Jpn J Hum Genet 1982;27:243–258.

58 Bernstein R, Pinto MR, Wallace C, Penfold G, Mendelow B: The incidence, type, and subsequent evolution of 14 variant Ph[1] translocations in 180 South African patients with Ph[1]-positive chronic myeloid leukemia. Cancer Genet Cytogenet 1984;12:225–238.

59 Gödde-Salz E, Schmitz N, Bruhn H-D: Philadelphia chromosome (Ph) positive chronic myelocytic leukemia (CML): Frequency of additional findings. Cancer Genet Cytogenet 1985;14:313–323.

60 GFCH (Groupe Français de Cytogénétique Hématologique): Unusual Ph translocations in the French prospective study on chronic myeloid leukemia. Cancer Genet Cytogenet 1985; 16:305–309.

61 Heim S, Billström R, Kristoffersson U, Mandahl N, Strömbeck B, Mitelman F: Variant Ph translocations in chronic myeloid leukemia. Cancer Genet Cytogenet 1985;18:215–227.

62 Sandberg AA: Chromosomes and causation of human cancer and leukemia. XL. The Ph[1] and other translocations in CML. Cancer 1980;46: 2221–2226.

63 De Braekeleer M: Variant Philadelphia translocations in chronic myeloid leukemia. Cytogenet Cell Genet 1987;44:215–222.

64 Verma RS, Macera MJ: Genomic diversity of Philadelphia-positive chronic myelogenous leukemia. Leuk Res 1987;11:833–842.

65 Yehuda O, Abeliovich D, Ben-Neriah S, Sverdlin I, Cohen R, Varadi G, Orr R, Ashkenazi YJ, Heyd J, Lugassy G, Ben Yehuda D: Clinical implications of fluorescence in situ hybridization analysis in 13 chronic myeloid leukemia cases: Ph-negative and variant Ph-positive. Cancer Genet Cytogenet 1999;114:100–107.

66 Hagemeijer A, Bartram CR, Smit EME, Van Agthoven AJ, Bootsma D: Is the chromosomal region 9q34 always involved in variants of the Ph[1] translocation? Cancer Genet Cytogenet 1984;13:1–16.

67 Hagemeijer A, de Klein A, Gödde-Salz E, Turc-Carel C, Smit EME, van Agthoven AJ, Grosveld GC: Translocation of c-abl to 'masked' Ph in chronic myeloid leukemia. Cancer Genet Cytogenet 1985;18:95–104.

68 Ishihara T, Minamihisamatsu M: The Philadelphia chromosome. Considerations based on studies of variant Ph translocations. Cancer Genet Cytogenet 1988;32:75–92.

69 Bernstein R: Cytogenetics of chronic myelogenous leukemia. Semin Hematol 1988;25:20–34.

70 Abe S, Minamihisamatsu M, Ishihara T, Sasaki M: Chromosomal in situ hybridization and Southern blot analyses using c-abl, c-sis, or bcr probe in chronic myelogenous leukemia cells with variant Philadelphia translocations. Cancer Genet Cytogenet 1989;38:61–74.

71 Calabrese G, Stuppia L, Guanciali Franchi P, Peila R, Morizio E, Liberati AM, Spadano A, Di Lorenzo R, Donti E, Antonucci A, Palka G: Complex translocations of the Ph chromosome and Ph negative CML arise from similar mechanisms, as evidenced by FISH analysis. Cancer Genet Cytogenet 1994;78:153–159.

72 Fitzgerald PH, Morris CM: Complex chromosomal translocations in the Philadelphia chromosome leukemias. Serial translocations or a concerted genomic rearrangement? Cancer Genet Cytogenet 1991;57:143–151.

73 McKeithan TW, Warshawsky L, Espinosa R III, Le Beau MM: Molecular cloning of the breakpoints of a complex Philadelphia chromosome translocation: Identification of a repeated region on chromosome 17. Proc Natl Acad Sci USA 1992;89:4923–4927.

74 Reddy KS, Sulcova V: A FISH study of variant Philadelphia rearrangements. Cancer Genet Cytogenet 2000;118:121–131.

75 Mitelman F, Levan G, Nilsson PG, Brandt L: Non-random karyotypic evolution in chronic myeloid leukemia. Int J Cancer 1976;18:24–30.

76 Sharp JC, Potter AM, Wood JK: Non-random and random chromosomal abnormalities in transformed chronic granulocytic leukaemia. Scand J Haematol 1976;16:5–12.

77 Lilleyman JS, Potter AM, Watmore AE, Cooke P, Sokol RJ, Wood JK: Myeloid karyotype and the malignant phase of chronic granulocytic leukaemia. Br J Haematol 1978;39:317–323.

78 Prigogina EL, Fleischman EW, Volkova MA, Frenkel MA: Chromosome abnormalities and clinical and morphologic manifestations of chronic myeloid leukemia. Hum Genet 1978; 41:143–156.

79 Stoll C, Oberling F: Non-random clonal evolution in 45 cases of chronic myeloid leukemia. Leuk Res 1978;3:61–66.
80 Carbonell F, Benitez J, Prieto F, Badia L, Sanchez-Fayos J: Chromosome banding patterns in patients with chronic myelocytic leukemia. Cancer Genet Cytogenet 1982;7:287–297.
81 Ishihara T, Sasaki M, Oshimura M, Kamada N, Yamada K, Okada M, Sakurai M, Sugiyama T, Shiraishi Y, Kohno S-I: A summary of cytogenetic studies on 534 cases of chronic myelocytic leukemia in Japan. Cancer Genet Cytogenet 1983;9:81–92.
82 Swolin B, Weinfeld A, Waldenström J, Westin J: Cytogenetic studies of bone marrow and extramedullary tissues and clinical course during metamorphosis of chronic myelocytic leukemia. Cancer Genet Cytogenet 1983;9:197–209.
83 Hashimoto T, Ohtaki M, Kamada N, Yamamoto H, Munaka M: Application of log-linear model in inference on karyotypic evolution in chronic myelocytic leukemia. Environ Health Perspect 1990;87:135–141.
84 Johansson B, Mertens F, Mitelman F: Primary vs. secondary neoplasia-associated chromosomal abnormalities – Balanced rearrangements vs. genomic imbalances? Genes Chromosomes Cancer 1996;16:155–163.
85 Berger R, Bernheim A, Daniel M-T, Flandrin G: t(15;17) in a promyelocytic form of chronic myeloid leukemia blastic crisis. Cancer Genet Cytogenet 1983;8:149–152.
86 Pintado T, Ferro MT, San Roman C, Mayayo M, Larana JG: Clinical correlations of the 3q21;q26 cytogenetic anomaly. A leukemic or myelodysplastic syndrome with preserved or increased platelet production and lack of response to cytotoxic drug therapy. Cancer 1985; 55:535–541.
87 Rubin CM, Larson RA, Bitter MA, Carrino JJ, Le Beau MM, Diaz MO, Rowley JD: Association of a chromosomal 3;21 translocation with the blast phase of chronic myelogenous leukemia. Blood 1987;70:1338–1342.
88 Sato Y, Abe S, Mise K, Sasaki M, Kamada N, Kouda K, Musashi M, Saburi Y, Horikoshi A, Minami Y, Miyakuni T, Yokoyama Y, Ishihara Y, Miura Y: Reciprocal translocation involving the short arms of chromosomes 7 and 11, t(7p-;11p+), associated with myeloid leukemia with maturation. Blood 1987;70:1654–1658.
89 Ferro MT, Steegman JL, Escribano L, Heiurichs B, Parada L, Garcia-Sagredo JM, Resino M, Cabello P, San Roman C: Ph-positive chronic myeloid leukemia with t(8;21) (q22;q22) in blastic crisis. Cancer Genet Cytogenet 1992;58:96–99.
90 Heim S, Egelund Christensen B, Fioretos T, Sörensen A-G, Tinggaard Pedersen N: Acute myelomonocytic leukemia with inv(16) (p13q22) complicating Philadelphia chromosome positive chronic myeloid leukemia. Cancer Genet Cytogenet 1992;59:35–38.
91 Heim S: Is cancer cytogenetics reducible to the molecular genetics of cancer cells? Genes Chromosomes Cancer 1992;5:188–196.
92 McCarthy DM, Rassool FV, Goldman JM, Graham SV, Birnie GD: Genomic alterations involving the c-myc proto-oncogene locus during the evolution of a case of chronic granulocytic leukaemia. Lancet 1984;ii:1362–1365.
93 Blick M, Romero P, Talpaz M, Kurzrock R, Shtalrid M, Andersson B, Trujillo J, Beran M, Gutterman J: Molecular characteristics of chronic myelogenous leukemia in blast crisis. Cancer Genet Cytogenet 1987;27:349–356.
94 Karasawa M, Okamoto K, Maehara T, Tsukamoto N, Morita K, Naruse T, Omine M: Detection of c-myc oncogene amplification in a CML blastic phase patient with double minute chromosomes. Leuk Res 1996;20:85–91.
95 Jennings BA, Mills KI: c-myc locus amplification and the acquisition of trisomy 8 in the evolution of chronic myeloid leukemia. Leuk Res 1998;22:899–903.
96 Virtaneva K, Wright FA, Tanner SM, Yuan B, Lemon WJ, Caligiuri MA, Bloomfield CD, de la Chapelle A, Krahe R: Expression profiling reveals fundamental biological differences in acute myeloid leukemia with isolated trisomy 8 and normal cytogenetics. Proc Natl Acad Sci USA 2001;98:1124–1129.
97 Ohmine K, Ota J, Ueda M, Ueno S, Yoshida K, Yamashita Y, Kirito K, Imagawa S, Nakamura Y, Saito K, Akutsu M, Mitani K, Kano Y, Komatsu N, Ozawa K, Mano H: Characterization of stage progression in chronic myeloid leukemia by DNA microarray with purified hematopoietic stem cells. Oncogene 2001;20:8249–8257.
98 Ravandi F, Hayes K, Cortes J, Albitar M, Glassman A, Talpaz M, Kantarjian HM: Translocation t(17;18)(q10;q10). A new nonrandom chromosomal translocation of clonal evolution in chronic myeloid leukemia. Cancer 2001;91:1704–1708.
99 Ahuja H, Bar-Eli M, Advani SH, Benchimol S, Cline MJ: Alterations in the p53 gene and the clonal evolution of the blast crisis of chronic myelocytic leukemia. Proc Natl Acad Sci USA 1989;86:6783–6787.
100 Kelman Z, Prokocimer M, Peller S, Kahn Y, Rechavi G, Manor Y, Cohen A, Rotter V: Rearrangements in the p53 gene in Philadelphia chromosome positive chronic myelogenous leukemia. Blood 1989;74:2318–2324.
101 Feinstein E, Cimino G, Gale RP, Alimena G, Berthier R, Kishi K, Goldman J, Zaccaria A, Berrebi A, Canaani E: p53 in chronic myelogenous leukemia in acute phase. Proc Natl Acad Sci USA 1991;88:6293–6297.
102 Foti A, Ahuja HG, Allen SL, Koduru P, Schuster MW, Schulman P, Bar-Eli M, Cline MJ: Correlation between molecular and clinical events in the evolution of chronic myelocytic leukemia to blast crisis. Blood 1991;77: 2441–2444.
103 Ahuja H, Bar-Eli M, Arlin Z, Advani S, Allen SL, Goldman J, Snyder D, Foti A, Cline M: The spectrum of molecular alterations in the evolution of chronic myelocytic leukemia. J Clin Invest 1991;87:2042–2047.
104 Mori N, Wada M, Yokota J, Terada M, Okada M, Teramura M, Masuda M, Hoshino S, Motoji T, Oshimi K, Mizoguchi H: Mutations of the p53 tumor suppressor gene in hematologic neoplasms. Br J Haematol 1992; 81:235–240.
105 Neubauer A, He M, Schmidt CA, Huhn D, Liu ET: Genetic alterations in the p53 gene in the blast crisis of chronic myelogenous leukemia: Analysis by polymerase chain reaction based techniques. Leukemia 1993;7:593–600.
106 Serra A, Guerrasio A, Gaidano G, Rosso C, Rege-Cambrin G, Petroni D, Mazza U, Saglio G: Molecular defects associated with the acute phase CML. Leuk Lymphoma 1993;11: 25–28.
107 Peller S, Yona R, Kopilova Y, Prokocimer M, Goldfinger N, Uysal A, Karabulut HG, Tukun A, Bokesoy I, Tuncman G, Rotter V: Molecular alterations in the *TP53* gene of peripheral blood cells of patients with chronic myeloid leukemia. Genes Chromosomes Cancer 1998;21:2–7.
108 Yamaguchi H, Inokuchi K, Sakuma Y, Dan K: Mutation of the p51/p63 gene is associated with blastic crisis in chronic myelogenous leukemia. Leukemia 2001;15:1729–1734.
109 Schütte J, Opalka B, Becher R, Bardenheuer W, Szymanski S, Lux A, Seeber S: Analysis of the p53 gene in patients with isochromosome 17q and Ph$^1$-positive or -negative myeloid leukemia. Leuk Res 1993;17:533–539.
110 Rege-Cambrin G, Gaidano G, Serra A, Scaravaglio P, Guglielmelli T, Guerrasio A, Giovinazzo B, Saglio G: Analysis of the p53 gene in myeloid malignancies associated with chromosomal abnormalities involving the short arm of chromosome 17. Leukemia 1994;8:23–26.
111 Fioretos T, Strömbeck B, Sandberg T, Johansson B, Billström R, Borg Å, Nilsson P-G, Van Den Berghe H, Hagemeijer A, Mitelman F, Höglund M: Isochromosome 17q in blast crisis of chronic myeloid leukemia and in other hematologic malignancies is the result of clustered breakpoints in 17p11 and is not associated with coding *TP53* mutations. Blood 1999;94:225–232.
112 Sill H, Goldman JM, Cross NCP: Homozygous deletions of the p16 tumor-suppressor gene are associated with lymphoid transformation of chronic myeloid leukemia. Blood 1995;85:2013–2016.
113 Serra A, Gottardi E, Della Ragione F, Saglio G, Iolascon A: Involvement of the cyclin-dependent kinase-4 inhibitor (CDKN2) gene in the pathogenesis of lymphoid blast crisis of chronic myelogenous leukaemia. Br J Haematol 1995;91:625–629.
114 Iolascon A, Della Ragione F, Giordani L, Serra A, Saglio G, Faienza MF: Expression of cell cycle regulatory genes in chronic myelogenous leukemia. Haematologica 1998;83:771–777.

115 Haidar MA, Cao XB, Manshouri T, Chan LL, Glassman A, Kantarjian HM, Keating MJ, Beran MS, Albitar M: p16INK4A and p15INK4B gene deletions in primary leukemias. Blood 1995;86:311–315.
116 Drexler HG: Review of alterations of the cyclin-dependent kinase inhibitor INK4 family genes p15, p16, p18 and p19 in human leukemia-lymphoma cells. Leukemia 1998; 12:845–859.
117 Towatari M, Adachi K, Kato H, Saito H: Absence of the human retinoblastoma gene product in the megakaryoblastic crisis of chronic myelogenous leukemia. Blood 1991; 78:2178–2181.
118 Gaidano G, Serra A, Guerrasio A, Rege-Cambrin G, Mazza U, Saglio G: Genetic analysis of p53 and RB1 tumor-suppressor genes in blast crisis of chronic myeloid leukemia. Ann Hematol 1994;68:3–7.
119 Hillion J, Larsen C-J, Berger R: Hemizygous deletion of the 3′ end of the RB1 gene in a case of Philadelphia positive-acute lymphoblastic leukemia. Cancer Genet Cytogenet 1991;54:197–203.
120 Ahuja HG, Jat PS, Foti A, Bar-Eli M, Cline MJ: Abnormalities of the retinoblastoma gene in the pathogenesis of acute leukemia. Blood 1991;78:3259–3268.
121 Beck Z, Kiss A, Toth FD, Szabo J, Bacsi A, Balogh E, Borbely A, Telek B, Kovacs E, Olah E, Rak K: Alterations of P53 and RB genes and the evolution of the accelerated phase of chronic myeloid leukemia. Leuk Lymphoma 2000;38:587–597.
122 Nakayama H, Ishimaru F, Avitahl N, Sezaki N, Fujii N, Nakase K, Ninomiya Y, Harashima A, Minowada J, Tsuchiyama J, Imajoh K, Tsubota T, Fukuda S, Sezaki T, Kojima K, Hara M, Takimoto H, Yorimitsu S, Takahashi I, Miyata A, Taniguchi S, Tokunaga Y, Gondo H, Niho Y, Nakao S, Kyo T, Dohy H, Kamada N, Harada M: Decreases in Ikaros activity correlate with blast crisis in patients with chronic myelogenous leukemia. Cancer Res 1999;59:3931–3934.
123 Mori N, Morosetti R, Lee S, Spira S, Ben-Yehuda D, Schiller G, Landolfi R, Mizoguchi H, Koeffler HP: Allelotype analysis in the evolution of chronic myelocytic leukemia. Blood 1997;90:2010–2014.
124 Mori N, Morosetti R, Spira S, Lee S, Ben-Yehuda D, Schiller G, Landolfi R, Mizoguchi H, Koeffler HP: Chromosome band 1p36 contains a putative tumor suppressor gene important in the evolution of chronic myelocytic leukemia. Blood 1998;92:3405–3409.
125 Hilgers W, Kern SE: Molecular genetic basis of pancreatic adenocarcinoma. Genes Chromosomes Cancer 1999;26:1–12.
126 Wada C, Shionoya S, Fujino Y, Tokuhiro H, Akahoshi T, Uchida T, Ohtani H: Genomic instability of microsatellite repeats and its association with the evolution of chronic myelogenous leukemia. Blood 1994;83:3449–3456.
127 Indraccolo S, Simon M, Hehlmann R, Erfle V, Chieco-Bianchi L, Leib-Moesch C: Genetic instability of a dinucleotide repeat-rich region in three hematologic malignancies. Leukemia 1995;9:1517–1522.
128 Silly H, Chase A, Mills KI, Apfelbeck U, Sormann S, Goldman JM, Cross NCP: No evidence for microsatellite instability or consistent loss of heterozygosity at selected loci in chronic myeloid leukaemia blast crisis. Leukemia 1994;8:1923–1928.
129 Mori N, Takeuchi S, Tasaka T, Lee S, Spira S, Ben-Yehuda D, Mizoguchi H, Schiller G, Koeffler HP: Absence of microsatellite instability during the progression of chronic myelocytic leukemia. Leukemia 1997;11:151–152.
130 Thijsen SFT, Schuurhuis GJ, van Oostveen JW, Ossenkoppele GJ: Chronic myeloid leukemia from basics to bedside. Leukemia 1999;13:1646–1674.
131 LeMaistre A, Lee MS, Talpaz M, Kantarjian HM, Freireich EJ, Deisseroth AB, Trujillo JM, Stass SA: Ras oncogene mutations are rare late stage events in chronic myelogenous leukemia. Blood 1989;73:889–891.
132 Collins SJ, Howard M, Andrews F, Agura E, Radich J: Rare occurrence of N-ras point mutations in Philadelphia chromosome positive chronic myeloid leukemia. Blood 1989;73: 1028–1032.
133 Cogswell PC, Morgan R, Dunn M, Neubauer A, Nelson P, Poland-Johnston NK, Sandberg AA, Liu E: Mutations of the Ras protooncogenes in chronic myelogenous leukemia: A high frequency of Ras mutations in bcr/abl rearrangement-negative chronic myelogenous leukemia. Blood 1989;74:2629–2633.
134 Watzinger F, Gaiger A, Karlic H, Becher R, Pillwein K, Lion T: Absence of N-ras mutations in myeloid and lymphoid blast crisis of chronic myeloid leukemia. Cancer Res 1994; 54:3934–3938.
135 Tanaka K, Takauchi K, Takechi M, Kyo T, Dohy H, Kamada N: High frequency of RAS oncogene mutation in chronic myeloid leukemia patients with myeloblastoma. Leuk Lymphoma 1994;13:317–322.
136 Steer EJ, Goldman JM, Cross NCP: Mutations of the transcription factor AML1/ CBFA2 are uncommon in blastic transformation of chronic myeloid leukemia. Leukemia 2001;15:476–477.
137 Collins SJ, Groudine MT: Chronic myelogenous leukemia: Amplification of a rearranged c-abl oncogene in both chronic phase and blast crisis. Blood 1987;69:893–898.
138 Andrews DF III, Collins SJ: Heterogeneity in expression of the bcr-abl fusion transcript in CML blast crisis. Leukemia 1987;1:718–724.
139 Gaiger A, Henn T, Hörth E, Geissler K, Mitterbauer G, Maier-Dobersberger T, Greinix H, Mannhalter C, Haas OA, Lechner K, Lion T: Increase of BCR-ABL chimeric mRNA expression in tumor cells of patients with chronic myeloid leukemia precedes disease progression. Blood 1995;86:2371–2378.
140 Johansson B, Moorman AV, Secker-Walker LM: Derivative chromosomes of 11q23-translocations in hematologic malignancies. Leukemia 1998;12:828–833.
141 Mitani K, Ogawa S, Tanaka T, Miyoshi H, Kurokawa M, Mano H, Yazaki Y, Ohki M, Hirai H: Generation of the *AML1-EVI-1* fusion gene in the t(3;21)(q26;q22) causes blastic crisis in chronic myelocytic leukemia. EMBO J 1994;13:504–510.
142 Nucifora G, Rowley JD: *AML1* and the 8;21 and 3;21 translocations in acute and chronic myeloid leukemia. Blood 1995;86:1–14.
143 Russell M, Thompson F, Spier C, Taetle R: Expression of the EVI1 gene in chronic myelogenous leukemia in blast crisis. Leukemia 1993;7:1654–1657.
144 Suzukawa K, Taki T, Abe T, Asoh H, Kamada N, Yokota J, Morishita K: Identification of translocational breakpoints within the intron region before the last coding exon (exon 12) of the *EVI1* gene in two cases of CML-BC with inv(3)(q21q26). Genomics 1997;42:356–360.
145 Ogawa S, Kurokawa M, Tanaka T, Tanaka K, Hangaishi A, Mitani K, Kamada N, Yazaki Y, Hirai H: Increased Evi-i expression is frequently observed in blastic crisis of chronic myelocytic leukemia. Leukemia 1996;10: 788–794.
146 Claxton DF, Liu P, Hsu HB, Marlton P, Hester J, Collins F, Deisseroth AB, Rowley JD, Siciliano MJ: Detection of fusion transcripts generated by the inversion 16 chromosome in acute myelogenous leukemia. Blood 1994;83: 1750–1756.
147 de Greef GE, Hagemeijer A, Morgan R, Wijsman J, Hoefsloot LH, Sandberg AA, Sacchi N: Identical fusion transcript associated with different breakpoints in the AML1 gene in simple and variant t(8;21) acute myeloid leukemia. Leukemia 1995;9:282–287.
148 Yamamoto K, Nakamura Y, Nakamura Y, Saito K, Furusawa S: Expression of the NUP98/HOXA9 fusion transcript in the blast crisis of Philadelphia chromosome-positive chronic myelogenous leukaemia with t(7;11)(p15;p15). Br J Haematol 2000;109: 423–426.
149 Fujino T, Suzuki A, Ito Y, Ohyashiki K, Hatano Y, Miura I, Nakamura T: Single-translocation and double-chimeric transcripts: Detection of NUP98-HOXA9 in myeloid leukemias with HOXA11 or HOXA13 breaks of the chromosomal translocation t(7;11)(p15;p15). Blood 2002;99:1428–1433.
150 Suzuki A, Ito Y, Sashida G, Honda S, Katagiri T, Fujino T, Nakamura T, Ohyashiki K: t(7;11)(p15;p15) chronic myeloid leukaemia developed into blastic transformation showing a novel NUP98/HOXA11 fusion. Br J Haematol 2002;116:170–172.
151 Höglund M, Johansson B, Pedersen-Bjergaard J, Marynen P, Mitelman F: Molecular characterization of 12p abnormalities in hematologic malignancies: Deletion of *KIP1*, rearrangement of *TEL*, and amplification of *CCND2*. Blood 1996;87:324–330.

152 Coignet LJA, Lima CSP, Min T, Streubel B, Swansbury J, Telford N, Swanton S, Bowen A, Nagai M, Catovsky D, Fonatsch C, Dyer MJS: Myeloid- and lymphoid-specific breakpoint cluster regions in chromosome band 13q14 in acute leukemia. Genes Chromosomes Cancer 1999;25:222–229.
153 Nishii K, Usui E, Sakakura M, Miyata E, Ridge SA, Ford AM, Masuya M, Chen F, Mitani H, Yamaguchi M, Katayama N, Kita K, Shiku H: Additional t(11;17)(q23;q21) in a patient with Philadelphia-positive mixed lineage antigen-expressing leukemia. Cancer Genet Cytogenet 2001;126:8–12.
154 Ahuja HG, Popplewell L, Tcheurekdjian L, Slovak ML: *NUP98* gene rearrangements and the clonal evolution of chronic myelogenous leukemia. Genes Chromosomes Cancer 2001; 30:410–415.
155 Barbouti A, Johansson B, Höglund M, Mauritzson N, Strömbeck B, Nilsson P-G, Tanke HJ, Hagemeijer A, Mitelman F, Fioretos T: Multicolor COBRA-FISH analysis of chronic myeloid leukemia reveals novel cryptic balanced translocations during disease progression. Genes Chromosomes Cancer 2002;35: 127–137.
156 Randhawa GS, Cui H, Barletta JA, Strichman-Almashanu LZ, Talpaz M, Kantarjian H, Deisseroth AB, Champlin RC, Feinberg AP: Loss of imprinting in disease progression in chronic myelogenous leukemia. Blood 1998;91:3144–3147.
157 Nelkin BD, Przepiorka D, Burke PJ, Thomas ED, Baylin SB: Abnormal methylation of the calcitonin gene marks progression of chronic myelogenous leukemia. Blood 1991;77:2431–2434.
158 Malinen T, Palotie A, Pakkala S, Peltonen L, Ruutu T, Jansson SE: Acceleration of chronic myeloid leukemia correlates with calcitonin gene hypermethylation. Blood 1991;77: 2435–2440.
159 Zion M, Ben-Yehuda D, Avraham A, Cohen O, Wetzler M, Melloul D, Ben-Neriah Y: Progressive de novo DNA methylation at the bcr-abl locus in the course of chronic myelogenous leukemia. Proc Natl Acad Sci USA 1994;91:10722–10726.
160 Asimakopoulos FA, Shteper PJ, Krichevsky S, Fibach E, Polliack A, Rachmilewitz E, Ben-Neriah Y, Ben-Yehuda D: ABL1 methylation is a distinct molecular event associated with clonal evolution of chronic myeloid leukemia. Blood 1999;94:2452–2460.
161 Schultheis B, Carapeti-Marootian M, Hochhaus A, Weisser A, Goldman JM, Melo JV: Overexpression of SOCS-2 in advanced stages of chronic myeloid leukemia: Possible inadequacy of a negative feedback mechanism. Blood 2002;99:1766–1775.
162 Tatematsu K, Nakayama J, Danbara M, Shionoya S, Sato H, Omine M, Ishikawa F: A novel quantitative 'stretch PCR assay', that detects a dramatic increase in telomerase activity during the progression of myeloid leukemias. Oncogene 1996;13:2265–2274.
163 Ohyashiki K, Ohyashiki JH, Iwama H, Hayashi S, Shay JW, Toyama K: Telomerase activity and cytogenetic changes in chronic myeloid leukemia with disease progression. Leukemia 1997;11:190–194.
164 Johansson B, Mertens F, Mitelman F: Geographic heterogeneity of neoplasia-associated chromosome aberrations. Genes Chromosomes Cancer 1991;3:1–7.
165 Alimena G, Brandt L, Dallapiccola B, Mitelman F, Nilsson PG: Secondary chromosome changes in chronic myeloid leukemia: Relation to treatment. Cancer Genet Cytogenet 1979;1:79–85.
166 Swolin B, Weinfeld A, Westin J, Waldenström J, Magnusson B: Karyotypic evolution in Ph-positive chronic myeloid leukemia in relation to management and disease progression. Cancer Genet Cytogenet 1985;18:65–79.
167 Nanjangud G, Kadam PR, Saikia T, Bhisey AN, Kumar A, Gopal R, Chopra H, Nair CN, Advani SH: Karyotypic findings as an independent prognostic marker in chronic myeloid leukaemia blast crisis. Leuk Res 1994; 18:385–392.
168 da Silva MAP, Heerema NA, Lowe LR, Tricot G, Jansen J: Cytogenetic patterns following bone marrow transplantation for chronic granulocytic leukemia. Cancer Genet Cytogenet 1988;36:35–44.
169 Calabrese G, Di Bartolomeo P, Stuppia L, Guanciali Franchi P, Parruti G, Ciancarelli M, Angrilli F, Geraci L, Palka G: Cytogenetics in patients with chronic myelogenous leukemia treated with bone marrow transplantation. Cancer Genet Cytogenet 1989;41:49–59.
170 Sessarego M, Frassoni F, Defferrari R, Bacigalupo A, Miceli S, Mareni C, Ajmar F: Cytogenetic follow-up after bone marrow transplantation for Philadelphia-positive chronic myeloid leukemia. Cancer Genet Cytogenet 1989;42:253–261.
171 Alimena G, De Cuia MR, Mecucci C, Arcese W, Mauro F, Screnci M, Mancini M, Cedrone M, Nanni M, Montefusco E, Mandelli F: Cytogenetic follow-up after allogeneic bone-marrow transplantation for $Ph^1$-positive chronic myelogenous leukemia. Bone Marrow Transplant 1990;5:119–127.
172 Offit K, Burns JP, Cunningham I, Jhanwar SC, Black P, Kernan NA, O'Reilly RJ, Chaganti RSK: Cytogenetic analysis of chimerism and leukemia relapse in chronic myelogenous leukemia patients after T cell-depleted bone marrow transplantation. Blood 1990;75: 1346–1355.
173 Sessarego M, Frassoni F, Defferrari R, Bacigalupo A, Fugazza G, Mareni C, Bruzzone R, Dejana A, Ajmar F: Karyotype evolution of Ph positive chronic myelogenous leukemia patients relapsed in advanced phases of the disease after allogeneic bone marrow transplantation. Cancer Genet Cytogenet 1991;57: 69–78.
174 Bilhou-Nabera C, Bernard Ph, Marit G, Viard F, Gharbi M-J, Wen Z, Lacombe F, Broustet A, Reiffers J: Serial cytogenetic studies in allografted patients with chronic myeloid leukemia. Bone Marrow Transplant 1992;9:263–268.
175 Shah NK, Wagner J, Santos G, Griffin CA: Karyotype at relapse following allogeneic bone marrow transplantation for chronic myelogenous leukemia. Cancer Genet Cytogenet 1992;61:183–192.
176 Chase A, Pickard J, Szydlo R, Coulthard S, Goldman JM, Cross NCP: Non-random involvement of chromosome 13 in patients with persistent or relapsed disease after bone-marrow transplantation for chronic myeloid leukemia. Genes Chromosomes Cancer 2000;27: 278–284.
177 Johansson B, Fioretos T, Billström R, Mitelman F: Aberrant cytogenetic evolution pattern of Philadelphia-positive chronic myeloid leukemia treated with interferon-alpha. Leukemia 1996;10:1134–1138.
178 Majlis A, Smith TL, Talpaz M, O'Brien S, Rios MB, Kantarjian HM: Significance of cytogenetic clonal evolution in chronic myelogenous leukemia. J Clin Oncol 1996;14: 196–203.
179 Fayad L, Kantarjian H, O'Brien S, Seong D, Albitar M, Keating M, Talpaz M: Emergence of new clonal abnormalities following interferon-alpha induced complete cytogenetic response in patients with chronic myeloid leukemia: Report of three cases. Leukemia 1997; 11:767–771.
180 Maloisel F, Uettwiller F, Laplace A, Lioure B, Herbrecht R, Mark M, Dufour P: Emergence of unusual cytogenetic abnormalities under interferon-alpha therapy in patients with chronic myelogenous leukemia. Cancer Genet Cytogenet 1999;113:172–176.
181 Gorre ME, Mohammed M, Ellwood K, Hsu N, Paquette R, Rao PN, Sawyers CL: Clinical resistance to STI-571 cancer therapy caused by *BCR-ABL* gene mutation or amplification. Science 2001;293:876–880.
182 O'Malley FM, Garson OM: Chronic granulocytic leukemia: Correlation of blastic transformation type with karyotypic evolution. Am J Hematol 1985;20:313–323.
183 Parreira L, Kearney L, Rassool F, Babapulle VB, Matutes E, Parreira A, Tavares de Castro J, Goldman JM, Catovsky D: Correlation between chromosomal abnormalities and blast phenotype in the blast crisis of Ph-positive CGL. Cancer Genet Cytogenet 1986;22:29–34.
184 Alimena G, De Cuia MR, Diverio D, Gastaldi R, Nanni M: The karyotype of blastic crisis. Cancer Genet Cytogenet 1987;26:39–50.
185 Diez-Martin JL, Dewald GW, Pierre RV: Possible cytogenetic distinction between lymphoid and myeloid blast crisis in chronic granulocytic leukemia. Am J Hematol 1988; 27:194–203.
186 Bernstein R, Gale RP: Do chromosome abnormalities determine the type of acute leukemia that develops in CML? Leukemia 1990;4: 65–68.

187 Kadam PR, Narjangud GJ, Advani SH, Nair C, Banavali S, Gopal R, Saikia T: Chromosomal characteristics of chronic and blastic phase of chronic myeloid leukemia. A study of 100 patients in India. Cancer Genet Cytogenet 1991;51:167–181.
188 Ch'ang H-J, Tien H-F, Wang C-H, Chuang S-M, Chen YC, Shen M-C, Lin D-T, Lin K-H: Comparison of clinical and biologic features between myeloid and lymphoid transformation of Philadelphia chromosome positive chronic myeloid leukemia. Cancer Genet Cytogenet 1993;71:37–93.
189 Murakami T, Ohyashiki K, Ohyashiki JH, Kawakubo K, Tauchi T, Nakazawa S, Kimura N, Toyama K: Cytogenetic and immunogenotypic alterations of blast crisis cells in chronic myelogenous leukemia independently linked to immunophenotypic expression. Cancer Genet Cytogenet 1994;72:48–54.
190 Secker-Walker LM, Mehta A, Bain B: Abnormalities of 3q21 and 3q26 in myeloid malignancy: A United Kingdom Cancer Cytogenetic Group study. Br J Haematol 1995;91:490–501.
191 Alimena G, Dallapiccola B, Gastaldi R, Mandelli F, Brandt L, Mitelman F, Nilsson PG: Chromosomal, morphological and clinical correlations in blastic crisis of chronic myeloid leukaemia. A study of 69 cases. Scand J Haematol 1982;23:103–117.
192 Sessarego M, Ajmar F: Correlation between acquired pseudo-Pelger-Huet anomaly and involvement of chromosome 17 in chronic myeloid leukemia. Cancer Genet Cytogenet 1987;25:265–270.
193 Hernandez-Boluda J-C, Cervantes F, Costa D, Carrio A, Montserrat E: Blast crisis of Ph-positive chronic myeloid leukemia with isochromosome 17q: Report of 12 cases and review of the literature. Leuk Lymphoma 2000:38:83–90.
194 Wong KF, So CC, Kwong YL: Cytogenetic divergence of the same blastic clone in transformed chronic granulocytic leukemia: No effect on morphologic and immunologic features. Cancer Genet Cytogenet 1999;113:54–57.
195 Anastasi J, Feng J, Le Beau MM, Larson RA, Rowley JD, Vardiman JW: The relationship between secondary chromosomal abnormalities and blast transformation in chronic myelogenous leukemia. Leukemia 1995;9:628–633.
196 Sercan HO, Sercan ZY, Kizildag S, Undar B, Soydan S, Sakizli M: Consistent loss of heterozygosity at 14Q32 in lymphoid blast crisis of chronic myeloid leukemia. Leuk Lymphoma 2000;39:385–390.
197 Ge X-Q, Tanaka K, Mansyur A, Tazawa H, Iwato K, Kyo T, Dohy H, Kamada N: Possible prediction of myeloid and lymphoid crises in chronic myelocytic leukemia at onset by determining the methylation status of the major breakpoint cluster region. Cancer Genet Cytogenet 2001;126:102–110.
198 Löwenberg B, Hagemeijer A, Abels J: Detection of the blastic crisis cell clone in chronic myeloid leukaemia. Br J Haematol 1985;59: 27–36.
199 Cervantes F, Ballesta F, Mila M, Rozman C: Cytogenetic studies in blast crisis of Ph-positive chronic granulocytic leukemia: Results and prognostic evaluation in 52 patients. Cancer Genet Cytogenet 1986;21:239–246.
200 Krulik M, Smadja N, de Gramont A, Gonzalez-Canali G, Audebert AA, Dray C, Brissaud P, Debray J: Sequential karyotype study on Ph-positive chronic myelocytic leukemia. Significance of additional chromosomal abnormalities during disease evolution. Cancer 1987;60:974–979.
201 Kantarjian HM, Keating MJ, Talpaz M, Walters RS, Smith TL, Cork A, McCredie KB, Freireich EJ: Chronic myelogenous leukemia in blast crisis. Analysis of 242 patients. Am J Med 1987;83:445–454.
202 Przepiorka D, Thomas ED: Prognostic significance of cytogenetic abnormalities in patients with chronic myelogenous leukemia. Bone Marrow Transplant 1988;3:113–119.
203 Mitelman F, Brandt L, Nilsson PG: Cytogenetic evidence for splenic origin of blastic transformation in chronic myeloid leukaemia. Scand J Haematol 1974;13:87–92.
204 Gomez G, Hossfeld DK, Sokal JE: Removal of abnormal clone of leukaemic cells by splenectomy. Br Med J 1975;ii:421–423.
205 Mitelman F, Nilsson PG, Brandt L: Abnormal clones resembling those seen in blast crisis arising in the spleen in chronic myelocytic leukemia. J Natl Cancer Inst 1975;54:1319–1321.
206 Sadamori N, Gomez GA, Sandberg AA: Therapeutic and prognostic value of initial chromosomal findings at the blastic phase of Ph1-positive chronic myeloid leukemia. Blood 1983;61:935–939.
207 Muehleck SD, McKenna RW, Arthur DC, Parkin JL, Brunning RD: Transformation of chronic myelogenous leukemia: Clinical, morphologic, and cytogenetic features. Am J Clin Pathol 1984;82:1–14.
208 Watmore AE, Potter AM, Sokol RJ, Wood JK: Value of cytogenetic studies in prediction of acute phase CML. Cancer Genet Cytogenet 1985;14:293–301.
209 Sokal JE, Gomez GA, Baccarani M, Tura S, Clarkson BD, Cervantes F, Rozman C, Carbonell F, Anger B, Heimpel H, Nissen NI, Robertson JE: Prognostic significance of additional cytogenetic abnormalities at diagnosis of Philadelphia chromosome-positive chronic granulocytic leukemia. Blood 1988; 72:294–298.
210 Borgström GH, Vuopio P, de la Chapelle A: Abnormalities of chromosome No. 17 in myeloproliferative disorders. Cancer Genet Cytogenet 1982;5:123–135.
211 Olah E, Rak K: Prognostic value of chromosomal findings in the blast phase of $Ph^1$-positive chronic myeloid leukaemia (CML). Int J Cancer 1981;27:287–295
212 Cervantes F, Rozman M, Rosell J, Urbano-Ispizua A, Montserrat E, Rozman C: A study of prognostic factors in blast crisis of Philadelphia chromosome-positive chronic myelogenous leukaemia. Br J Haematol 1990;76: 27–32.
213 Griesshammer M, Heinze B, Hellmann A, Popp C, Anger B, Heil G, Bangerter M, Heimpel H: Chronic myelogenous leukemia in blast crisis: Retrospective analysis of prognostic factors in 90 patients. Ann Hematol 1996;73:225–230.

Dr. Bertil Johansson
Department of Clinical Genetics
University Hospital
SE–221 85 Lund (Sweden)
E-Mail bertil.johansson@klingen.lu.se

Bain BJ (ed): Chronic Myeloproliferative Disorders.
Basel, Karger, 2003, pp 62–68

# The 8p11 Myeloproliferative Syndrome: A Distinct Clinical Entity Caused by Constitutive Activation of *FGFR1*

Donald Macdonald[a] Andreas Reiter[b] Nicholas C.P. Cross[c,d]

[a]Department of Haematology, Faculty of Medicine, Imperial College of Science, Technology and Medicine, Charing Cross Hospital, London, UK; [b]III Medizinische Universitätsklinik, Klinikum Mannheim, Germany; [c]Wessex Regional Genetics Laboratory, Salisbury, [d]Human Genetics Division, University of Southampton School of Medicine, Southampton, UK

**Key Words**
*FGFR1* · 8p11 · Myeloproliferative disorder · Eosinophilia

**Abstract**
Several recurrent translocations that involve chromosome band 8p11 have been described in myeloid malignancies. These translocations target two distinct genes: (1) *FGFR1*, a receptor tyrosine kinase for fibroblast growth factors, and (2) *MOZ*, a putative histone acetyltransferase whose precise function remains to be defined. Disruption of *FGFR1* is associated with a disease entity known as the 8p11 myeloproliferative syndrome (EMS)/stem cell leukaemia-lymphoma syndrome (SCLL), a chronic myeloproliferative disorder that frequently presents with eosinophilia and associated T-cell lymphoblastic lymphoma. The disease is aggressive and rapidly transforms to acute leukaemia, usually of myeloid phenotype. Currently, only allogeneic stem cell transplantation appears to be effective in eradicating or suppressing the malignant clone. To date, four gene fusions associated with distinct translocations have been described in EMS: the t(8;13)(p11;q12), t(8;9)(p11;q33), t(6;8)(q27;p11) and t(8;22)(p11q22) fuse *ZNF198*, *CEP110*, *FOP* and *BCR*, respectively, to *FGFR1*. The resulting fusion proteins have constitutive tyrosine kinase activity and activate multiple signal transduction pathways. These pathways and the fusion proteins are attractive targets for targeted signal transduction therapy.

This article is based on a review first published in *Acta Haematologica* 2002;107:101–107.

### Introduction

The term myeloproliferative disorder (MPD) was originally proposed to describe the interrelationship between a group of blood disorders characterised by abnormal proliferation of one or more cell lineages derived from a common haemopoietic stem cell. The term also recognises the common strands in the natural history of these disorders, specifically the observation that all MPDs may evolve into a terminal myelofibrotic phase, or undergo transformation into an acute leukaemia. Currently, the term MPD includes the blood diseases polycythaemia vera (PV), essential thrombocythaemia, idiopathic myelofibrosis, and chronic myeloid leukaemia (CML). Each of these syndromes has a set of defined clinical and laboratory features which are often set out as diagnostic criteria. Consequently, an individual case can be compared against these criteria and, if particular features are present, an appropriate diagnosis made. Whilst most cases of MPD fit into one of the four categories above, there remain a miscellaneous group of conditions which appear to overlap these well-defined diseases. Such cases have tended to be grouped together under diagnostic labels such as atypical MPD or Philadelphia (*BCR-ABL*)-negative CML [1, 2]. Some of these cases also have an elevated monocyte count and thus overlap with chronic myelomonocytic leukaemia (CMML) [3].

In 1995 we published a review article in which we speculated on the existence, within this miscellaneous group, of a novel, distinct myeloproliferative syndrome with eosinophilia, generalised lymphadenopathy and, unusually, a high incidence of T-cell non-Hodgkin's lymphoma (NHL) [4]. The common finding in all cases was an acquired reciprocal translocation involving chromosome band 8p11 and a range of chromosome partners. For this reason we suggested the name '8p11 myeloproliferative syndrome' (EMS) for the disease, but it has also been referred to as 'stem cell leukaemia/lymphoma' (SCLL) [5]. Subsequent publications have confirmed the original hypothesis and elucidated the molecular basis of EMS/SCLL.

## Clinical Features

In the original description of EMS we identified 13 published case reports. We are now aware of 29 published cases of a myeloproliferative disorder associated with one of the 3 chromosomal translocations that were initially associated with EMS: t(8;13)(p11;q12) (n = 19); t(8;9)(p11;q34) (n = 7); and t(6;8)(q27;p11) (n = 4) [6–31]. Moreover, since the molecular basis of EMS has been elucidated, 7 further cases with novel translocations have been described; 2 cases with a t(8;22)(p11;q11) [32] and one each with a t(8;11)(p11;p15), t(8;12)(p11;q15), ins(12;8)(p11;p11p21), t(8;17)(p11;q25) or t(8;19)(p12;q13) [33, 34] (table 1).

Of the cases with the three most common translocations, the median age at presentation was 32 years, but the disorder has been described in all age groups (range 3–84 years). Indeed, of the 29 published cases, 7 were aged 18 or less at presentation. There is a slight male predominance of 1.5:1. Frequent presenting features included symptoms of hypermetabolism, lethargy, weight loss, fevers and night sweats. Of the cases for whom information was available, more than 60% had splenomegaly. Of interest, 2 of the patients with a t(6;8) were diagnosed initially as having PV [28] and 3 patients presented with acute leukaemia [16, 18, 30].

**Table 1.** Translocations reported to disrupt *FGFR1*

| Fusion gene | Translocation | Reference |
|---|---|---|
| *ZNF198-FGFR1* | t(8;13)(p11;q12) | 35–38 |
| *FOP-FGFR1* | t(6;8)(q27;p11) | 28 |
| *CEP110-FGFR1* | t(8;9)(p11;q33) | 29 |
| *BCR-FGFR1* | t(8;22)(p11;q11) | 32 |
| *?-FGFR1* | t(8;19)(p12;q13) | 33 |
| *?-FGFR1* | t(8;17)(p11;q25) | 34 |
| *?-FGFR1* | t(8;11)(p11;p15) | 34 |
| *?-FGFR1* | t(8;12)(p11;q15) | 34 |
| *?-FGFR1* | ins(12;8)(p11;p11p21) | 34 |

As for other blood disorders, some cases were identified by abnormalities seen on routine full blood counts. The appearance of the peripheral blood at presentation is in some respects similar to CML. Marked leukocytosis was found in all but 1 patient with the predominant cell types being neutrophils, metamyelocytes and myelocytes. However, the degree of left shift and the percentage of blasts varied and in some cases the appearances were of blastic transformation. A distinguishing feature of EMS is the finding of a marked eosinophilia, which is seen in 90% of patients in either the bone marrow, peripheral blood or both. In 14 cases, the median eosinophil count in the peripheral blood was 4 × $10^9$/l (range 1.2–40). The leucocyte alkaline phosphatase (LAP) score was reported for 6 cases only and found to be low in 5. The bone marrow showed myeloid hyperplasia in all cases. However, thrombocytosis and basophilia were found rarely. In some cases, particularly those associated with the t(8;9), thrombocytopenia and monocytosis were seen, and thus the disease resembled CMML but without major dysplastic signs in either lineage.

## Lymphadenopathy and Non-Hodgkin's Lymphoma

One of the most striking finding of EMS is the high frequency of lymphoblastic lymphoma, which is uncommon in other MPDs, but which is seen in more than three quarters of EMS cases. In common with CML, for which lymphoid transformation is well recognised, this suggests that the target for transformation in EMS is a lymphoid/myeloid stem cell. In some cases, the lymphoma is present at diagnosis whereas in others it appears during the course of the disease.

Lymph node biopsy was performed in 19 cases: the histology was T-cell NHL or T-cell lymphoblastic lymphoma (T-LBL) in 16 and myeloid metaplasia in 3. Of the 7 cases without lymphadenopathy, 2 presented as B-lineage acute lymphoblastic leukaemia (B-lineage ALL) and 5 had the clinical and laboratory features of a CML-like MPD (n = 3) or PV (n = 2). Two of the MPD cases subsequently progressed to B-lineage ALL. The observation of 16 cases of T-NHL and 4 cases of B-lineage ALL in a group of 29 patients with an MPD is remarkable and puts the incidence of high-grade lymphoid malignancy in EMS at approximately 60–70%.

All the NHLs seen in EMS are high-grade T-LBLs, diagnosed on the basis of morphology and immunophenotyping. Gene rearrangement analysis of the immunoglobulin *(IgH)* and T cell receptor *(TCR)* loci was reported for 6 cases. In 3 cases, only germline configurations of both the TCR and Ig loci were seen. In 2 cases, either the TCR-γ or -β loci were rearranged and in 1 case *IgH* was rearranged and the *TCR* was in the germline configuration. These variable results are consistent with the notion that the malignant transformation giving rise to the T-NHL occurs in a precursor cell prior to rearrangement of the TCR. Six patients had successful cytogenetics on the lymphoma (n = 5) or the tonsils (n = 1) and in each case the same 8p11 translocation that was seen in the bone marrow was found [5, 8, 11, 19, 22]. Of note, the incidence of T-NHL was higher in cases that presented with a t(8;13) compared to patients that presented with variant translocations: 13/18 patients with a t(8;13) had T-NHL compared to 3/11 patients with a t(6;8) or t(8;9).

## Natural History and Treatment

The natural history of EMS has some similarities to CML in that there is usually a chronic phase that is characterised by myeloid hyperplasia and overproduction of myeloid cells which retain the capacity to differentiate into mature cells. In the absence of effective treatment, most patients progress to acute myeloid leukaemia (AML) or, less commonly, B-lineage ALL. Transplantation and early death from complications associated with the lymphoma are likely explanations for the case reports where disease progression was not recorded.

It seems likely that the molecular events underlying EMS constitute potent stimuli to disease progression since evolution often occurs rapidly. Whilst it is impossible from the available case reports to give a precise estimate of the duration of chronic phase, it is apparent that almost all patients will progress within 1 or 2 years of diagnosis. The median time to transformation is probably between 6 and 9 months. This contrasts with CML where the chronic phase typically lasts between 2 and 7 years. Nevertheless, exceptions to the short chronic phase have been described; in particular 1 case had a chronic phase of 7 years before transformation to AML [28]. The most common cytogenetic abnormality associated with transformation is +21, which was seen in 9 patients who progressed to AML and 1 patient with lymphoma.

As EMS is a rare MPD, the ideal treatment is not established. It would appear from published reports that, similarly to CML, the malignant clone cannot be eradicated by conventional chemotherapy. In the absence of stem cell transplantation (SCT), disease progression was observed in the vast majority of cases. Despite aggressive chemotherapy with either AML or NHL treatment protocols, most patients died from resistant disease or early relapse within 1.5 years of diagnosis. One case presented with B-ALL, received ALL chemotherapy and re-entered a chronic myeloproliferative phase with t(8;13)-positive cells in the bone marrow. At the time of reporting, this chronic phase had been maintained for 4 years [18]. One patient who developed T-NHL received ALL chemotherapy followed by autologous SCT. At the time of reporting, this patient was in complete remission 21 months after diagnosis and 11 months after autograft [17]. In addition, we are aware of 1 case who achieved complete cytogenetic remission for several months with interferon-α (IFN) plus cytosine arabinoside, but who then transformed to AML [J.M. Goldman, pers. commun.].

It appears that only allogeneic SCT offers the prospect of cure. Seven cases have been reported who received chemotherapy followed by allogeneic SCT. Of these, 3 died of transplant-related complications and 4 survived in complete remission. One of the survivors had been managed on IFN for 1 year during chronic phase and achieved a partial cytogenetic response prior to undergoing SCT [26].

## Molecular Characterisation of the t(8;13)

Cloning of the t(8;13)(p11;q12) was reported by four groups in 1998 [35–38]. This translocation fuses a novel gene called *ZNF198* (also referred to as *RAMP* or *FIM*) at 13q11–12 to the fibroblast growth factor receptor 1 (*FGFR1*), the gene encoding a receptor tyrosine kinase that was already known to map to the 8p11–12 region. It was particularly striking that EMS and CML were not

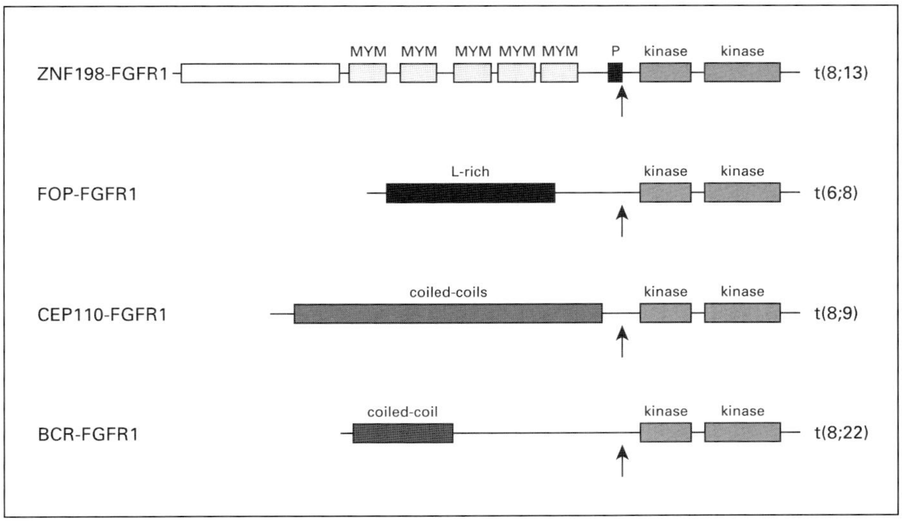


**Fig. 1.** Structures of the four known fusion proteins in EMS. MYM = Zinc-finger-related MYM domain; P = proline-rich domain; kinase = tyrosine kinase domain; L-rich = leucine rich region. The arrow indicates the points of fusion.

only similar clinically, but that they also shared a common molecular pathogenesis in that they both involve tyrosine kinase fusion genes.

The full length *ZNF198* gene is widely expressed and is predicted to encode a protein of 1,377 amino acids [36]. The structure of this gene was somewhat confused initially as two of the groups who characterised the t(8;13) inadvertently described truncated versions of *ZNF198* that were almost certainly cloning artefacts [37, 38]. The complete *ZNF198* protein is predicted to contain a series of novel zinc finger-like motifs which can be viewed as either 10 repeats of a 4 × cysteine motif [35] or 5 repeats of an 8 × cysteine motif [36, 37]. Presumably, structural analysis will eventually determine the correct size of the functional unit. *ZNF198* normally localises principally to the nucleolus and it has been suggested that it might play a role in the control or execution of ribosomal RNA transcription [39].

The t(8;13) fuses about two thirds of the *ZNF198* coding sequence to the entire kinase domain of *FGFR1*, starting just downstream of the transmembrane domain. The reciprocal *FGFR1-ZNF198* fusion is not detectable in all patients and is therefore probably pathogenetically unimportant. All cases thus far reported share a common mRNA fusion in which exon 17 of *ZNF198* is fused to exon 9 of *FGFR1* which can be conveniently detected by RT-PCR analysis. However, at the molecular level, the t(8;13) appears to be a complex rearrangement and so cytogenetic or fluorescence in situ hybridisation (FISH) analysis alone may lead to diverse assignments of the position of the translocation breakpoints on chromosome 13 [36, 40, 41]. FISH has been used to detect probable *FGFR1* rearrangements in patients with other translocations of 8p11–12, including 2 patients who presented with AML and 1 patient who presented with systemic mastocytosis, who subsequently progressed to Ph-negative CML [33, 34].

## The Molecular Basis of the Variant Translocations

The t(8;9)(p11;q33) and t(6;8)(q27;p12) have been cloned and shown to fuse *CEP110* and *FOP*, respectively, to *FGFR1* [28–31]. The demonstration that *FGFR1* was also involved in these cases confirmed our original postulate that the t(8;13), t(8;9) and t(6;8) constitute a discrete clinical entity. To date, mRNA fusions have been reported for 4 cases with a t(6;8), FOP exon 5 (n = 1), FOP exon 6 (n = 2) or FOP exon 7 (n = 1) was fused to *FGFR1* exon 9 [28, 34]. Of 4 cases reported with a t(8;9), CEP110 exon 15 was fused to *FGFR1* exon 9 in each case [29–31, 34]. Thus all EMS patients reported to date are characterised by a chimeric mRNA consisting of a partner gene spliced to *FGFR1* exon 9 (fig. 1).

*FOP* is widely expressed and predicted to encode an alternatively spliced 339 amino acid protein of unknown function that contains a novel leucine-rich repeat [28]. *CEP110* is predicted to encode a 994-amino-acid coiled-coil protein with 4 consensus leucine zipper motifs that was first cloned as a protein that cross-reacted with sera from a patient with autoimmune disease [29–31]. CEP110 is a widely expressed centrosome protein that may play a function in centriole maturation. The 3 principal *FGFR1* partner genes in EMS are entirely unrelated in sequence but they do share 2 features in common. First, they are all widely expressed and are therefore presumably expressed in haemopoietic stem cells. This is to be expected since the elements that control the expression of the partner gene will almost certainly also control expression of the *FGFR1* fusion gene. Second, they all contain putative dimerisation domains that are likely to be crucial for the activation of the *FGFR1* tyrosine kinase (see below).

Recently, we have identified 2 patients with a t(8;22)(p11;q12) and a *BCR-FGFR1* fusion [32]. Although it is difficult to generalise from just 2 cases, these 2 patients had a clinical and morphological picture that was essentially indistinguishable from typical, *BCR-ABL*-positive CML. Both patients had basophilia, a feature that is uncommon in *BCR-ABL*-negative MPDs and, as noted above, is rare in EMS. Both patients had eosinophilia, although the levels were within the range seen in CML, and 1 patient had an unusual marrow lymphocytosis. Thus these individuals appeared to resemble classical CML more closely than individuals with other *FGFR1* fusion proteins. One possible explanation for this is that the *BCR* moiety contributes to specific features of the disease phenotype.

### Mechanism and Consequences of FGFR1 Activation

Tyrosine kinases are normally tightly regulated signalling proteins that impact on proliferation, apoptosis and differentiation [42]. Normal FGFR1 is a trans-plasma membrane protein with an extracellular ligand binding domain, a transmembrane domain and a cytoplasmic tyrosine kinase domain. FGFR1 and its relatives FGFRs 2–4 play important roles in early development, in conjunction with their ligands, the fibroblast growth factors (FGFs), of which there are currently more than 20 members [43]. FGFRs may also impact on haemopoiesis, although their role in this context has not been clearly defined.

In their inactive state, receptors such as FGFR1 are thought to exist as monomers in the plasma membrane. Binding of FGF ligands induces dimerisation, which juxtaposes the two catalytic domains, inducing a conformational change which partially activates the enzymatic activity. This leads to transphosphorylation of a key tyrosine residue in the activation loop resulting in an increase in enzymatic activity, phosphorylation of additional tyrosines and subsequently phosphorylation and/or recruitment of target substrates [42, 43]. Normal FGFRs activate multiple signalling pathways including those involving Ras/MAPK, PI3K, PLCγ and STAT proteins. It is likely that these pathways are also activated by FGFR1 fusion proteins in EMS [44].

It is well established that tyrosine kinase fusion proteins such as BCR-ABL and ETV6-PDGFRB exist as dimers or multimers as a result of a homotypic interaction between specific domains of the partner gene, e.g. the BCR coiled-coil domain and the ETV6 pointed domain [45, 46]. Dimerisation mimics the initial stage of normal tyrosine kinase activation and thus the fusion proteins have constitutive enzymatic activity. Similarly, the FGFR1 fusion proteins in EMS have constitutive enzymatic activity [39, 44]. For ZNF198-FGFR1, the dimerisation domain is an unusual proline-rich region within the ZNF198 moiety, immediately proximal to the point of fusion with FGFR1 [47]. Dimerisation domains within CEP110 and FOP have not been defined experimentally but are likely to be composed of one or more coiled-coil or leucine-rich domains.

### Targeted Therapy for 8p11 Myeloproliferative Syndrome?

Recently, there has been considerable excitement over the initial dramatic results of treatment of CML patients with the tyrosine kinase inhibitor imatinib mesylate (STI571) (Glivec™) [48]. STI571 is inactive against FGFR1 and so it is very unlikely that this compound will benefit EMS patients. Other reagents, however, have been developed that do have anti-FGFR1 activity, for example SU5404, SU6668 and PD173074 [49, 50]. Paralleling the effect of STI571 on the BCR-ABL protein, these compounds block the FGFR1 ATP binding site and are able to specifically inhibit the growth of *ZNF198-FGFR1*-transformed cells [32]. Although no patient with EMS has yet been treated with a specific inhibitor, the remarkable effects of STI571 in CML suggest that this approach may also be beneficial for patients with EMS.

## References


1 Shepherd PCA, Ganesan TS, Galton DA: Haematological classification of the chronic myeloid leukaemias Baillières Clin Haematol 1987;1:887–906.
2 Oscier D: Atypical chronic myeloid leukemias. Pathol Biol 1997;45:587–593.
3 Bain BJ: The relationship between the myelodysplastic syndromes and the myeloproliferative disorders. Leuk Lymphoma 1999;34:443–449.
4 Macdonald D, Aguiar RCT, Mason PJ, Goldman JM, Cross NCP: A new myeloproliferative disorder associated with chromosomal translocations involving 8p11: A review. Leukemia 1995;9:1628–1630.
5 Inhorn RC, Aster JC, Roach SA, Slapak CA, Soiffer R, Tantravahi R, Stone RM: A syndrome of lymphoblastic lymphoma, eosinophilia, and myeloic hyperplasia/malignancy associated with t(8;13)(p11;q11): Description of a distinctive clinicopathologic entity. Blood 1195;85:1881–1887.
6 Friedhoff F, Rajendra B, Moody R, Alapatt T: Novel reciprocal translocation between chromosomes 8 and 9 found in a patient with myeloproliferative disorder. Cancer Genet Cytogenet 1983;9:391–394.
7 Lewis JP, Jenks H, Lazerson J: Philadelphia chromosome-negative chronic myelogenous leukemia in a child with t(8;9)(p11 or 12;q34). Am J Pediatr Hematol Oncol 1983;5:265–269.
8 Vannier JP, Bizet M, Bastard C, Bernard A, Ducastelle T, Tron P: Simultaneous occurrence of a T-cell lymphoma and a chronic myelogenous leukemia with an unusual karyotype. Leuk Res 1984;8:647–657.
9 Abruzzo LV, Jaffe ES, Cotelingam JD, Whang-Peng J, Del DV Jr, Medeiros LJ: T-cell lymphoblastic lymphoma with eosinophilia associated with subsequent myeloid malignancy. Am J Surg Pathol 1992;16:236–245.
10 Jotterand BM, Muhlematter D, Wicht M, Delacretaz F, Schmidt PM: t(8;9)(p11;q32) in atypical chronic myeloid leukaemia: A new cytogenetic-clinicopathologic association? Br J Haematol 1992;81:307–308.
11 Rao PH, Cesarman G, Coleman M, Acaron S Verma RS: Cytogenetic evidence for extramedullary blast crisis with t(8;13)(q11;p11) in chronic myelomonocytic leukemia. Acta Haematol 1992;88:201–203.
12 Fagan K, Hyde S, Harrison P: Translocation (8;13) and T-cell lymphoma. A case report. Cancer Genet Cytogenet 1993;65:71–73.
13 Oscier DG, Gardiner A, Dyer M: t(8;9) in chronic myeloid leukaemia. Br J Haematol 1993;83:176–177.
14 Leslie J, Barker T, Glancy M, Jennings B, Pearson J: t(8;13) (p11;q12) translocation in a myeloproliferative disorder associated with a T-cell non-Hodgkin lymphoma. Br J Haematol 1994; 85:876–878.
15 Macdonald D, Sheerin SM, Cross NCP, Spencer A, Goldman JM: An atypical myeloproliferative disorder with a t(8;13)(p11;q12): A third case. Br J Haematol 1994;86:879–880.
16 Elsner S, Martin H, Rode C, Wassmann B, Ganser A, Hoelzer D: An uncommon translocation t(6;8) associated with atypical acute myeloid leukaemia/myeloproliferative disease detected by fluorescence in-situ hybridization. Br J Haematol 1994;87:124.
17 Behringer D, Schaefer HE, Kunzmann R, Mertelsmann R, Dolken G: Translocation t(8;13) in a patient with T cell lymphoma and features of a myeloproliferative syndrome. Leukemia 1995;9:988–992.
18 Michaux L, Mecucci C, Pereira Velloso ER, Dierlamm J, Criel A, Louwagie A, van Orshoven A, Van den BH: About the t(8;13) (p11;q12) clinico-pathologic entity. Blood 1996;87:1658–1659.
19 Nakayama H, Inamitsu T, Ohga S, Kai T, Suda T, Suda M, Matsuzaki A, Ueda K: Chronic myelomonocytic leukaemia with t(8;9)(p11; q34) in childhood: An example of the 8p11 myeloproliferative disorder? Br J Haematol 1996;92:692–695.
20 van den Berg H, Kroes W, van der Schoot CE, Dee R, Pals ST, Bouts TH, Slater RM: A young child with acquired t(8;9)(p11;q34). Additional proof that 8p11 is involved in mixed myeloid/ T lymphoid malignancies. Leukemia 1996;10: 1252–1253.
21 Aguiar RCT, Chase A, Coulthard S, Macdonald DH, Carapeti M, Reiter A, Sohal J, Lennard A, Goldman JM, Cross NCP: Abnormalities of chromosome band 8p11 in leukemia: Two clinical syndromes can be distinguished on the basis of MOZ involvement. Blood 1997; 90:3130–3135.
22 Somers GR, Slater H, Rockman S, Ekert H, Southey MC, Chow CW, Armes JE, Venter DJ: Coexistent T-cell lymphoblastic lymphoma and an atypical myeloproliferative disorder associated with t(8;13)(p21;q14). Pediatr Pathol Lab Med 1997;17:141–158.
23 Still IH, Chernova O, Hurd D, Stone RM, Cowell JK: Molecular characterization of the t(8;13)(p11;q12) translocation associated with an atypical myeloproliferative disorder: Evidence for three discrete loci involved in myeloid leukemias on 8p11. Blood 1997;90:3136–3141.
24 Chaffanet M, Popovici C, Leroux D, Jacrot M, Adelaide J, Dastugue N, Gregoire MJ, Hagemeijer A, Lafage-Pochitaloff M, Birnbaum D, Pebusque MJ: t(6;8), t(8;9) and t(8;13) translocations associated with stem cell myeloproliferative disorders have close or identical breakpoints in chromosome region 8p11–12. Oncogene 1998;16:945–949.
25 Chernova O, Still I, Kalaycio M, Hoeltge G, Cowell JK: Characterization of the breakpoints in a t(8;13)(p11;q12) translocation from a patient with myeloproliferative disease using fluorescence in situ hybridization. Genes Chromosomes Cancer 1998;21:160–165.
26 Martinez-Climent JA, Vizcarra E, Benet I, Marugan I, Terol MJ, Solano C, Arbona C, Tormo M, Comes AM, Garcia-Conde J: Cytogenetic response induced by interferon alpha in the myeloproliferative disorder with eosinophilia, T cell lymphoma and the chromosomal translocation t(8;13)(p11;q12). Leukemia 1998;12: 999–1000.
27 Matsumoto K, Morita K, Takada S, Sakura T, Shiozaki H, Murakami H, Miyawaki S: A chronic myelogenous leukemia-like myeloproliferative disorder accompanied by T-cell lymphoblastic lymphoma with chromosome translocation t(8;13)(p11;q12): A Japanese case. Int J Hematol 1999;70:278–282.
28 Popovici C, Zhang B, Gregoire MJ, Jonveaux P, Lafage-Pochitaloff M, Birnbaum D, Pebusque MJ: The t(6;8)(q27;p11) translocation in a stem cell myeloproliferative disorder fuses a novel gene, FOP, to fibroblast growth factor receptor 1. Blood 1999;93:1381–1389.
29 Guasch G, Mack GJ, Popovici C, Dastugue N, Birnbaum D, Rattner JB, Pebusque MJ: FGFR1 is fused to the centrosome-associated protein CEP110 in the 8p12 stem cell myeloproliferative disorder with t(8;9)(p12;q33). Blood 2000;95:1788–1796.
30 Roy S, Szer J, Campbell LJ, Juneja S: Sequential transformation of the t(8;13) related disease. Acta Haematol 2002;107:95–97.
31 Al-Obaidi MJ, Rymes N, White P, Pomfret M, Smith H, Starczynski J, Johnson R: A fourth case of 8p11 Myeloproliferative disorder transforming to B-lineage acute lymphoblastic leukaemia. Acta Haematol 2002;107:98–100.
32 Demiroglu A, Steer EJ, Heath C, Taylor K, Bentley M, Allen SL, Koduru P, Brody JP, Hawson G, Rodwell R, Doody M-L, Carnicero F, Reiter A, Goldman JM, Melo JV, Cross NCP: The t(8;22) in chronic myeloid leukemia fuses BCR to FGFR1: Transforming activity and specific inhibition of FGFR1 fusion proteins. Blood 2001;98:3778–3783.
33 Mugneret F, Chaffanet M, Maynadie M, Guasch G, Favre B, Casasnovas O, Birnbaum D, Pebusque MJ: The 8p12 myeloproliferative disorder. t(8;19)(p12;q13.3): A novel translocation involving the FGFR1 gene. Br J Haematol 2000;111:647–649.
34 Sohal J, Chase A, Mould S, Corcoran M, Oscier D, Iqbal S, Parker S, Welborn J, Harris I, Martinelli G, Montefusco V, Sinclair P, Wilkins BS, van den Berg H, Vanstraelen D, Goldman JM, Cross NCP: Identification of four new translocations involving FGFR1 in myeloid disorders. Genes Chromosomes Cancer 2001; 32:155–163.
35 Popovici C, Adelaide J, Ollendorff V, Chaffanet M, Guasch G, Jacrot M, Leroux D, Birnbaum D, Pebusque MJ: Fibroblast growth factor receptor 1 is fused to FIM in stem-cell myeloproliferative disorder with t(8;13). Proc Natl Acad Sci USA 1998;95:5712–5717.

36 Reiter A, Sohal J, Kulkarni S, et al: Consistent fusion of ZNF198 to the fibroblast growth factor receptor-1 in the t(8;13)(p11;q12) myeloproliferative syndrome. Blood 1998;92:1735–1742.
37 Smedley D, Hamoudi R, Clark J, Warren W, Abdul-Rauf M, Somers G, Venter D, Fagan K, Cooper C, Shipley J: The t(8;13)(p11;q11–12) rearrangement associated with an atypical myeloproliferative disorder fuses the fibroblast growth factor receptor 1 gene to a novel gene RAMP. Hum Mol Genet 1998;7:637–642.
38 Xiaou S, Nalabolu SR, Aster JC, et al: FGFR1 is fused with a novel zinc-finger gene, ZNF198, in the t(8;13) leukaemia/lymphoma syndrome. Nat Genet 1998;18:84–87.
39 Ollendorff V, Guasch G, Isnardon D, Galindo R, Birnbaum D, Pebusque MJ: Characterization of FIM-FGFR1, the fusion product of the myeloproliferative disorder-associated t(8;13) translocation. J Biol Chem 1999;274:26922–26930.
40 Smedley D, Somers G, Venter D, Chow CW, Cooper C, Shipley J: Characterization of a t(8;13)(p11;q11–12) in an atypical myeloproliferative disorder. Genes Chromosomes Cancer 1998;21:70–73.
41 Kempski H, Macdonald D, Michalski AJ, Roberts T, Goldman JM, Cross NCP, Cowell JK: Localization of the 8;13 translocation breakpoint associated with myeloproliferative disease to a 1.5 Mbp region of chromosome 13. Genes Chromosomes Cancer 1995;12:283–287.
42 Jiang G, Hunter T: Receptor signaling: when dimerization is not enough. Curr Biol 1999;9:R568–R571.
43 Mason IJ: The ins and outs of fibroblast growth factors. Cell 1994;78:547–552.
44 Smedley D, Demiroglu A, Abdul-Rauf M, Heath C, Cooper C, Shipley J, Cross NCP: ZNF198-FGFR1 transforms Ba/F3 cells to growth factor independence and results in high level tyrosine phosphorylation of STATs 1 and 5. Neoplasia. 1999;1:349–355.
45 McWhirter JR, Galasso DL, Wang JY: A coiled-coil oligomerization domain of Bcr is essential for the transforming function of Bcr-Abl oncoproteins. Mol Cell Biol 1993;13:7587–7595.
46 Carroll M, Tomasson MH, Barker GF, Golub TR, Gilliland DG: The TEL/platelet-derived growth factor beta receptor (PDGF beta R) fusion in chronic myelomonocytic leukemia is a transforming protein that self-associates and activates PDGF beta R kinase-dependent signaling pathways. Proc Natl Acad Sci USA 1996;93:14845–14850.
47 Xiao S, McCarthy JG, Aster JC, Fletcher JA: ZNF198-FGFR1 transforming activity depends on a novel proline-rich ZNF198 oligomerization domain. Blood 2000;96:699–704.
48 Druker BJ, Sawyers CL, Kantarjian H, Resta DJ, Reese SF, Ford JM, Capdeville R, Talpaz M: Activity of a specific inhibitor of the BCR-ABL tyrosine kinase in the blast crisis of chronic myeloid leukemia and acute lymphoblastic leukemia with the Philadelphia chromosome. N Engl J Med 2001;344:1038–1042.
49 Mohammadi M, McMahon G, Sun L, Tang C, Hirth P, Yeh BK, Hubbard SR, Schlessinger J: Structures of the tyrosine kinase domain of fibroblast growth factor receptor in complex with inhibitors. Science 1997;276:955–960.

Prof. N.C.P. Cross
Wessex Regional Genetics Laboratory
Salisbury District Hospital
Salisbury SP2 8BJ (UK)
E-Mail ncpc@soton.ac.uk

Bain BJ (ed): Chronic Myeloproliferative Disorders.
Basel, Karger, 2003, pp 69–79

# Myeloproliferative Disorders with Translocations of Chromosome 5q31–35: Role of the Platelet-Derived Growth Factor Receptor Beta

E. Joanna Steer Nicholas C.P. Cross

Wessex Regional Genetics Laboratory, Salisbury, and Human Genetics Division, University of Southampton School of Medicine, Southampton, UK

**Key Words**
PDGFRB · 5q33 · 5q31 · Myeloproliferative disorder · Eosinophilia

**Abstract**
Acquired reciprocal chromosomal translocations that involve chromosome bands 5q31–33 are associated with a significant minority of patients with *BCR-ABL*-negative chronic myeloid leukemias. The most common abnormality is the t(5;12)(q33;p13), which fuses the *ETV6/TEL* gene to the platelet-derived growth factor receptor-β *(PDGFRB)*, a receptor tyrosine kinase that maps to 5q33. *PDGFRB* is disrupted by other translocations and to date four additional partner genes *(H4, HIP1, CEV14* and *Rab5)* have been reported. Clinically, most patients present with a myeloproliferative disorder (MPD) with eosinophilia, eosinophilic leukemia or chronic myelomonocytic leukemia and thus fall into the broader category of myelodysplastic syndromes/myeloproliferative disorders (MDS/MPD). With the advent of targeted signal transduction therapy, patients with rearrangement of *PDGFRB* might be better classified as a distinct subgroup of MDS/MPD.

This article is based on a review first published in *Acta Haematologica* 2002;107:113–122.

## Introduction

*t(5;12)*

To our knowledge, the association between the t(5;12)(q33;p13) and a chronic myeloid leukemia (CML)-like disorder with eosinophilia was first made by Keene et al. [1] in 1987. Currently, at least 34 cases with a related disorder and a t(5;12)(q31–33;p12–13) have been reported in the literature [1–20]. A distinct t(5;12)(q31q33;p12) has been reported in association with pre-B acute lymphoblastic leukemia [21]. The myeloid t(5;12) was cloned in 1994 by Golub et al. [5] and shown to fuse a novel gene called *TEL* (subsequently renamed *ETV6*) at 12p13 to the platelet-derived growth factor receptor-β *(PDGFRB)* gene at 5q33. ETV6 is an ets-related transcription factor that plays an essential role in bone marrow hemopoiesis [22]. PDGFRB is a receptor tyrosine kinase which in adults plays an important role in wound healing and other processes. The ETV6-PDGFRB fusion protein generated as a consequence of the t(5;12) contains the pointed (helix-loop-helix) domain of ETV6 plus the entire tyrosine kinase and transmembrane domains of PDGFRB (fig. 1).

*Other Translocations That Involve PDGFRB*

In addition to the t(5;12), several other cases with translocations of 5q31–5q33 have been described [10, 23–40]. To date three other partner genes have been

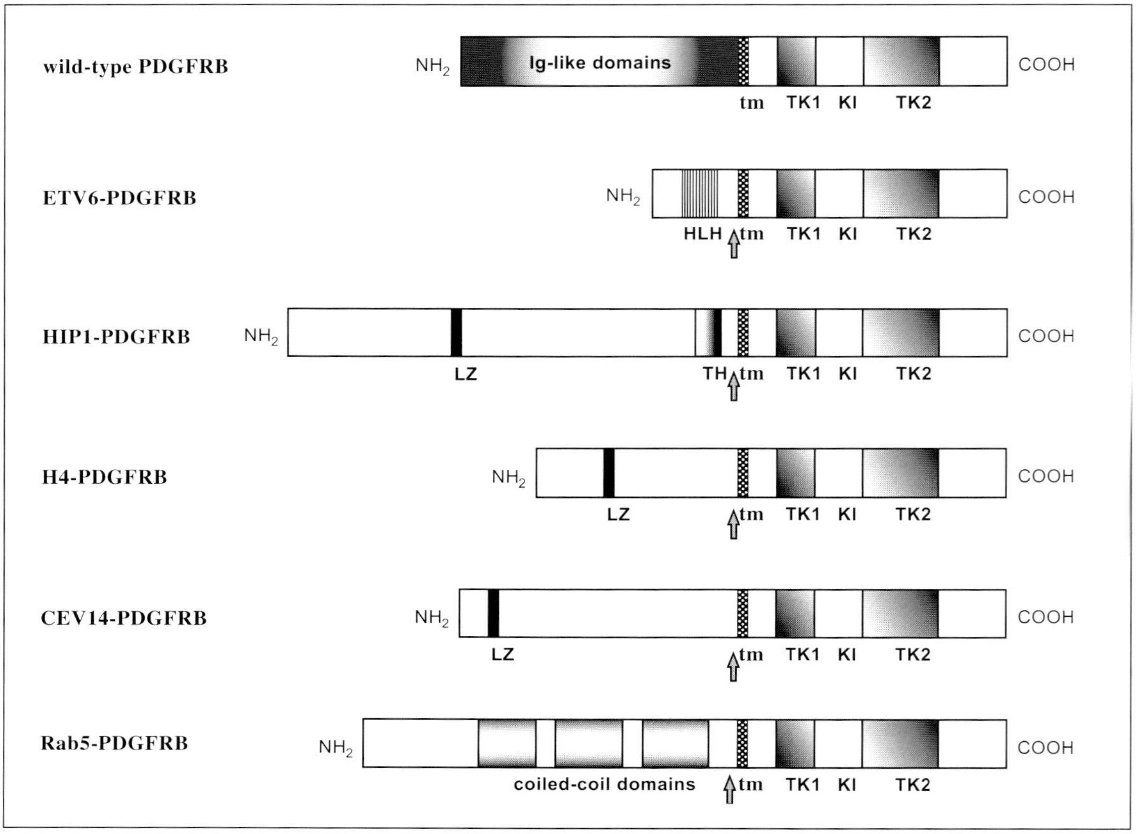


**Fig. 1.** Structure of PDGFRB fusion proteins. tm = Transmembrane domain; TK1 and TK2 = tyrosine kinase domains; KI = kinase insert; HLH = helix-loop-helix; LZ = leucine zipper; TH = talin homology domain. The arrows indicate the point of fusion in each case.

cloned from patients who presented with a CML/CMML-like disease: the t(5;7)(q33;q11), t(5;10)(q33;q21.2) and t(5;17)(q33;p13) fuse *HIP1*, *H4/D10S170* and *Rab5* to *PDGFRB* (fig. 1) [30, 31, 39]. A fourth partner gene, *CEV14*, was described in a patient who acquired a t(5;14)(q33;q32) as a secondary abnormality at relapse, having initially presented with AML. However, it is notable that the acquisition of the t(5;14) was associated with the appearance of marked eosinophilia and hepatosplenomegaly, i.e. MPD-like features [36, 41]. The four partner genes are unrelated in sequence, but they are all widely expressed and contain dimerization domains that are critical to the aberrant function of the fusion proteins (see below).

Unpublished results from our laboratory have shown that other translocations target *PDGFRB* and suggest that several partner genes remain to be characterized. We have also found that translocations assigned by cytogenetics to either 5q31 or 5q33 may disrupt *PDGFRB*. However, fluorescence in situ hybridization (FISH) analysis alone may be misleading: the chromosome 5 breakpoint in a patient with a t(5;10) appeared to be distal to *PDGFRB* by FISH, but a rearrangement of this gene was found by Southern blotting and a *H4-PDGFRB* fusion mRNA was detected by RT-PCR. The reason for this discrepancy was that a complex rearrangement had occurred with a small inversion in addition to the translocation [31]. Complex rearrangements appear to be a common feature of rare translocations and so a combination of approaches may be necessary to determine or exclude gene involvement in any new case.

*Alternative Target Genes at 5q31–35*

Currently, it is not clear what proportion of CML-like MPDs with a translocation of 5q31–33 actually involve *PDGFRB*. Certainly, there are other genes in the region

**Table 1.** 5q31–5q34 translocations in leukaemia that do not involve PDGFRB

| Translocation | Phenotype | 5q partner | Fusion | Ref. |
|---|---|---|---|---|
| t(3;5)(q25;q34) | trilineage MDS/AML | *MLF1* | *NPM-MLF1* | 42 |
| t(5;11)(q31;q23) | JMML | *GRAF* | *MLL-GRAF* | 43 |
| ins(5;11)(q31;q13q23) | pre-B ALL | *AF5q31* | *MLL-AF5q31* | 44 |
| t(5;11)(q35;p15.5) | AML | *NSD1* | *NUP98-NSD1* | 45 |
| t(5;12)(q31q33;p12) | pre B-ALL | ? | ? | 21 |
| t(5;12)(q31;p13) | MDS/AML | *ACS2* | *ETV6-ACS2* | 46 |
| t(5;14)(q31;q32) | B-ALL | *IL-3* | *IgH-IL-3* | 47 |

that are targeted by translocations in hematological disorders, for example *MLF1, GRAF, AF5q31, NSD1, ACS2* and *IL-3* (table 1) [21, 42–47]. However, only one of these fusions has been reported in an MPD – a single case with juvenile myelomonocytic leukemia (JMML) with a t(5;11)(q31;q23) and an *MLL-GRAF* fusion [43]. Cases with a t(3;5)(q25;q34) or t(5;12)(q35;p15) typically present with AML, or trilineage MDS that rapidly progresses to AML, and the remaining translocations are associated with lymphoid malignancies [44, 47]. Three MDS/AML patients were reported with a t(5;12)(q31;p13) and involvement of *ACS2*. Interestingly, 1 of these cases had acute eosinophilic leukemia, 1 had AML with eosinophilia and 1 had refractory anemia with excess blasts with basophilia, i.e. features that might be considered as similar to those seen in patients with *PDGFRB* fusions. However, the fusion mRNAs reported in these cases were very unusual, either being out of frame or involving a noncoding exon [46]. It is possible therefore that the true oncogenic lesion in these cases was not identified.

### Clinical Features of Patients with a t(5;12)

*Myeloid Disorders with a t(5;12)*

MPDs with 5q31–33 abnormalities are uncommon and it is not clear what range of clinical features they are associated with. We have therefore undertaken a literature review of all reports of patients carrying a t(5;12)(q31 ~ 33;p13) in association with a myeloid malignancy. Due to the problem of complex translocations referred to above, we included cases for whom FISH analysis apparently excluded involvement of *ETV6* or *PDGFRB*. We excluded all cases with lymphoid leukemias since these probably do not involve *PDGFRB* [21]. Overall we identified 34 patients for analysis (table 2).

*Initial Diagnosis*

The most common diagnoses (n = 30) were MPDs: CMML, eosinophilic leukemia, aCML, Philadelphia or *BCR-ABL*-negative CML. Nine of these cases were reported as an abstract by Hanson et al. [20] with clinical details limited to median values and ranges. We have used this group for comparison with all other patients. Two of the remaining patients had MDS and 2 had AML. Of the 2 MDS patients, 1 had eosinophilia with a rearrangement of *PDGFRB;* no clinical details were given for the second case. Both of the AML cases presented with de novo acute leukemia and the t(5;12) was noted at relapse. It was not stated if this translocation was also present at earlier stages of the disease. We included these as it is well established that MPDs can progress to AML and may occasionally present in advanced phase. Only 4 patients were known to have the *ETV6-PDGFRB* fusion and a further 7 cases had *ETV6* or *PDGFRB* rearrangements as determined by FISH or Southern blotting. The median age for t(5;12) patients was 42 years (range 8–80; n = 22), which is comparable with the Hanson series (median 58, range 42–67, n = 9).

*Eosinophilia*

Of the 25 patients for whom some information was available, only 3 had no indication of elevated eosinophil counts in either the peripheral blood or bone marrow. In the 15 cases for whom it was reported, the median percentage of eosinophils in the peripheral blood was 11 (range 3–94). However, blood eosinophilia was not observed in any of the 9 cases reviewed by Hanson et al. [20].

*Splenomegaly and Monocytosis*

Splenomegaly was reported in 19 of 25 patients and hepatomegaly was also described in a minority of cases. Of 23 cases for whom information was provided on monocyte counts, 12 were described as having monocyto-

**Table 2.** Cases in the literature with t(5;12) in myeloid leukaemias

| Patient | Sex | Age | Diagnosis | Translocation | Survival m | Splenomegaly | WBC | Neutrophils % | Eosinophils % | Monocytes % | Notes | Ref. |
|---|---|---|---|---|---|---|---|---|---|---|---|---|
| 1 | M | 42 | aCML | t(5;12)(q33;p13) | 72 | N | 26.5 | 73 | | 1 | | 2 |
| 2 | M | 60 | Eo CMPD | t(5;12)(q33;p13) | 48d | N | 14.6 | 55 | 7.5 | 4 | transformed after 4 yrs, died 1 month later | 3 |
| 3 | M | 80 | Eo CMPD | t(5;12)(q33;p13) | 2d | N | 10.7 | 14 | 23 | 0 | | 3 |
| 4 | M | 16 | CMML | t(5;12)(q31;p12) | 156 | 8 cm | 46.4 | 61 | 6 | 5 | transformed after 12 yrs, then BMT | 4 |
| 5 | M | | CMML | t(5;12) | | | | | | | transformed to AML, acquired t(8;21) | 5 |
| 6 | M | | CMML | t(5;12) | | | | | | | | 5 |
| 7 | M | | CMML | t(5;12) | | | | | | | | 5 |
| 8 | M | 23 | CEoL | t(5;12)(q33;p13) | 14 | Y | 14 | 44 | 11 | 1 | skin lesions at presentation *PDGFRB* rearranged by FISH | 6 |
| 9 | M | 72 | MDS | t(5;12)(q32-3;p12) | 12d | | | | | | *ETV6* rearranged by FISH. Died of ALS | 7 |
| 10 | M | | CMML | t(5;12) | | | | | | | | 8 |
| 11 | M | 59 | EoL | t(5;12)(q31;p13) | 3d | Y | | | 94 | | died of haemorrhage and heart failure | 1 |
| 12 | M | 23 | EoL | t(5;12)(q33;p13) | 5 | 8 cm | | | 29 | 5 | schistosomiasis and residual bilharzia | 1 |
| 13 | M | 52 | CMML | t(5;12)(q31;p12.3) | 34 | N | 13.6 | | | 10 | | 9 |
| 14 | M | 34 | Ph- CML | t(5;12)(q31;p12-3) | 14 | | | | | | | 10 |
| 15 | F | 8 | MDS with Eo. | t(5;12)(q31;p13) | 168 | Y | 33.7 | 49 | 6 | | *PDGFRB* rearranged by Southern blot; breakpoint tel to *ETV6* by FISH | 11 |
| 16 | M | 25 | MPD | t(5;12)(q33;p12) | | | | | | | *ETV6* rearranged by Southern blot | 12 |
| 17 | M | 60 | MPD | t(5;12)(q33;p12-13) | | | | | | | *ETV6* rearranged by Southern blot | 12 |
| 18 | F | 41 | AML M4 R | t(5;12)(q33;p13) | | | | | | | additional t(11;19)(q23;p13.1) breakpoint cen. to *ETV6* by FISH | 13 |
| 19 | M | 41 | CMML | t(5;12)(q31;p12) | 31 | moderate | 32.4 | 40 | 11 | 7 | Kras overexpression | 14 |
| 20 | M | 53 | CML | t(5;12)(q31;p12) | 9d | slight | 32 | 17 | 3 | 6 | died of sepsis in BC | 15 |
| 21 | M | 40 | CML | t(5;12)(q31;p12) | 14d | N | 46.9 | 26 | 12 | 2 | died post BMT due to leukaemia | 15 |
| 22 | M | | MDS/CMML | t(5;12)(q33;p13) | | | | | | | | 16 |
| 23 | M | 67 | CMML | t(5;12) | 20d | 5 cm | 38.6 | 56 | 5 | 10 | *ETV6* rearranged by FISH | 17 |
| 24 | M | 44 | CMML | t(5;12) | 3d | 6 cm | 87 | 5 | 6 | 9 | *ETV6* rearranged by FISH | 17 |
| 25 | M | 63 | CMML | t(5;12) | 5 | 6 cm | 70 | 40 | 27 | 10 | *ETV6-PDGFRB* positive by RT-PCR | 17 |
| 26 | M | 30 | AML M2 2nd R | t(5;12)(q31;p12) | 1d | | 41.1 | | 42 | | died of sepsis | 18 |
| 27 | M | 59 | EoL | t(5;12)(q31;p12) | 3d | moderate | 20.9 | 68 | 10 | 2 | additional t(19;21) | 19 |
| Median | | 42 | | | 20 | | 32.4 | 44 | 11 | 5 | | 1–19 |
| 28–37 | | med: 58 | CMPD | t(5;12) | | 8 of 9 patients | med: 52.6 | med: 82 | 0/9 | 3 of 9 patients | BM eosinophilia in 7/9, med: 8% | 20 |

Age = age at diagnosis
ALS = amyotrophic lateral sclerosis
aCML = atypical chronic myeloid leukaemia
CMML = chronic myelomonocytic leukaemia
MPD = myeloproliferative disorder
MDS = myelodysplastic syndrome
(C)EoL = (chronic) eosinophilic leukaemia
EoCMPD = eosinophilic chronic myeloproliferative disorder
JCML = juvenile chronic myeloid leukaemia

RAEB = refractory anaemia with excess of blasts
RA = refractory anaemia
R = relapse
BC = blast crisis
d = died
med = median
cen = centromeric
tel = telomeric

sis. For the 14 patients with available data, the median percentage of monocytes was 5 (range 0–10). Apart from 1 case (and the 2 patients with AML) peripheral blood blasts were not elevated. The median basophil count was 1.5% (range 0–7; n = 12) but basophilia was not seen in any of the 9 Hanson cases.

*Treatment*

The treatment regimen was reported for 17 cases, of whom 1 patient received no therapy. For the 15 patients who were in a chronic phase at presentation, patients were treated with hydroxyurea, busulfan or interferon-alpha, in some cases with the addition of cytosine arabinoside or other cytoxic agents. Of the 3 patients treated with interferon, 1 was described as having a good response but stopped treatment due to adverse side effects [4]. No information was provided for the other 2 cases. One patient underwent allogeneic bone marrow transplantation but relapsed and died without further treatment. Nine patients were treated for leukemia from diagnosis, the other cases began therapy between 3 and 21 months after presentation. Three patients were initially treated for other disorders (schistosomiasis; skin nodules caused by eosinophilic infiltration that was initially diagnosed as contact dermatitis; reactive arthritis) before antileukemia treatment.

*Extreme Male Bias*

Perhaps the most remarkable feature of the t(5;12) cases is that virtually all are male. Of the 34 patients, only 2 were female, 1 of whom was a child and the other had relapsed AML, with the translocation present at relapse, but it was not stated whether it was present previously. This is an extraordinary male bias not seen in any other leukemia, although an excess of males is seen in patients with idiopathic eosinophilia [48] and a very small but significant male excess is seen in CML. The reason for this phenomenon is completely unclear but possibilities include: (a) a male-specific mechanism for inducing a double-stranded break in *PDGFRB* as the initial event in the formation of the translocation; (b) a female-specific immune response against cells expressing *ETV6-PDGFRB;* (c) a female-specific hormonal mechanism that in some way blocks the expansion of the *ETV6-PDGFRB*-positive clone, and (d) that the development of the disease requires two hits: *ETV6-PDGFRB* plus an unknown X-linked gene, mutations of which might be present in the general population.

*Disease Progression and Survival*

The median follow up of cases was highly variable and in some cases was very short. We therefore wrote to the original authors to enquire about the course of the disease in each patient alive at reporting. Not surprisingly many patients were lost to follow-up but additional information was obtained for 15 cases, the majority of whom were still alive.

Only 5 patients (16% of the total) who presented in chronic phase transformed to acute leukemia. However, because of the variable follow-up this is likely to be an underestimate of the true probability of transformation. Nevertheless, these figures do suggest that transformation is infrequent relative to other similar diseases. For example, it has been estimated that one third to one half of patients with CMML will progress to acute leukemia [10]. The time to transformation was highly variable and ranged from 9 months to 12 years. Of the 10 patients who were known to have died, the cause of death was infection (n = 2), leukemia (n = 5), amyotropic lateral sclerosis (n = 1) and not specified (n = 2). For the 19 evaluable patients the 2-year survival was 47%. The median survival for all patients was 20 months.

**Clinical Features of Patients with Variant Translocations**

*Patients with 5q31–35 Translocations*

A review of patients carrying possible variant translocations that might involve *PDGFRB* is hampered by the fact that there are numerous cases of MDS or AML in the literature with abnormalities of chromosome 5q. Many of these might involve the uncharacterized target of recurrent 5q deletions rather than *PDGFRB* [49]. We have included here patients with myeloid disorders and a reciprocal translocation that targeted 5q31–35, unless this was a recurrent translocation that is known not to involve *PDGFRB*, such as the t(3;5)(q25;q34). However, we did include patients with similar but cytogenetically distinct translocations, such as the t(3;5)(q25;q33). We also included 1 patient with eosinophilic leukemia and a t(2;5)(p23;q35), a translocation that is normally associated with the *NPM-ALK* fusion in T-lineage anaplastic large cell lymphoma [50]. A total of 20 cases were identified in the literature for analysis (table 3).

*Clinical Features*

In this group of patients there is a greater diversity of phenotypes and less information available than in cases with a t(5;12)(q31 ~ q33;p13). Patient diagnosis is much more variable, probably because only a subset of cases actually involved *PDGFRB;* however, 5 cases were definitely known to involve this gene. This group included more children and the median age at presentation was 38 years. There were 15 males and 5 females. Splenomegaly was seen in 10 of 13 cases but monocytosis was rare. The 2-year survival was 76% and the median survival was not reached.

Despite the likely heterogeneity of this group, it is interesting to note that many had eosinophilia (11 of 11 cases where information was given). However, in 1 case it was established that the eosinophils did not carry a translocation, and in some other reports the eosinophils were described as fully mature with normal morphology [34]. This suggests that the increased eosinophil count in some cases may be reactive. Possibly relevant to this is the fact that the gene encoding IL-5, the cytokine that is required for eosinophil differentiation, is located at 5q31 and therefore might be upregulated by translocations that disrupt this region. In addition, a family with familial eosinophilia has been described in which the disease is linked to 5q31–33, but which appears not to involve the genes for IL-3, IL-5 or GM-CSF. Unless a subtle mutation has been missed, these findings suggest that there may be a novel

**Table 3.** Cases in the literature with translocations involving 5q31–5q35 in myeloid leukaemias

| Patient | Sex | Age | Diagnosis | Translocation | Partner | Survival m | Splenomegaly | WBCs | Neutrophils % | Eosinophils % | Monocytes % | Notes | Ref. |
|---|---|---|---|---|---|---|---|---|---|---|---|---|---|
| 37 | F | 5 m | MPD | t(1;5)(q23;q33) | | 9 d | marked | 48.3 | 27 | 30 | 3 | died of infection | 23 |
| 38 | M | 7 m | MPD | t(1;5)(q23;q33) | | 18 d | marked | 42.1 | | 13 | | moderate dysmyelopoiesis | 23 |
| 39 | M | 44 | EoL | t(2;5)(p23;q35) | | 6 | mild | 55 | 5 | 88.5 | 1.5 | pustular folliculitis at presentation | 24 |
| 40 | M | 17 | Ph-ve CML 2R | t(3;5)(p21;q35) | | 31 | N | 0.9 | | | | | 25 |
| 41 | M | 38 | MDS/RAEB | t(3;5)(q21;q31) | | 15 | N | 5.2 | 60 | | | transformed after 11m | 26 |
| 42 | M | Child | MDS | t(3;5)(q25;q33) | | 72 d | | | | | | BMT relapse after 5 years, died 1 year later | 27 |
| 43 | F | 38 | MDS | t(3;5)(q25;q33) | | 18 | | 22.7 | | | | developed RAEB after 18m | 28 |
| 44 | F | 2 | JCML | t(3;5)(q27;q33) | | 60 | 4 cm | 29 | 24 | 16 | 1 | no treatment for 4 yrs, lymphoma | 29 |
| 45 | M | 54 | CMML | t(5;7)(q33;q11.2) | *HIP1* | | Y | | | raised | raised | dysplastic monocytes | 30 |
| 46 | M | 48 | MPD | t(5;10)(q33;q21.2) | *H4* | 24+ | 20 cm | 94 | | 7 | | splenectomy and BMT | 31 |
| 47 | M | 49 | aCML | t(5;10)(q33;q22) | *H4* | | N | 33 | 30 | 6 | 3 | industrial exposure | 32 |
| 48 | M | 57 | CEoL | t(5;12)(q31;q13) | | 12 | Y | | | 22 | | skin lesions at presentation | 33 |
| 49 | M | 7 | MDS | t(5;12)(q33;q13) | | 156 | N | | | 40 | | generalized skin rash at presentation | 34 |
| 50 | M | 35 | Ph-CML | t(5;13)(q33;q14) | | 11 d | 10 cm | | | | | died in BC | 35 |
| 51 | F | 18 | AML M2 R | t(5;14)(q33;q32) | *CEV14* | 5 d | Y | 150 | NA | 36 | | itchy skin and erythema | 36 |
| 52 | M | 60 | AML M1 R | t(5;16)(q33;q22) | | 2 d | Y | 98.1 | 2 | 16.5 | 6.5 | died from cerebral thrombosis | 37 |
| 53 | M | 10 m | AML M4 | t(5;16)(q33;q22) | | 190 | | | | 0 | | 33% eosinophils in BM | 38 |
| 54 | M | | CMML | t(5;17)(q33;q13) | *Rab5* | 24+ | | | | | | PBSCT, still has translocation | 39 |
| 55 | M | 48 | Ph-CML | t(5;17)(q33;p11) | | 54 | | | | | | | 10 |
| 56 | F | 73 | MDS/CMML | t(X;5)(q13;q33) | | | | | | | | translocation in 20% of cells | 40 |
| Median | | 38 | | | | | | 42.1 | 27 | 16 | 3 | | |

BM = Bone marrow; BMT = bone marrow transplant; CML = chronic myeloid leukaemia; PBSCT = peripheral blood stem cell transplant. Other abbreviations, see legend of table 2.

gene in this region that is involved in eosinophil development and which could also be affected by 5q31–33 translocations [51].

*Skin Involvement*

Of note are 2 patients with a t(5;12)(q31 ~ q33;q13), a translocation that involves an unknown target gene at 12q13. In both cases there were severe skin rashes and lesions at presentation in addition to prominent eosinophilia. Although it is unclear if these abnormalities share a common pathology, it is conceivable that they arise as a consequence of aberrant PDGFRB signaling since this receptor is normally involved in wound healing and repair of skin lesions. Alternatively, the skin lesions may be a consequence of eosinophilic infiltrations. Why this should occur in association with this particular translocation is unclear, but it is possible that the gene on 12q13 activates *PDGFRB* to a greater degree than *ETV6*, or that it might have a role in maturation and function of eosinophils. However, relatively mild skin involvement is also seen in other cases, suggesting that the severe lesions in these 2 cases is an exaggerated feature of a process that is common to cases with diverse 5q31–33 translocations. This is an area which deserves further investigation.

## The Normal Role of *PDGFRB*

*Structure and Function of PDGFRB*

The *PDGFRB* gene is located at chromosome band 5q33. It is transcribed and translated into a 5.45-kb mRNA and 1,106-amino-acid protein, respectively [52]. The 190-kD protein consists of 5 extracellular immunoglobulin-like structures, a transmembrane domain, a protein-protein interacting WW-like domain in the juxtamembrane region, an ATP-binding site and split tyrosine kinase domain in the intracellular portion of the protein [53]. The mature receptor is a 121-kD protein which is formed after cleavage of the signal peptide encoded by the first 33 amino acids. It normally acts as a receptor for platelet-derived growth factor (PDGF) ligands, of which there are four isoforms called PDGF A–D. PDGFRB is capable of binding strongly to PDGF-BB homodimers, weakly to AB heterodimers and not at all to the PDGF-AA form [54]. PDGFRA, which is highly related to PDGFRB, binds to A, B and C PDGF homodimers and AB heterodimers with roughly equal affinity but scarcely to D homodimers. The β-receptor is more specific and binds only PDGF-B and D homodimers [55–57]. In other words, the two receptors are functionally distinct.

PDGFRB is a member of the type III receptor tyrosine kinase family that includes PDGFRA, c-kit (SCFR), flt-1, flt-3, KDR (flk-1) and CSF1R (c-fms) [58, 59]. These receptors are all characterized by the presence of an intracellular tyrosine kinase domain that is interrupted by a kinase insert.

*Function of PDGFRB*

In the hemopoietic system *PDGFRB* is normally expressed in erythroid and myeloid precursors in the bone marrow. In addition, it is expressed by mature monocytes, megakaryocytes, fibroblasts, endothelial cells and osteoblasts [60]. Expression of *PDGFRB* in monocytes suggests a role in monocytic differentiation, possibly activating macrophages once matured. The protein is also involved in vascular repair through expression in smooth muscle cells, receiving PDGF-BB released into the blood vessel. It is also possible that the receptor plays a role in the processes of inflammation and regulation of tissue interstitial fluid pressure by mediating the involvement of monocytes, macrophages and platelets in inflamed tissue [61, 62].

*Signaling by PDGFRB*

Activation of *PDGFRB* results in mitogenesis, cytoskeletal rearrangements and, in some circumstances, chemotaxis. PDGFRB and other tyrosine kinase receptors are believed to exist in the plasma membrane as monomers. Binding of PDGF ligand dimers induces dimerization of the receptor, a critical event that activates the PDGFRB tyrosine kinase activity. Several tyrosine residues in the juxtamembrane, kinase insert domain and C-tail of the receptor are autophosphorylated and these serve as specific docking sites for downstream signaling proteins such as PI3K, PLCγ, Ras-GAP, Src, various SH2 domain proteins and adaptors such as Grb2, Shc and Nck (reviewed by Claesson-Welsh [54] and Heldin et al. [62]). Site-directed mutagenesis and domain deletion experiments have shown that certain tyrosine residues are necessary for signaling via specific proteins. Two downstream targets that are believed to be important intermediates in the mitogenic signal are Src and STAT5 [63, 64]. By analogy with other tyrosine kinase fusion proteins such as BCR-ABL, it is also likely that the Ras and PI3K pathways play an important role in transformation mediated by PDGFRB fusion proteins [65].

The PDGFRs are known to be involved in other malignancies; for example, PDGFRB is expressed with constitutively phosphorylated Y751 in human meningiomas. This results in mitogenesis and migration through constitutive activation of the PI3K pathway [66]. Some pancreatic cancers have a seven-fold increase in the mRNAs of both PDGFRs [67], and there is amplification of the alpha receptor in some forms of advanced brain tumour [68]. Recently, we cloned the t(4;22)(q12;q11) in two patients with a CML-like disease, and found that this translocation results in the fusion of BCR to PDGFRA, the first translocation known to involve this tyrosine kinase [69].

PDGFRB may be involved in the pathogenesis of chronic disease; for example, overexpression or activation of PDGFRB may play a role in the deposition of atherosclerotic plaques in the arteries of patients with atherosclerosis. This is a result of increased proliferation in cells of the vessel wall [70]. It is also thought that it might be involved in the development of germ cell aplasia and chronic pancreatitis [71, 72].

## Transformation by *PDGFRB* Fusion Genes

*Constitutive Activity of Fusion Proteins*

PDGFRB fusion proteins in MDS/MPD are no longer responsive to PDGF ligands as they are cytoplasmic proteins and the ligand binding domain of the receptor is missing. Instead, the fusion proteins are constitutively active and can, for example, transform IL-3-dependent cell lines to growth factor independence. As mentioned above, the partner proteins contain a specific dimerization motif: this enables the fusion proteins to self-associate, mimicking the normal process of receptor activation after binding of ligand.

The fusion that has been analyzed in the greatest detail is *ETV6-PDGFRB*. In mouse models this chimeric gene induces an MPD that is characterized by an excess of neutrophils and megakaryocytes in the bone marrow and splenomegaly, with extramedullary hemopoiesis [73, 74]. When expression of the fusion gene is driven by a lymphoid-specific promoter, the mice succumb to lymphoblastic lymphomas [75]. The mechanism of transformation is believed to involve STATs 1 and 5, signaling proteins that are also activated by several other tyrosine kinase fusion proteins, e.g., BCR-ABL, ETV6-ABL, ETV6-JAK2 and ZNF198-FGFR1 [76, 77]. As expected, transformation is absolutely dependent on a functional PDGFRB tyrosine kinase domain and the presence of the dimerization motif encoded by the partner gene.

*Potential Targeted Therapy of MDS/MPD with PDGFRB Involvement*

Theoretically, any agent that is able to block the kinase activity of PDGFRB or relevant downstream signalling pathways might be useful for the treatment of patients with disease arising from the presence of a PDGFRB fusion protein. This reasoning has been dramatically exemplified by the demonstration of the marked anti-leukemic activity of imatinib mesylate (STI571) in CML [78]. STI571 acts by occupying and blocking the ABL kinase ATP binding site and thereby inhibits its enzymatic activity. As might be expected from its mode of action, STI571 is not completely specific for the ABL kinase and it turns out that it also inhibits KIT and PDGFRB [79–81]. Indeed, STI571 was able to reverse the leukemic phenotype in a murine model of PDGFRB. Encouraging clinical trials with the drug for patients with a known rearrangement of PDGFRB indicate a rapid response in 3 out of 4 patients, with a normalised eosinophil and white cell count after just 1 week [82]. One patient in this study (corresponding to patient 49 in this review) had extensive and disfiguring skin lesions prior to therapy, but after only 5 days of imatinib treatment, the lesions started to flatten. This reduction led to the patient being able to have skin grafts for the first time in several years.

Recently, murine studies and clinical trials have shown that imatinib also has efficacy in the treatment of some patients with dermatofibrosarcoma protuberans (DFSP) and hypereosinophilic syndrome (HES). DFSP is a recurrent skin tumour characterised in most cases by the presence of a t(17;22) translocation which fuses the gene encoding the major component of collagen, COL1A1, to PDGFB. PDGFRB is constitutively activated in DFSP cells, and mouse tumours derived from primary human DFSP cells show a reduction in tumour size by up to one third compared to controls after only 9 days of treatment with imatinib [83, 84]. Patients with HES present with persistent eosinophilia of an unknown origin and often have splenomegaly. Of 8 patients thus far reported, 7 responded to imatinib and 6 of these showed improvement after only 1 week of treatment [85–87]. Patients were maintained at a drug concentration ten times lower than that in patients with CML, suggesting that the target protein is more sensitive to imatinib inhibition than BCR-ABL. Currently, there are no known molecular lesions in HES to explain this responsiveness.

Because tyrosine kinase fusion genes utilize the Ras pathway and because activated Ras is only functional when attached to the inner surface of the plasma membrane by a prenyl group, any agent that inhibits farnesylation might interrupt the transforming signal. A number of different farnesyl transferase inhibitors (FTIs) have been tested in preclinical work and one has shown activity against a BCR-ABL transfected cell line in a murine model system [88]. Almost certainly the anti-leukemic properties of FTIs are due to their effect on multiple, largely unknown farnesylated targets and not only on Ras. Since diverse tumours depend of Ras activation, FTIs are likely to find wide application in the treatment of malignancy. Importantly, model systems have shown that FTIs may be active against cell lines that have acquired resistance to STI571 suggesting that they may play a role in future combination therapy [88]. Other pathways, such as those involving PI3K or STAT proteins are also attractive targets for therapeutic intervention and it is likely that candidate inhibitors will become available in the near future. Such compounds might also be useful for treatment of *PDGFRB*-rearranged MDS/MPD patients.

## Conclusions

Reciprocal translocations of 5q31–33 are associated with a distinct MDS/MPD that is characterized by splenomegaly, monocytosis and eosinophilia. Transformation to acute leukemia occurs in a minority of cases with variable latency, but death may occur from other complications. Molecular analysis should be undertaken to determine whether *PDGFRB* is involved. Although FISH is informative in many cases, Southern blotting or RT-PCR is necessary to detect the involvement of this gene in some cases.

## Acknowledgments

We are very grateful to the following for providing additional patient information: L. Baranger (Angers, France), M. Magnusson (Bethesda, Md., USA), K. Bhambhani (Detroit, Mich., USA), Y. Neumann (Tel-Hashomer, Israel), P.J. Darbyshire (Birmingham, UK), W.G. Woods (Atlanta, Ga., USA), N. Dastugue (Toulouse, France), J. Chessells (London, UK), B. Johansson (Lund, Sweden), C. Hanson (Rochester, Minn., USA) and M. Gotic (Belgrade, Serbia). This work was supported by the Leukemia Research Fund.

## References


1 Keene P, Mendelow B, Pinto MR, Bezwoda W, MacDougall L, Falkson G, Ruff P, Bernstein R: Abnormalities of chromosome 12p13 and malignant proliferation of eosinophils: A non-random association. Br J Haematol 1987;67:25–31.

2 Aurich J, Duchayne E, Huguet-Rigal F, Bauduer F, Navarro M, Perel Y, Pris J, Caballin MR, Dastugue N: Clinical, morphological, cytogenetic and molecular aspects of a series of Ph-negative chronic myeloid leukemias. Hematol Cell Ther 1998;40:149–158.

3 Baranger L, Szapiro N, Gardais J, Hillion J, Derre J, Francois S, Blanchet O, Boasson M, Berger R: Translocation t(5;12)(q31-q33;p12-p13): A non-random translocation associated with a myeloid disorder with eosinophilia. Br J Haematol 1994;88:343–347.

4 Berkowicz M, Rosner E, Rechavi G, Mamon Z, Neuman Y, Ben Bassat I, Ramot B: Atypical chronic myelomonocytic leukemia with eosinophilia and translocation (5;12). A new association. Cancer Genet Cytogenet 1991;51:277–278.

5 Golub TR, Barker GF, Lovett M, Gilliland DG: Fusion of PDGF receptor beta to a novel ets-like gene, tel, in chronic myelomonocytic leukemia with t(5;12) chromosomal translocation. Cell 1994;77:307–316.

6 Granjo E, Lima M, Lopes JM, Dória S, Orfão A, Ying S, Barata LT, Miranda M, Cross NCP, Bain BJ: Chronic eosinophilic leukaemia presenting with erythroderma, mild eosinophilia and hyper-IgE: Clinical, immunological and cytogenetical features and therapeutic approach. Acta Haematol 2002;107:108–112.

7 Hoglund M, Johansson B, Pedersen-Bjergaard J, Marynen P, Mitelman F: Molecular characterization of 12p abnormalities in hematologic malignancies: Deletion of KIP1, rearrangement of TEL, and amplification of CCND2. Blood 1996;87:324–330.

8 Horiike S, Misawa S, Nakai H, Kaneko H, Yokota S, Taniwaki M, Yamane Y, Inazawa J, Abe T, Kashima K: N-ras mutation and karyotypic evolution are closely associated with leukemic transformation in myelodysplastic syndrome. Leukemia 1994;8:1331–1336.

9 Lerza R, Castello G, Sessarego M, Cavallini D, Pannacciulli I: Myelodysplastic syndrome associated with increased bone marrow fibrosis and translocation (5;12)(q33;p12.3). Br J Haematol 1992;82:476–477.

10 Martiat P, Michaux JL, Rodhain J: Philadelphia-negative (Ph-) chronic myeloid leukemia (CML): comparison with Ph+ CML and chronic myelomonocytic leukemia. The Groupe Français de Cytogénétique Hématologique. Blood 1991;78:205–211.

11 Pellier I, Le Moine PJ, Rialland X, Francois S, Baranger L, Blanchet O, Larget-Piet L, Ifrah N: Myelodysplastic syndrome with t(5;12)(q31;p12-p13) and eosinophilia: A pediatric case with review of literature. J Pediatr Hematol Oncol 1996;18:285–288.

12 Romana SP, Mauchauffe M, Le Coniat M, Chumakov I, Le Paslier D, Berger R, Bernard OA: The t(12;21) of acute lymphoblastic leukemia results in a tel-AML1 gene fusion. Blood 1995;85:3662–3670.

13 Sato Y, Bohlander SK, Kobayashi H, Reshmi S, Suto Y, Davis EM, Espinosa R, Hoopes R, Montgomery KT, Kucherlapati RS, Le Beau MM, Rowley JD: Heterogeneity in the breakpoints in balanced rearrangements involving band 12p13 in hematologic malignancies identified by fluorescence in situ hybridization: TEL (*ETV6*) is involved in only one half. Blood 1997;90:4886–4893.

14 Srivastava A, Boswell HS, Heerema NA, Nahreini P, Lauer RC, Antony AC, Hoffman R, Tricot GJ: KRAS2 oncogene overexpression in myelodysplastic syndrome with translocation 5;12. Cancer Genet Cytogenet 1988;35:61–71.

15 Wessels JW, Fibbe WE, van der Keur D, Landegent JE, van der Plas, den Ottolander GJ, Roozendaal KJ, Beverstock GC: t(5;12)(q31;p12). A clinical entity with features of both myeloid leukemia and chronic myelomonocytic leukemia. Cancer Genet Cytogenet 1993;65:7–11.

16 Willman CL, Sever CE, Pallavicini MG, Harada H, Tanaka N, Slovak ML, Yamamoto H, Harada K, Meeker TC, List AF: Deletion of IRF-1, mapping to chromosome 5q31.1, in human leukemia and preleukemic myelodysplasia. Science 1993;259:968–971.

17 Wlodarska I, Mecucci C, Marynen P, Guo C, Franckx D, La Starza R, Aventin A, Bosly A, Martelli MF, Cassiman JJ: TEL gene is involved in myelodysplastic syndromes with either the typical t(5;12)(q33;p13) translocation or its variant t(10;12)(q24;p13). Blood 1995;85:2848–2852.

18 Yahata N, Ohyashiki K, Ohyashiki JH, Kimura Y, Miyazawa K, Kodama A, Fukutake K, Toyama K: Late appearance of t(5;12)(q31;p12) in acute myeloid leukemia associated with eosinophilia. Cancer Genet Cytogenet 1998;107:147–150.

19 Yates P, Potter MN: Eosinophilic leukaemia with an abnormality of 5q31, the site of the IL-5 gene. Clin Lab Haematol 1991;13:211–215.

20 Hanson CA, Persons DL, Kurtin PJ, Dewald GW: t(5;12)(q31;p13): A distinct chronic myeloproliferative disorder (CMPD) with marrow eosinophilia. Mod Pathol 1995;8, 112A.

21 Wlodarska I, Aventin A, Ingles-Esteve J, Falzetti D, Criel A, Cassiman JJ, Mecucci C, Van den Bergh H, Marynen P: A new subtype of pre-B acute lymphoblastic leukemia with t(5;12)(q31q33;p12), molecularly and cytogenetically distinct from t(5;12) in chronic myelomonocytic leukemia. Blood 1997;89:1716–1722.

22 Wang LC, Swat W, Fujiwara Y, Davidson L, Visvader J, Kuo F, Alt FW, Gilliland DG, Golub TR, Orkin SH: The TEL/*ETV6* gene is required specifically for hematopoiesis in the bone marrow. Genes Dev 1998;12:2392–2402.

23 Darbyshire PJ, Shortland D, Swansbury GJ, Sadler J, Lawler SD, Chessells JM: A myeloproliferative disease in two infants associated with eosinophilia and chromosome t(1;5) translocation. Br J Haematol 1987;66:483–486.

24 Sato H, Danbara M, Tamura M, Morita M: Eosinophilic leukaemia with a t(2;5) (p23;q35) translocation. Br J Haematol 1994;87:404–406.

25 De Braekeleer M, Vekemans M: A t(3;5) in blastic phase of a Philadelphia chromosome-negative chronic myeloid leukemia. Cancer Genet Cytogenet 1989;37:163–168.

26 Sharp RA, Robertson J, Heppleston AD: t(3;5)(q21;q31) in a myelodysplastic syndrome. Leuk Res 1987;11:629–633.

27 Barnard DR, Kalousek DK, Wiersma SR, Lange BJ, Benjamin DR, Arthur DC, Buckley JD, Kobrinsky N, Neudorf S, Sanders J, Miller LP, Shina DC, Hammond GD, Woods WG: Morphologic, immunologic, and cytogenetic classification of acute myeloid leukemia and myelodysplastic syndrome in childhood: A report from the Children's Cancer Group. Leukemia 1996;10:5–12.

28 Daly HA, Williams J, Bond LR: t(3;5)(q25;q33) in myelodysplastic syndrome. Cancer Genet Cytogenet 1991;57:233–234.

29 Stollmann B, Fonatsch C, Havers W: Persistent Epstein-Barr virus infection associated with monosomy 7 or chromosome 3 abnormality in childhood myeloproliferative disorders. Br J Haematol 1985;60:183–196.

30 Ross TS, Bernard OA, Berger R, Gilliland DG: Fusion of Huntingtin interacting protein 1 to platelet-derived growth factor beta receptor (PDGFbetaR) in chronic myelomonocytic leukemia with t(5;7)(q33;q11.2). Blood 1998;91:4419–4426.

31 Kulkarni S, Heath C, Parker S, Chase A, Iqbal S, Pocock CF, Kaeda J, Cwynarski K, Goldman JM, Cross NC: Fusion of H4/D10S170 to the platelet-derived growth factor receptor beta in BCR-ABL-negative myeloproliferative disorders with a t(5;10)(q33;q21). Cancer Res 2000;60:3592–3598.

32 Schwaller J, Anastasiadou E, Cain D, Kutok J, Wojiski S, Williams IR, LaStarza R, Crescenzi B, Sternberg DW, Andreasson P, Schiavo R, Siena S, Mecucci C, Gilliland DG: H4(D10S170), a gene frequently rearranged in papillary thyroid carcinoma, is fused to the platelet-derived growth factor receptor beta gene in atypical chronic myeloid leukemia with t(5;10)(q33;q22). Blood 2001;97:3910–3918.

33 Gotic M, Radosevic-Radojkovic N, Ludoski M, Spasic P, Basara N, Avramovic D, Boskovic D, Rolovic Z: Eosinophilic leukemia with chromosomal aberration t(5;12)(q31;q13): An evidence of the existence of chronic eosinophilic leukemia as a distinct myeloproliferative entity. Br J Haematol 1994;87(suppl 1):168.

34 Jani K, Kempski HM, Reeves BR: A case of myelodysplasia with eosinophilia having a translocation t(5;12) (q31;q13) restricted to myeloid cells but not involving eosinophils. Br J Haematol 1994;87:57–60.
35 Rolovic Z, Ciric M, Mijovic A: A case of chronic myeloid leukemia without Ph1 translocation. Cancer Genet Cytogenet 1983;8:75–79.
36 Abe A, Tanimoto M, Towatari M, Matsuoka A, Kitaori K, Kato H, Toyozumi H, Takeo T, Adachi K, Emi N: Acute myeloblastic leukemia (M2) with translocation (7;11) followed by marked eosinophilia and additional abnormalities of chromosome 5. Cancer Genet Cytogenet 1995;83:37–41.
37 Sanada I, Asou N, Kojima S, Kawano F, Shido T, Takatsuki K: Acute myelogenous leukemia (FAB M1) associated with t(5;16) and eosinophilia. Report of an additional case. Cancer Genet Cytogenet 1989;43:139–141.
38 Bhambhani K, Inoue S, Tyrkus M, Gohle N: Acute myelomonocytic leukemia type M4 with bone marrow eosinophilia and t(5;16)(q33;q22). Cancer Genet Cytogenet 1986;20:187–188.
39 Magnusson MK, Meade KE, Brown KE, Arthur DC, Krueger LA, Barrett AJ, Dunbar CE: Rabaptin-5 is a novel fusion partner to platelet-derived growth factor beta receptor in chronic myelomonocytic leukemia. Blood 2001;98:2518–2525.
40 Chen Z, Richkind K, Roherty S, Velasco J, Lytle C, Brothman AR, Sandberg AA: A group of previously not recognized cytogenetic abnormalities in myeloid hematological malignancies. Cancer Genet Cytogenet 1999;113:162–165.
41 Abe A, Emi N, Tanimoto M, Terasaki H, Marunouchi T, Saito H: Fusion of the platelet-derived growth factor receptor beta to a novel gene CEV14 in acute myelogenous leukemia after clonal evolution. Blood 1997;90:4271–4277.
42 Yoneda-Kato N, Look AT, Kirstein MN, Valentine MB, Raimondi SC, Cohen KJ, Carroll AJ, Morris SW: The t(3;5)(q25.1;q34) of myelodysplastic syndrome and acute myeloid leukemia produces a novel fusion gene, NPM-MLF1. Oncogene 1996;12:265–275.
43 Borkhardt A, Bojesen S, Haas OA, Fuchs U, Bartelheimer D, Loncarevic IF, Bohle RM, Harbott J, Repp R, Jaeger U, Viehmann S, Henn T, Korth P, Scharr D, Lampert F: The human GRAF gene is fused to MLL in a unique t(5;11)(q31;q23) and both alleles are disrupted in three cases of myelodysplastic syndrome/acute myeloid leukemia with a deletion 5q. Proc Natl Acad Sci USA 2000;97:9168–9173.
44 Taki T, Kano H, Taniwaki M, Sako M, Yanagisawa M, Hayashi Y: AF5q31, a newly identified AF4-related gene, is fused to MLL in infant acute lymphoblastic leukemia with ins(5;11)(q31;q13q23). Proc Natl Acad Sci USA 1999;96:14535–14540.
45 Jaju RJ, Fidler C, Haas OA, Strickson AJ, Watkins F, Clark K, Cross NCP, Cheng J-F, Aplan PD, Kearney L, Boultwood J, Wainscoat JS: A novel gene, NSD1, is fused to NUP98 in the t(5;11)(q35;p15.5) in de novo childhood acute myeloid leukemia. Blood 2001 98:1264–1267.
46 Yagasaki F, Jinnai I, Yoshida S, Yokoyama Y, Matsuda A, Kusumoto S, Kobayashi H, Terasaki H, Ohyashiki K, Asou N, Murohashi I, Bessho M, Hirashima K: Fusion of TEL/*ETV6* to a novel ACS2 in myelodysplastic syndrome and acute myelogenous leukemia with t(5;12)(q31;p13). Genes Chromosomes Cancer 1999;26:192–202.
47 Grimaldi JC, Meeker TC: The t(5;14) chromosomal translocation in a case of acute lymphocytic leukemia joins the interleukin-3 gene to the immunoglobulin heavy chain gene. Blood 1989;73:2081–2085.
48 Bain BJ: Eosinophilic leukaemias and the idiopathic hypereosinophilic syndrome. Br J Haematol 1996;95:2–9.
49 Boultwood J, Fidler C, Strickson AJ, Watkins F, Kostrzewa M, Jaju RJ, Muller U, Wainscoat JS: Transcription mapping of the 5q– syndrome critical region: Cloning of two novel genes and sequencing, expression, and mapping of a further six novel cDNAs. Genomics 2000;66:26–34.
50 Morris SW, Kirstein MN, Valentine MB, Dittmer K, Shapiro DN, Look AT, Saltman DL: Fusion of a kinase gene, ALK, to a nucleolar protein gene, NPM, in non-Hodgkin's lymphoma. Science 1995;267:316–317.
51 Rioux JD, Stone VA, Daly MJ, Cargill M, Green T, Nguyen H, Nutman T, Zimmerman PA, Tucker MA, Hudson T, Goldstein AM, Lander E, Lin AY: Familial eosinophilia maps to the cytokine gene cluster on human chromosomal region 5q31-q33. Am J Hum Genet 1998;63:1086–1094.
52 Gronwald RG, Grant FJ, Haldeman BA, Hart CE, O'Hara PJ, Hagen FS, Ross R, Bowen-Pope DF, Murray MJ: Cloning and expression of a cDNA coding for the human platelet-derived growth factor receptor: Evidence for more than one receptor class. Proc Natl Acad Sci USA 1988;85:3435–3439.
53 Irusta PM, DiMaio D: A single amino acid substitution in a WW-like domain of diverse members of the PDGF receptor subfamily of tyrosine kinases causes constitutive receptor activation. EMBO J 1998;17:6912–6923.
54 Claesson-Welsh L: Platelet-derived growth factor receptor signals. J Biol Chem 1994;269:32023–32026.
55 Li X, Ponten A, Aase K, Karlsson L, Abramsson A, Uutela M, Backstrom G, Hellstrom M, Bostrom H, Li H, Soriano P, Betsholtz C, Heldin CH, Alitalo K, Ostman A, Eriksson U: PDGF-C is a new protease-activated ligand for the PDGF alpha-receptor. Nat Cell Biol 2000;2:302–309.
56 Bergsten E, Uutela M, Li X, Pietras K, Ostman A, Heldin CH, Alitalo K, Eriksson U: PDGF-D is a specific, protease-activated ligand for the PDGF beta-receptor. Nat Cell Biol 2001;3:512–516.
57 LaRochelle WJ, Jeffers M, McDonald WF, Chillakuru RA, Giese NA, Lokker NA, Sullivan C, Boldog FL, Yang M, Vernet C, Burgess CE, Fernandes E, Deegler LL, Rittman B, Shimkets J, Shimkets RA, Rothberg JM, Lichenstein HS: PDGF-D, a new protease-activated growth factor. Nat Cell Biol 2001;3:517–521.
58 Andre C, Martin E, Cornu F, Hu WX, Wang XP, Galibert F: Genomic organization of the human c-kit gene: Evolution of the receptor tyrosine kinase subclass III. Oncogene 1992;7:685–691.
59 Matthews W, Jordan CT, Gavin M, Jenkins NA, Copeland NG, Lemischka IR: A receptor tyrosine kinase cDNA isolated from a population of enriched primitive hematopoietic cells and exhibiting close genetic linkage to c-kit. Proc Natl Acad Sci USA 1991;88:9026–9030.
60 Yoon SY, Tefferi A, Li CY: Cellular distribution of platelet-derived growth factor, transforming growth factor-beta, basic fibroblast growth factor, and their receptors in normal bone marrow. Acta Haematol 2000;104:151–157.
61 Heuchel R, Berg A, Tallquist M, Ahlen K, Reed RK, Rubin K, Claesson-Welsh L, Heldin CH, Soriano P: Platelet-derived growth factor β receptor regulates interstitial fliud homeostasis through phosphatidylinositol-3′ kinase signaling. Proc Natl Acad Sci USA 1999;96:11410–11425.
62 Heldin CH, Ostman A, Ronnstrand L: Signal transduction via platelet-derived growth factor receptors. Biochim Biophys Acta 1998;1378:F79–113.
63 Valgeirsdottir S, Paukku K, Silvennoinen O, Heldin CH, Claesson-Welsh L: Activation of Stat5 by platelet-derived growth factor (PDGF) is dependent on phosphorylation sites in PDGF beta-receptor juxtamembrane and kinase insert domains. Oncogene 1998;16:505–515.
64 Sachsenmaier C, Sadowski HB, Cooper JA: STAT activation by the PDGF receptor requires juxtamembrane phosphorylation sites but not Src tyrosine kinase activation. Oncogene 1999;18:3583–3592.
65 Deininger MW, Goldman JM, Melo JV: The molecular biology of chronic myeloid leukemia. Blood 2000;96:3343–3356.
66 Shamah SM, Alberta JA, Giannobile WV, Guha A, Kwon YK, Carroll RS, Black PM, Stiles CD: Detection of activated platelet-derived growth factor receptors in human meningioma. Cancer Res 1997;57:4141–4147.
67 Ebert M, Yokoyama M, Friess H, Kobrin MS, Büchler MW, Korc M: Induction of platelet-derived growth factor receptor A and B chains and over-expression of their receptors in human pancreatic cancer. Int J Cancer 1995;62:529–535.
68 Muleris M, Almeida A, Dutrillaux AM, Pruchon E, Vega F, Delattre JY, Poission V, Malfoy B, Dutrillaux B: Oncogene amplification in human gliomas: A molecular cytogenetic analysis. Oncogene 1994;9:2717–2722.

69 Baxter EJ, Hochhaus A, Bolufer P, Reiter A, Fernandez JM, Senent L, Cervera J, Moscardo F, Sanz MA, Cross NC: The t(4;22)(q12;q11) in atypical chronic myeloid leukaemia fuses BCR to PDGFRA. Hum Mol Genet 2002;11:1391–1397.

70 Sirois MG, Simons M, Edelman ER: Antisense oligonucleotide inhibition of PDGFR-beta receptor subunit expression directs suppression of intimal thickening. Circulation 1997;95: 669–676.

71 Basciani S, Mariani S, Arizzi M, Ulisse S, Rucci N, Jannini EA, Rocca CD, Manicone A, Carani C, Spera G, Gnessi L: Expression of platelet-derived growth factor-A (PDGF-A), PDGF-B and PDGF receptor-alpha and -beta during human testicular development and disease. J Clin Endocrinol Metab 2002;87:2310–2319.

72 Ebert M, Kasper H-U, Hernberg S, Friess H, Büchler MW, Roessner A, Korc M, Malfertheiner P: Overexpression of platelet-derived growth factor (PDGF) B chain and type β PDGF receptor in human chronic pancreatitis. Dig Dis Sci 1998;43:567–574.

73 Ritchie KA, Aprikyan AA, Bowen-Pope DF, Norby-Slycord CJ, Conyers S, Bartelmez S, Sitnicka EH, Hickstein DD: The Tel-*PDGFRBeta* fusion gene produces a chronic myeloproliferative syndrome in transgenic mice. Leukemia 1999;13:1790–1803.

74 Tomasson MH, Sternberg DW, Williams IR, Carroll M, Cain D, Aster JC, Ilaria RL Jr, Van Etten RA, Gilliland DG: Fatal myeloproliferation, induced in mice by TEL/PDGFbetaR expression, depends on PDGFbetaR tyrosines 579/581. J Clin Invest 2000;105:423–432.

75 Tomasson MH, Williams IR, Hasserjian R, Udomsakdi C, McGrath SM, Schwaller J, Druker B, Gilliland DG: TEL/PDGFbetaR induces hematologic malignancies in mice that respond to a specific tyrosine kinase inhibitor. Blood 1999;93:1707–1714.

76 Wilbanks AM, Mahajan S, Frank DA, Druker BJ, Gilliland DG, Carroll M: TEL/PDGFbetaR fusion protein activates STAT1 and STAT5: A common mechanism for transformation by tyrosine kinase fusion proteins. Exp Hematol 2000;28:584–593.

77 Smedley D, Demiroglu A, Abdul-Rauf M, Heath C, Cooper C, Shipley J, Cross NC: ZNF198-FGFR1 transforms Ba/F3 cells to growth factor independence and results in high level tyrosine phosphorylation of STATS 1 and 5. Neoplasia 1999;1:349–355.

78 Druker BJ, Tamura S, Buchdunger E, Ohno S, Segal GM, Fanning S, Zimmermann J, Lydon NB: Effects of a selective inhibitor of the Abl tyrosine kinase on the growth of Bcr-Abl positive cells. Nat Med 1996;2:561–566.

79 Heinrich MC, Griffith DJ, Druker BJ, Wait CL, Ott KA, Zigler AJ: Inhibition of c-kit receptor tyrosine kinase activity by STI 571, a selective tyrosine kinase inhibitor. Blood 2000; 96:925–932.

80 Carroll M, Ohno-Jones S, Tamura S, Buchdunger E, Zimmermann J, Lydon NB, Gilliland DG, Druker BJ: CGP 57148, a tyrosine kinase inhibitor, inhibits the growth of cells expressing BCR-ABL, TEL-ABL, and TEL-PDGFR fusion proteins. Blood 1997;90:4947–4952.

81 Buchdunger E, Zimmermann J, Mett H, Meyer T, Muller M, Druker BJ, Lydon NB: Inhibition of the Abl protein-tyrosine kinase in vitro and in vivo by a 2-phenylaminopyrimidine derivative. Cancer Res 1996;56:100–104.

82 Apperley JF, Gardembas M, Melo JV, Russell-Jones R, Bain BJ, Baxter EJ, Chase A, Chessells JM, Colombat M, Dearden CE, Dimitrijevic S, Mahon FX, Marin D, Niklova Z, Olavarria E, Silberman S, Schultheis B, Cross NCP, Goldman JM: Response to imatinib mesylate in patients with chronic myeloproliferative diseases with rearrangements of the platelet-derived growth factor receptor beta. N Engl J Med 2002;347:481–487.

83 Simon M-P, Navarro M, Roux D, Pouysségur J: Structural and functional analysis of a chimeric protein COL1A1-PDGFB generated by the translocation t(17;22)(q22;q13.1) in dermatofibrosarcoma protuberans (DP). Oncogene 2001;20:2965–2975.

84 Sjoblom T, Shimizu A, O'Brien KP, Pietras K, Dal Cin P, Buchdunger E, Dumanski JP, Östman A, Heldin C-H: Growth inhibition of dermatofibrosarcoma protuberans tumors by the platelet-derived growth factor receptor antagonist STI571 through induction of apoptosis. Cancer Res 2001;61:5778–5783.

85 Gleich GJ, Leiferman KM, Pardanani A, Tefferi A, Butterfield JH: Treatment of hypereosinophilic syndrome with imatinib mesilate. Lancet 2002;359:1577–1578.

86 Schaller JL, Burkland GA: Case report: Rapid and complete control of idiopathic hypereosinophilia with imatinib mesylate. MedGenMed 2001;3:9.

87 Severino SA, Pane E, Anghel G, Ruscio C, Petti N, Della Seta R, Majolino I: Glivec (STI 571) in idiopathic hypereosinophilic syndrome (IHS) resistant to conventional treatment. Hematol J 2002;3(suppl 1):307.

88 Peters DG, Hoover RR, Gerlach MJ, Koh EY, Zhang H, Choe K, Kirschmeier P, Bishop WR, Daley GQ: Activity of the farnesyl protein transferase inhibitor SCH66336 against BCR/ABL-induced murine leukemia and primary cells from patients with chronic myeloid leukemia. Blood 2001;97:1404–1412.



Prof. N.C.P. Cross
Wessex Regional Genetics Laboratory
Salisbury District Hospital
Salisbury SP2 8BJ (UK)
E-Mail ncpc@soton.ac.uk

Bain BJ (ed): Chronic Myeloproliferative Disorders.
Basel, Karger, 2003, pp 80–89 / Acta Haematol 2002;108:171–179

# Cytogenetic and Molecular Genetic Aspects of Childhood Myeloproliferative/Myelodysplastic Disorders

Georgina W. Hall

Paediatric Haematology/Oncology Unit, John Radcliffe Hospital, Oxford, UK

**Key Words**
Juvenile myelomonocytic leukaemia · Transient abnormal myelopoiesis · Childhood

**Abstract**
Of the myeloproliferative/myelodysplastic disorders (MPD/MDS) that occur in childhood most, regarding the cytogenetic and molecular genetic basis, is known about the two purely paediatric disorders: juvenile myelomonocytic leukaemia (JMML) and transient myeloproliferative disorder (TMD). Although much has been published about these two disorders, their aetiology is by no means fully established. It would appear, however, that in this paediatric subset of MPDs a stage/developmentally specific vulnerability for proliferation and transformation exists. The study of the molecular basis of many other MPD-like syndromes that also occur in childhood, has been greatly accelerated by the identification of rare, but recurring, cytogenetic abnormalities involving 8p11 and 5q31–33. Good collaborative studies could result in similar progress being made in the understanding of JMML and TMD.

## Introduction

There are two paediatric conditions included in the recent World Health Organisation (WHO) classification [1] of myeloproliferative disorders (MPDs), the 'myelodysplastic/myeloproliferative' disorder (MDS/MPD), juvenile myelomonocytic leukaemia (JMML), and transient myeloproliferative disorder (TMD) or transient abnormal myelopoiesis (TAM) which is seen in neonates with Down's syndrome. These two disorders occur only in childhood/infancy.

Of the adult 'chronic MPDs' (CMPDs): chronic myeloid leukaemia (CML), essential thrombocythaemia (ET), polycythaemia vera (PV) and chronic idiopathic myelofibrosis (CIMF), and the MDS/MPD atypical CML (aCML), all are known to occur in childhood. There are several other unclassified overlapping syndromes which occur in all age groups. Some may be variants of JMML, aCML or chronic myelomonocytic leukaemia (CMML) and these include the 5q31–33 syndromes and the 8p11 myeloproliferative syndrome (EMS). The cytogenetics and molecular genetics of the latter group of disorders are discussed in detail in accompanying chapters [2–5], and will only be touched on briefly here with reference to their occurrence in children. What is known of the cytogenetics and molecular genetics of the two primary paediatric disorders, JMML and TMD, is discussed in more detail.

## Chronic Myeloproliferative Diseases

It would appear that the cytogenetic abnormalities and molecular genetics of most chronic myeloproliferative disorders that occur in childhood are no different from those in adults, but this has yet to be confirmed.

*Chronic Myeloid Leukaemia*

It is well known that Philadelphia chromosome-positive (Ph+)/*BCR/ABL*-positive CML occurs in childhood although the exact incidence needs clarification. European databases and the new UK Paediatric CML registry set up by the Medical Research Council (MRC) Childhood Leukaemia Working Party on Rare Diseases should help to establish both incidence and natural history and in time allow the development of an appropriate paediatric prognostic scoring system. As yet there is no evidence to suggest that the disease, its molecular/cytogenetic basis or behaviour is any different in childhood. References 2–5 deal, in great detail, with the various cytogenetic and molecular genetic aspects of CML.

*Essential Thrombocythaemia, Polycythaemia vera, Chronic Idiopathic Myelofibrosis*

The remaining chronic MPDs: ET, PV and CIMF are known to occur in childhood [6–8] although the true incidence and natural history are not well established. Cases and reviews of cases reported in the literature indicate that the natural history, certainly in ET and PV, may be similar to that seen in adults [6, 9, 10]. The UK Childhood Myeloproliferative Diseases Registry will try to ascertain the incidence and natural history of these disorders in childhood. The more common cytogenetic abnormalities seen occasionally in adults with ET and PV, i.e. trisomy 8, trisomy 9, del (13q) and del (20q) have not been found in reported childhood cases nor so far in any of the UK registry patients.

## Myelodysplastic/Myeloproliferative Diseases

*Atypical Chronic Myeloid Leukaemia*

According to the recent WHO classification, atypical, *BCR/ABL* fusion gene negative CML (aCML) occurs predominantly in the 7th or 8th decade of life and has an incidence of 1 or 2 for every 100 cases of Ph+ CML [1]. A significant minority of *BCR/ABL*-negative CML patients and certain patients with MPD-like features, especially those with eosinophilia, have translocations involving 5q31–33. Many of these translocations disrupt the platelet-derived growth factor beta receptor gene *(PDGFRB)* [11–13]. Various 5q31 translocations have been identified in all age groups including a t(5;11)(q31;q23) *(MML/GRAF)* (not involving *PDGFRB*) in a child with JMML [14], an 8-year-old with 'MDS/eosinophilia' and t(5;12)(q31;p13) [15] and a 16-year-old with 'CMML' and t(5;12)(q31;p12) [16]. The t(5;12)(q31;q12) translocation [17] is now well described and results in the *TEL/PDGFRB* fusion gene and is most commonly associated with a CMML phenotype. Hence there appears to be overlapping clinical syndromes of MPD associated with the disruption of *PDGFRB* in particular. The striking feature of eosinophilia in many of these cases is interesting because some children with JMML and eosinophilia have been found to have translocations involving 5q31–33. A specific eosinophil/MDS category used to exist in the modified French-American-British (FAB) Classification for paediatric MDS produced in 1996 [18] which included all children with MDS who had >25% eosinophils in the bone marrow. Patients classified in this category often had translocations involving 5q31 or 5q33 breakpoints, i.e. t(5;12)(q31;q13), hepatosplenomegaly and skin infiltration and were felt by some to have a variant of JMML [19].

MPDs with 5q31–35 translocations are discussed in detail in Steer and Cross [20].

*8p11 Myeloproliferative Syndrome*

The term EMS, coined by Inhorn et al. [21], describes a myeloproliferative syndrome characterized by eosinophilia, lymphadenopathy and a high incidence of T cell-lymphoblastic non-Hodgkin lymphoma (NHL) with progression to acute myeloid leukaemia (AML) in association with 8p11 translocations. EMS, also called 'stem cell leukaemia/lymphoma' (SCLL) is not recognized in the WHO classification. A sizeable number of children have been reported with EMS. The first report of this breakpoint t(8;9)(p11 or 12;q34) was in a 3-year-old [22] with aCML. The authors where initially interested in the new translocation because of the 9q34 breakpoint involving *ABL* and the possibility of a new gene fusion partner producing a similar CML-like phenotype. However, further reports of patients with a myeloproliferative disorder and lymphoid malignancies were all associated with an acquired reciprocal translocation involving chromosome band 8p11. Many cases were children including a 13-year-old child with a myeloproliferative disorder who developed T cell-NHL (T-NHL) in association with t(6;8)(q27;p12) [23], a 10-year-old with 'CMML', tonsillar lymphoma and rapid leukaemic transformation with t(8;9)(p11;q34) [24] and an 18-year-old female with MPD/eosinophilia and

**Table 1.** 1996 Modified FAB Classification for paediatric MDS [18]

| Category | Definition |
|---|---|
| RA | as for FAB unless fits *Imo7* or Eos |
| RARS | as for FAB |
| RAEB | as for FAB unless fits *Imo7* or Eos |
| RAEB-t | as for FAB unless fits *Imo7* or Eos |
| JCML | CMML with Hb F>10% *and no monosomy 7* |
| *Imo7* | MDS, *age<4 years and monosomy 7* |
| Other CMML | CMML but not JCML or *Imo7* |
| Eos | MDS with >25% eosinophils in bone marrow |

*Imo7* = a term abandoned in 2000, i.e. no longer used.

T-NHL in association with t(8;13)(p11;q11–12) [21]. To date, four different gene fusions associated with EMS have been described; t(8;13)(p11;q12), t(8;9)(p11;q33), t(6;8)(q27;p11) and t(8;22)(p11;q22) in which *ZNF198, CEP110, FOP* and *BCR* all fuse to the entire tyrosine kinase domain of fibroblast growth factor receptor-1 gene *(FGFR1)*. The resulting fusion proteins all have constitutive tyrosine kinase activity and affect various signal transduction pathways [25, 26]. Of the 27 published cases of EMS, 7 were children. A more detailed account of the molecular and cytogenetic basis of EMS is to be found in Macdonald et al. [26].

*Juvenile Myelomonocytic Leukaemia*

JMML, previously called 'juvenile chronic myeloid leukaemia' (JCML), bears little relationship to CML morphologically or cytogenetically, although it is clearly a myeloproliferative disorder. Its morphological features are however, more similar to those of the myelodysplastic syndrome CMML, which has now been reclassified as an MDS/MPD. Although myelodysplasia is a diagnostic criterion for CMML, it is not a striking feature of JMML.

*Clinical/Diagnostic Criteria.* The annual incidence of JMML in northern Europe and Canada is estimated to be in the region of 1 per million, although data from the UK Paediatric Myelodysplasia Registry indicate a slightly lower rate of 0.6/million in the UK, which may be due to a classification bias. JMML probably accounts for only 2–3% of all childhood leukaemias and shows a predominance in males, young children under the age of 2 years and an association with neurofibromatosis type 1 (NF1). Up to 15% of children with JMML have clinical signs of NF1 [27, 28] and children with NF1 are reported to have a 200- to 500-fold increased risk of myeloid malignancies, specifically JMML [29, 30]. Typically, patients with JMML present with massive hepatosplenomegaly, lymphadenopathy, an erythematous skin rash – classically facial with a butterfly distribution, bleeding problems secondary to thrombocytopenia, fever, pallor and marked wasting [28]. The blood film morphology is similar to that of adult CMML, according to the FAB Classification, with myelomonocytic proliferation. Occasionally, children will have a predominance of erythroblasts in their peripheral blood with an erythroleukaemia (AML-M6)-like picture whilst in others it is difficult to differentiate between progressive JMML and acute myelomonocyte leukaemia (FAB AML-M4). Very few patients will progress to a CML-like blast crisis which, when it occurs, is almost never lymphoid. JMML has never been adequately described by any adult MDS classification and in 1996 a modified FAB Classification for paediatric MDS was proposed [18] (table 1). This classification suggested, however, that only patients without monosomy 7 could be classified as having JCML and created a separate category of 'infantile monosomy 7 syndrome' (*Imo7*), which has since been abandoned.

The most recent WHO classification [1] is based on the criteria of Niemeyer et al. [31] and requires the following for diagnosis: (1) peripheral blood monocytosis $>1 \times 10^9$/l; (2) blasts (including promomocytes) <20% of peripheral blood leucocytes and/or nucleated bone marrow cells; (3) no Ph chromosome or BCR/ABL fusion gene; (4) plus two or more of the following: Hb F increased for age, immature granulocytes in the peripheral blood, WBC count $>10 \times 10^9$/l, clonal chromosomal abnormality (which may be monosomy 7) or GM-CSF hypersensitivity of myeloid progenitors in vitro.

The similarity with CMML is evident, especially as the first three main diagnostic criteria of CMML (according to the WHO classification [1]) are the same as for JMML. It is likely that children who have in the past been reported as having CMML, had in fact JMML but were insufficiently investigated to confirm the diagnosis.

Essential in the diagnosis of JMML is the exclusion of chronic/persistent Epstein-Barr virus infection [32] although various other infections can also mimic JMML in young children, namely, infection by human herpesvirus 6, cytomegalovirus, mycobacteria, *Histoplasma capsulatum* and *Toxoplasma carinii*.

*Clonality.* Although JMML has been shown to be clonal in approximately 50% of patients studied, only 35% have a detectable cytogenetic abnormality [28]. Clonal X chromosome inactivation studies using the HUMARA assay have been reported in some girls with JMML [33] and the loss of the normal *NF1* allele in CD34+ cells has been reported in other JMML patients [34]. Activating *Ras* mutations are found in approximately 20–30% of JMML cases [35, 36]. *TP53* (previously *p53*) mutations are rare in JMML and even in the few patients that progress to blast crisis there is no evidence of increased *TP53* mutations occurring [37]. Other investigations have revealed that not all lineages are involved, and JMML is probably a heterogeneous disorder with respect to clonality in different lineages [38].

*Cytogenetics* The most common cytogenetic abnormality at presentation is monosomy 7 (table 2) although this defect can be acquired later in the course of the disease. The incidence at presentation varies from study to study, from 6% [39] to 33% [40] and most likely reflects differences in the definition of JMML (table 1). Approximately half of the other cytogenetic abnormalities identified in JMML patients involved 7q including deletions (7q–), something more commonly seen in secondary MDS in adults. Speculation as to whether defects or loss of 7q were important in the development of JMML led to the hunt for cryptic deletions of 7q in cytogenetically normal bone marrow from patients with JMML, but none was found [41].

**Table 2.** Cytogenetic abnormalities found in JMML

| |
|---|
| Monosomy 7 |
| Del (7q) |
| Trisomy 8 |
| Trisomy 21 |
| t(13;14) |
| t(1;3)(p13;p21) [46] |
| der (7)t(7;12)(q21;q13) |
| t(3;12)(q21;p13) |
| t(3;15)(q13.1;q26) |
| 5q31–33 abnormalities |
| t(5;12)(q31;q13)[1] |
| t(1;5)(q22;q33)[1] |

[1] These cases had eosinophilia [19].

Trisomy 8 and 21 are the next most frequently seen abnormality in JMML. Interestingly, 2 cases, 1 CMML (probably JMML) and 1 JMML, have been reported in children with constitutional trisomy 8 mosaicism [42, 43]. This is of interest because of a possible association with EMS; perhaps cryptic translocations involving 8p11 or disomic homozygosity in these trisomy 8 patients results in these MPDs.

Less commonly reported cytogenetic findings include t(13;14) (q12.?22; q32.?3) [44] and defects of 3q; t(3;12) (q21;p13) and t(3;15) (q13.1;q26), including an unbalanced der (12)t(3;12)(q21 ~ 22;p13.33) resulting in a partial trisomy of 3q [45].

Children with NF1 who develop MDS/AML have an increased incidence of monosomy 7 [47] and those NF1 children, particularly boys, with JMML may acquire monosomy 7 during the course of their disease [48].

*Pathobiology/Molecular Genetics.* The classic signs of aggressive JMML include fevers and profound wasting, now known to be due to the unceasing production of tumour necrosis factor-alpha from overactivated monocytes. The massive proliferation of myeloid precursors and particularly monocytes in JMML appears to be due to an exquisite hypersensitivity to granulocyte/macrophage colony-stimulating factor (GM-CSF) [49]. JMML cells have a normal response to interleukin-3 and granulocyte colony-stimulating factor (G-CSF), however. There is no evidence of a defect or mutation in the GM-CSF receptor [33, 50] and the problem appears to reside further down the signal transduction pathway of GM-CSF which is Ras dependent. There are two distinct signalling pathways triggered by the binding of GM-CSF to its receptor, the Ras-mediated signal pathway and another via the activation of Jak2 tyrosine kinase (fig. 1). In the former, once GM-CSF has bound to its receptor, Shc is phosphorylated, which increases the amount of GTP-bound Ras, which in turn activates the mitogen-activated (MAP) kinases, and the signal continues on down the pathway. Activating point mutations in the *Ras* genes have been identified which result in elevated levels of Ras-GTP and hence constitutive activation of the GM-CSF signal transduction pathway in 30% of JMML patients' cells [36, 51]. Another mechanism by which increased levels of Ras-GTP occurs is when there is a lack of its natural inhibitor neurofibromin. Neurofibromin normally downregulates Ras-GTP by hydrolysing Ras from its active GTP-bound state to its inactive GDP-bound state. Neurofibromin is encoded by the neurofibromatosis type 1 tumour suppressor gene (*NF1*) and is a GTPase-activating protein (GAP) [52]. Leukaemic cells from children with NF1 have a selective decrease in NF1 GAP-like activity for Ras and raised Ras-GTP levels [53]. Similar changes have been reported in *Nf1–/–* knockout mice [54, 55]. These data support a specific role for neurofibromin in negatively regulating GM-CSF signalling through Ras.

As mentioned earlier, 15% of children with JMML have a clinical diagnosis of NF1 and hence a reduction of

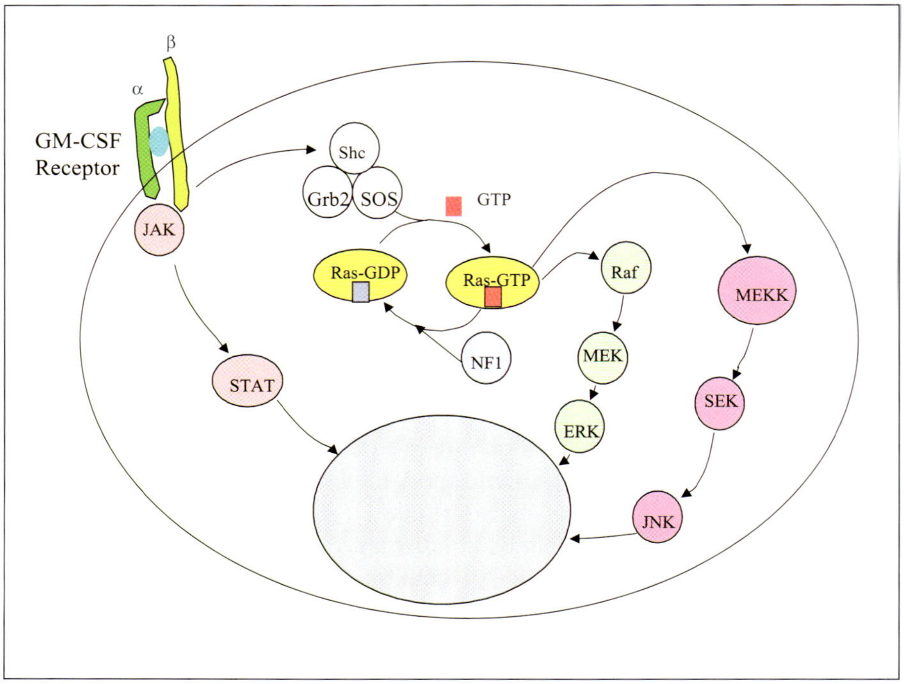


**Fig. 1.** Schematic diagram [modified from ref. 93] showing the two signal transduction pathways of GM-CSF. The first via Shc phosphorylation, following the binding of GM-CSF to the α- and β-subunits of its receptor, increases GTP-bound Ras and the activation of mitogen-activated (MAP) kinases. *NF1* codes for neurofibromin which has GTPase-activating protein (GAP)-like activity and hydrolyses Ras-GTP to inactive Ras-GDP, hence negatively regulates the pathway. Activating *Ras* mutations or loss of *NF1* both result in 'constitutive' activation of the pathway, provided GM-CSF is first bound.

neurofibromin production. Both alleles of the *NF1* gene are inactivated in leukaemic cells in patients with NF1 and this loss of heterozygosity suggests that *NF1* is functioning as a tumour suppressor gene in immature myeloid cells [56]. Mutations in *NF1* have also been found in the leukaemic cells of approximately 15% of JMML cases with no clinical sign of NF1 [35]. In addition *Ras*-activating mutations have been identified in another (mutually exclusive) 30% of JMML patients, so it would seem that disruption of the GM-CSF pathway is central to the pathogenesis of JMML. However, defects in this pathway are not present in all cases of JMML. Activation of Jak2 tyrosine kinase or overexpression of Shc could also produce a similar phenotype, leading to GM-CSF hypersensitivity, but there is as yet no evidence of disruption in this particular pathway in JMML cells. Some work has been done on disturbing the normal functioning of the Ras pathway in an attempt to inhibit the spontaneous growth of JMML cells. To function properly, Ras needs to be prenylated, and farnesyl transferase inhibitors, which are compounds developed specifically to block the prenylation of Ras, appear to have a significant growth-inhibitory effect on JMML cells in culture [57]. Recent work on human JMML and murine *Nf1–/–* deficient cells indicates that the presence of GM-CSF is essential for maintaining the MPD phenotype [58]. Hence blocking the binding of GM-CSF to its receptor would be another way to reduce signal transduction. A modified GM-CSF called E21R, which selectively binds to the GM-CSF receptor, has been shown to result in a transient clinical and haematological response in a child with end stage JMML [59]. Clearly, GM-CSF hypersensitivity is a central feature of JMML, however, apart from the inevitable myelomonocytic proliferation, the erythroid lineage is also affected. Clinically, there is a switch to fetal type erythropoiesis with high levels of Hb F and 'i' antigen expression as well as, in some cases, an erythroleukaemia-like picture on the blood film. There is also increased expression of GATA 1, the erythropoietin receptor and α- and γ-globin in JMML cells [60].

Cytogenetic clonal analysis of burst-forming units-erythroid (BFU-E) from JMML patients suggests that the disease arises from an early progenitor capable of differentiating into both monocyte-macrophages and erythroid cells. In addition, loss of the *NF1* allele has been observed in erythroblasts harvested from BFU-E from some JMML patients [34].

How can we explain the association with myeloid leukaemia in NF1 patients, which is restricted to the first few years of life and not present in adulthood? If the defects and mutations identified in children with JMML are still present in adulthood (should they reach it), and in NF1 adults who did not develop JMML in childhood, then it would appear that some stage or developmentally specific component, 'the predisposing vulnerability' has been switched off or is no longer important.

### Transient Myeloproliferative Disorder

TMD, classified by the WHO as a variant of acute megakaryoblastic leukaemia [1], is also known as TAM, or transient leukaemia (TL). TMD is a myeloproliferative phenomenon seen in the neonatal period almost exclusively in Down's syndrome. This disorder is self-limiting and regresses in infancy usually by 3 months of age. It occurs in an estimated 10% of neonates with Down's syndrome [60] but has been reported in normal newborns who are constitutional mosaics for trisomy 21 [62] and in infants in whom trisomy 21 is present only in the 'leukaemic' clone [63, 64]. TMD has also been reported in a phenotypically normal newborn infant with trisomy 18 and 21 and complex intrachromosomal rearrangements of chromosome 21 on bone marrow examination [65]. TMD is usually an incidental finding as the majority of patients are well, although most do have hepatosplenomegaly. Ten percent is probably an underestimate, as clinically well neonates with TMD may be missed. The WBC is usually raised to greater than 50 × $10^9$/l, and significant numbers of blasts are present in the peripheral blood, often 50% of the total WBC (generally a higher percentage than is seen in the bone marrow). The rest of the blood count is generally preserved, and although thrombocytopenia is commonly seen, profound cytopenias are rare. The proliferation is clonal [66] and the blasts are cytochemically and immunophenotypically megakaryoblasts or, occasionally, erythroblasts and are derived from a common erythroid and megakaryoblastic progenitor. Acquired chromosomal abnormalities occur in less than a third of patients with TMD. Tetrasomy 21, t(21;21) and i21 are the most common abnormalities seen [67]. Trisomy 12, 14 and additional +21 has been reported in 1 child [65]. Although such changes may raise the possibility of 'true', i.e. 'non-transient' leukaemia, they do not predict an aggressive clinical course [63]. Indeed spontaneous and complete haematological recovery, typical of TMD, occurs despite the presence of cytogenetic abnormalities [68]. TMD is essentially a TL as it is alternatively known, which remits in weeks to months, rarely requiring other than supportive therapy. However, there are those in whom the disease is severe and ultimately fatal. Such patients can present moribund at birth, if not stillborn [69], with severe hydrops fetalis (probably secondary to organ, particularly cardiopulmonary infiltration, rather than anaemia). More commonly, these severe cases develop progressive hepatic fibrosis [70]. Liver histology (biopsy or autopsy) shows infiltration of the liver with megakaryoblasts [71]. These blasts have been shown to express high levels of platelet-derived growth factor (PDGF) and transforming growth factor-β (TGFβ1), both of which, but particularly PDGF, are known to induce hepatic fibrosis [72]. A similar process is thought to result in the myelofibrosis associated with acute megakaryoblastic leukaemia (AMKL, AML-M7) [73]. There is a suggestion that the clonal population of megakaryoblasts seen in TMD is derived from fetal haemopoietic cells in the liver rather than the bone marrow [74]. If so, this might explain the spontaneous resolution of the disease as haemopoiesis switches from hepatic to medullary. Presumably, whatever triggers the switch triggers the end of the proliferation. A recent review of patients from the Paediatric Oncology Group Study indicates that these severe, life-threatening forms of TMD respond well to low-dose cytarabine therapy [70].

It is estimated that in 20–30% of cases of TMD the 'leukaemia', which is nearly always AMKL (AML-M7), recurs within the first 3 years of life (median age of 16 months) usually preceded by a myelodysplastic phase [62, 63, 75]. At this time, treatment with cytotoxic therapy is definitely required and these Down's syndrome children do better with standard chemotherapy than non-Down's syndrome children with AML, although they do suffer increased toxicity due to their increased sensitivity to cytarabine [76]. It is also estimated that approximately 20% of Down's syndrome children who develop MDS/AMKL had TMD as neonates [77–79] and this leukaemia appears to be derived from the original TMD. Occasionally, AML which is not AMKL, occurs later in childhood usually after the age of 5 years. In those patients known to have had TMD as neonates, the new leukaemia appears to be clonally unrelated to the original TMD, i.e. is due to the development of a new myeloid leukaemia rather than a clonal evolution of residual TMD cells [80, 81].

*Pathobiology/Molecular Genetics.* It is assumed that TMD is related to the presence of trisomy 21, but is the increased risk of (or predisposition for) developing TMD/AMKL simply due to a gene dosage effect (an extra chromosome 21) or perhaps a double dose of a mutated tumour suppressor gene (disomic homozygosity via the non-disjoined chromosomes from one parent). Cryptic deletions of chromosome 21, confined to leukaemic cells of Down's syndrome children with leukaemia, have been identified which suggest a loss of function of a critical gene on chromosome 21 is associated with leukaemogenesis [82]. There appears to be instability around the 21q11 locus in Down's syndrome which may be crucial in the development of AMKL [65] although such instability was detected in very few of the clonal cells in a case of TMD. Disomic homozygosity of DNA markers in the pericentromeric 21q11 region has repeatedly been found in Down's syndrome children with TMD and AMKL compared to Down's syndrome children without leukaemia [83–85]. These data would suggest that the myeloproliferation seen in TMD and AML is based on a gene dosage effect due to a predisposing allele located very close to the centromere of chromosome 21 [86]. There are as yet no obvious candidate genes at the 21q11 locus, although in another candidate region, 21q22, there are several: *AML1* or *AML1-CBFB* transcription factor genes, IFN-α/B receptor, CRF2–4, GART and SON [87]. As TMD predominantly involves the liver and the spontaneous regression of TMD mimics the normal spontaneous regression of liver haemopoiesis, the *AML1* gene locus is an attractive candidate as AML1 is essential for normal fetal liver haemopoiesis in a mouse model [88]. Immediately distal to the *AML1* gene is the cystithionine-B-synthetase gene *(CBS),* whose abnormal activity may account for the increased sensitivity to cytarabine of Down's syndrome [89] patients but would not obviously explain the pathogenesis of TMD/AMKL.

Still more needs to be understood about the relationship between TMD and AMKL. Are they the same? Is one just a transient fledgling form of the more inevitably lethal full-blown version? A study of the telomerase activity, as a marker of malignancy, indicates that TMD, with virtually no telomerase activity, is the benign version of the malignant AMKL where telomerase activity is frequently seen [90]. So what makes TMD eventually turn 'truly' malignant? One seems to evolve into the other, but why not straight away? Why the inevitable regression and then reappearance after a relatively predictable delay in a significant minority of patients? A 'second hit' may be required. Overexpression of BCL-2 protein has been found in TMD cells, which may explain their prolonged survival although expression of Fas and p53 was normal [91].

Are the rarer, life-threatening forms of TMD actually the same as the later-onset AMKL? Certainly these more severe cases do have telomerase activity [90] and require cytotoxic treatment with cytarabine [70].

Collaborative, multi-institutional collection and investigation of these cases is essential in order to answer these questions. TMD is a fascinating paradigm in which to study possible stage-specific mechanisms in developmental haemopoiesis and myeloproliferation.

## Conclusion

In childhood myeloproliferative disorders we have the opportunity to study the mechanisms integral to myeloid proliferation, differentiation and leukaemogenesis. There are early signs of common threads running through these rare and intriguing disorders such as the increased expression and disruption of *PDGF* and *PDGFRB*, respectively (TMD and aCML, JMML) and the overexpression of *TGFB1* and downregulation/mutation of *TGFBRII* (TMD, AMKL and CIMF [92]). Why is there a predisposition for myeloid malignancy, namely JMML, in NF1 children but not adults? And why is the occurrence per se of JMML limited to childhood? Is Ras no longer important in the GM-CSF pathway or do other adult mechanisms/pathways come into play/switch on, obviating or neutralizing the effect of *NF1* loss? Regarding TMD, what makes a leukaemia regress and hence transient? And what then allows its reappearance and persistence? Once elucidated, the molecular basis of these fascinating disorders should give insight into not only the mechanisms of myeloproliferation and transformation, but perhaps also normal developmental haemopoiesis.

## References


1 Jaffe ES, Harris NL, Stein H, Vardiman JW (eds): Pathology and Genetics of Tumours of Haematopoietic and Lymphoid Tissues, ed 1. World Health Organization Classification of Tumours, edited by Kleihues P, Sobin L. Lyon, IARC Press, 2001.

2 Bain BJ: An overview of translocation-related oncogenesis in chronic myeloid leukaemias; in Bain BJ (ed): Chronic Myeloproliferative Disorders. Cytogenetic and Molecular Genetic Abnormalities. Basel, Karger, 2003, pp 1–8.

3 Barnes DJ, Melo JV: Molecular basis of chronic myeloid leukaemia; in Bain BJ (ed): Chronic Myeloproliferative Disorders. Cytogenetic and Molecular Genetic Abnormalities. Basel, Karger, 2003, pp 9–31.

4 Kaeda J, Chase A, Goldman JM: Cytogenetic and molecular monitoring of residual disease in chronic myeloid leukaemia; in Bain BJ (ed): Chronic Myeloproliferative Disorders. Cytogenetic and Molecular Genetic Abnormalities. Basel, Karger, 2003, pp 32–43.

5 Johansson B, Fioretos T, Mitelman F: Cytogenetic and molecular genetic evolution of Philadelphia-chromosome-positive chronic myeloid leukaemia; in Bain BJ (ed): Chronic Myeloproliferative Disorders. Cytogenetic and Molecular Genetic Abnormalities. Basel, Karger, 2003, pp 44–61.

6 Dror Y: Essential thrombocythaemia in children. Br J Haematol 1999;107:691–698.

7 Heilmann E, Klein CE, Beck JD: Primary polycythaemia in childhood and adolescence. Folia Haematol 1983;110:935–941.

8 Sekhar M, Prentice HG, Popat U, Anderson D, Janmohammed R, Roberts I, Britt RP: Idiopathic myelofibrosis in children. Br J Haematol 1996;93:2:394–397.

9 Michiels JJ, Van Genderen PJ: Essential thrombocythemia in childhood. Semin Thromb Hemost 1997;23:295–301.

10 Yang RC, Qian LS: Essential thrombocythaemia in children: A report of nine cases. Br J Haematol 2000;110:1002–1014.

11 Siena S, Sammarelli G, Grimoldi MG, Schiavo R, Nozza A, Roncalli M, Mecucci C, Santoro A, Carlo-Stella C: New reciprocal translocation t(5;10)(q33;q22) associated with atypical chronic myeloid leukemia. Haematologica 1999;84:369–372.

12 Kulkarni S, Heath C, Parker S, Chase A, Iqbal S, Pocock CF, Kaeda J, Cwynarski K, Goldman JM, Cross NC: Fusion of H4/D10S170 to the platelet-derived growth factor receptor beta in BCR-ABL-negative myeloproliferative disorders with a t(5;10)(q33;q21). Cancer Res 2000;60:3592–3598.

13 Keene P, Mendelow B, Pinto MR, Bezwoda W, MacDougall L, Falkson G, Ruff P, Bernstein R: Abnormalities of chromosome 12q13 and malignant proliferation of eosinophils: A non-random association. Br J Haematol 1987;67:25–31.

14 Borkhardt A, Bojesen S, Haas OA, Fuchs U, Bartelheimer D, Loncarevic IF, Bohle RM, Harbott J, Repp R, Jaeger U, Viehmann S, Henn T, Korth P, Scharr D, Lampert F: The human GRAF gene is fused to MLL in a unique t(5;11)(q31;q23) and both alleles are disrupted in three cases of myelodysplastic syndrome/acute myeloid leukemia with a deletion 5q. Proc Natl Acad Sci USA 2000;97:9168–9173.

15 Berkowicz M, Rosner E, Rechavi G, Mamon Z, Neuman Y, Ben-Bassat I, Ramot B: Atypical chronic myelomonocytic leukemia with eosinophilia and translocation (5;12). A new association. Cancer Genet Cytogenet 1991;51:277–278.

16 Pellier I, LeMoine PJ, Rialland X, François S, Baranger L, Blanchet O, Larget-Piet L, Ifrah N: Myelodysplastic syndrome with t(5;12)(q31;p12–p13) and eosinophilia: A pediatric case with review of literature. J Pediatr Hematol Oncol 1996;18:285–983.

17 Golub TR, Barker GF, Lovett M, Gilliland DG: Fusion of PDGF receptor beta to a novel ets-like gene, tel, in chronic myelomonocytic leukemia with t(5;12) chromosomal translocation. Cell 1994;77:307–316.

18 Passmore SJ, Hann IM: Paediatric myelodysplasia. Br Med Bull 1996;52:778–786.

19 Jani K, Kempski HM, Reeves BR: A case myelodysplasia with eosinophilia having a translocation t(5;12)(q31;q13) restricted to myeloid cells but not involving eosinophils. Br J Haematol 1994;87:57–60.

20 Steer EJ, Cross NCP: Myeloproliferative disorders with translocations of chromosome 5q31–35: Role of the platelet-derived growth factor receptor beta; in Bain BJ (ed): Chronic Myeloproliferative Disorders. Cytogenetic and Molecular Genetic Abnormalities. Basel, Karger, 2003, pp 69–79.

21 Inhorn RC, Aster JC, Roach SA, Slapak CA, Sniffer R, Tantravahi R, Stone RM: A syndrome of lymphoblastic lymphoma, eosinophilia, and myeloid hyperplasia/malignancy associated with t(8;13)(p11;q11): Description of a distinctive clinicopathologic entity. Blood 1995;85:1881–1887.

22 Lewis JP, Jenks H, Larzerson J: Philadelphia chromosome negative chronic myelogenous leukemia in a child with t(8;9)(p11 or 12;q34).Am J Pediatr Hematol Oncol 1983;5:265–269.

23 Vannier JP, Bizet M, Bastard C, Bernard A, Ducastelle T, Tron P: Simultaneous occurrence of a T-cell lymphoma and a chronic myelogenous leukemia with an unusual karyotype. Leuk Res 1984;8:647–657.

24 Nakayama H, Inamitsu T, Ohga S, Kai T, Suda M, Matsuzaki A, Ueda K: Chronic myelomonocytic leukaemia with t(8;9)(p11;q34) in childhood: An example of the 8p11 myeloproliferative disorder? Br J Haematol 1996;92:692–695.

25 Sohal J, Chase A, Mould S, Corcoran M, Oscier D, Iqbal S, Parker S, Welborn J, Harris RI, Martinelli G, Montefusco V, Sinclair P, Wilkins BS, van den Berg H, Vanstraelen D, Goldman JM, Cross NC: Identification of four new translocations involving FGFR1 in myeloid disorders. Genes Chromosomes Cancer 2001;32:155–163.

26 Macdonald D, Reiter A, Cross NCP: The 8p11 myeloproliferative syndrome: A distinct clinical entity caused by constitutive activation of *FGFR1*; in Bain BJ (ed): Chronic Myeloproliferative Disorders. Cytogenetic and Molecular Genetic Abnormalities. Basel, Karger, 2003, pp 62–68.

27 Shannon KM, O'Connell P, Martin GA, Paderanga D, Olson K, Dinndor P, McCormick F: Loss of the normal NF1 allele from the bone marrow of children with type 1 neurofibromatosis and malignant myeloid disorders. N Engl J Med 1994;330:597–601.

28 Niemeyer CM, Arico M, Basso G, Biondi A, Cantu Rajnoldi A, Creutzig U, Hass O, Harbott J, Hasle H, Kerndrup G, Locatelli F, Mann G, Stollmann-Gibbels B, van't Veer-Korthof ET, van Wering E, Zimmermann M: Chronic myelomonocytic leukemia in childhood: A retrospective analysis of 110 cases. European Working Group on Myelodysplastic Syndromes in Childhood (EWOG-MDS) (see comments). Blood 1997;89:3534–3543.

29 Suda T, Miura Y, Mizoguchi H, Ijima H, Eguchi M, Kaku H, Ide C: Characterization of hemopoietic precursor cells in juvenile-type chronic myelocytic leukemia. Leuk Res 1982;6:43–53.

30 Stiller CA, Chessells JM, Fitchett M: Neurofibromatosis and childhood leukaemia/lymphoma: A population-based UKCCSG study (see comments). Br J Cancer 1994;70:969–972.

31 Niemeyer CM, Fenu S, Hasle H, Mann G, Stary J, van Wering E: Differentiating juvenile myelomonocytic leukemia from infectious disease. Blood 1998;91:365–366.

32 Herrod H, Dow L, Sullivan J: Persistent Epstein-Barr virus infection mimicking juvenile chronic myelogenous leukemia: Immunologic and hematologic studies. Blood 1983;61:1098–1104.

33 Busque L, Gilliland DG, Prachal JT, Sieff CA, Weinstein HJ, Sokol JM, Belickova M, Wayne AS, Zuckerman KS, Sokol L, Castleberry RP, Emanuel PD: Clonality in juvenile chronic myelogenous leukemia. Blood 1995;85:21–30.

34 Miles DK, Freedman MH, Stephens K, Pallavicini M, Sievers EL, Weaver M, Grunberger T, Thompson P, Shannon KM: Pattern of hematopoietic lineage involvement in children with neurofibromatosis, type 1 and malignant myeloid disorders. Blood 1996;88:4314–4320.

35 Kalra R, Dale D, Freedman M, Bonilla MA, Weinblatt M, Ganser A, Bowman P, Abish S, Priest J, Oseas RS, et al: Monosomy 7 and activating RAS mutations accompany malignant transformation in patients with congenital neutropenia. Blood 1995;86:4579–4586.

36 Miyauchi J, Asada M, Sasaki M, Tsunematsu Y, Kojima S, Mizutani S: Mutations of the N-ras gene in juvenile chronic myelogenous leukemia. Blood 1994;83:2248–2254.
37 Miyauchi J, Asada M, Tsunematsu Y, Kaneko Y, Kojima S, Mizutani S: Abnormalities of the p53 gene in juvenile myelomonocytic leukaemia. Br J Haematol 1999;106:980–986.
38 Flotho C, Valcamonica S, Mach-Pascual S, Schmahl G, Corral L, Ritterbach J, Hasle H, Arico M, Biondi A, Niemeyer CM: RAS mutations and clonality analysis in children with juvenile myelomonocytic leukemia (JMML). Leukemia 1999;13:32–37.
39 Luna-Fineman S, Shannon KM, Lange BJ: Childhood monosomy 7: Epidemiology, biology and mechanistic implications. Blood 1995; 85:1985–1999.
40 Luna-Fineman S, Shannon KM, Atwater SK, Davis J, Masterson M, Ortega J, Sanders J, Steinherz P, Weinberg V, Lange BJ: Myelodysplastic and myeloproliferative disorders of childhood: A study of 167 patients. Blood 1999;93:459–466.
41 Le Beau MM, Espinosa R 3rd, Davis EM, Eisenbart JD, Larson RA, Green ED: Cytogenetic and molecular delineation of a region of chromosome 7 commonly deleted in malignant myeloid disease. Blood 1996;88:1930–1935.
42 Brady AF, Waters CS, Pocha MJ, Brueton LA: Chronic myelomonocytic leukaemia in a child with constitutional partial trisomy 8 mosaicism. Clin Genet 2000;58:142–146.
43 Palmar CG, Profisor AJ, Weaver DD, Hodes ME, Heerema N: Juvenile chronic granulocytic leukemia in a patient with trisomy 8, neurofibromatosis and prolonged Epstein-Barr virus infection. J Pediatr 1983;102:888–892.
44 Bown N, Yule SM, Evans J, Kernahan J, Reid MM: Chronic myelomonocytic leukemia with t(13;14) in a child. Cancer Genet Cytogenet 1992;60:190–192.
45 Tosi S, Mosna G, Cazzaniga G, Giudici G, Kearney L, Biondi A, Privitera E: Unbalanced t(3;12) in a case of juvenile myelomonocytic leukemia (JMML) results in partial trisomy of 3q as defined by FISH. Leukemia 1997;11: 1465–1468.
46 Passmore SJ, Hann IM, Stiller CA, Ramani P, Swansbury GJ, Gibbons B, Reeves BR, Chessells JM: Pediatric myelodysplasia: A study of 68 children and a new prognostic scoring system. Blood 1995;85:1742–1750.
47 Shannon KM, Watterson J, Johnson P, O'Connell P, Lange B, Shah N, Steinherz P, Kann YW, Priest JR: Monosomy 7 myeloproliferative disease in children with neurofibromatosis, type 1: Epidemiology and molecular analysis. Blood 1992;79:1311–1318.
48 Kaneko Y, Maseki N, Sakurai M, Shibuya A, Shinohara T, Fujimoto T, Kanno H, Nishikawa A: Chromosome pattern in juvenile chronic myelogenous leukemia, myelodysplastic syndrome, and acute leukemia associated with neurofibromatosis. Leukemia 1989;3:36–41.
49 Emanuel P, Bates LJ, Castleberry RP, Gaultieri RJ, Zuckerman KS: Selective hypersensitivity to granulocyte-macrophage colony-stimulating factor by juvenile chronic myeloid leukaemia hematopoietic progenitors. Blood 1991;77: 925–929.
50 Freeburn RW, Gale RE, Wagner HM, Linch DC: Analysis of the coding sequence for the GM-CSF receptor alpha and beta chains in patients with juvenile chronic myeloid leukemia (JCML). Exp Hematol 1997;25:306–311.
51 Kalra R, Paderanga DC, Olson K, Shannon KM: Genetic analysis is consistent with the hypothesis that NF1 limits myeloid cell growth through p21ras. Blood 1994;84:3435–3439.
52 Brodeur GM: The NF1 gene in myelopoiesis and childhood myelodysplastic syndromes. N Engl Med 1994;330:637–639.
53 Bollag G, Clapp DW, Shih S, Adler F, Zhang YY, Thompson P, Lange BJ, Freedman MH, McCormick F, Jacks T, Shannon K: Loss of NF1 results in activation of the Ras signaling pathway and leads to aberrant growth in haematopoietic cells. Nat Genet 1996;12:144–148.
54 Largaespada DA, Brannan CI, Jenkins NA, Copeland NG: Nf1 deficiency causes Ras-mediated granulocyte/macrophage colony stimulating factor hypersensitivity and chronic myeloid leukaemia. Nat Genet 1996;12:137–143.
55 Side LE, Emanuel PD, Taylor B, Franklin J, Thompson P, Castlebery RP, Shannon KM: Mutations of the NF1 gene in children with juvenile myelomonocytic leukemia without clinical evidence of neurofibromatosis, type 1. Blood 1998;92:267–272.
56 Side L, Taylor B, Cayouette M, Conner E, Thompson P, Luce M, Shannon K: Homozygous inactivation of the NF1 gene in bone marrow cells from children with neurofibromatosis type 1 and malignant myeloid disorders. N Engl J Med 1997;336:1713–1720.
57 Emanuel PD, Snyder RC, Wiley T, Gopurala B, Castleberry RP: Inhibition of juvenile myelomonocytic leukemia cell growth in vitro by farnesyltransferase inhibitors. Blood 2000;95: 639–645.
58 Birnbaum RA, O'MarcaighA, Wardak Z, Zhang YY, Dranoff G, Jacks F, Clapp DW, Shannon K: Nf1 and Gmcsf interact in myeloid leukemogenesis. Mol Cell 2000;5:189–195.
59 Bernard F, Thomas C, Emile JF, Hercus T, Cassinat B, Chomienne C, Donadieu J: Transient hematologic and clinical effect of E21R in a child with end-stage juvenile myelomonocytic leukemia. Blood 2002;99:2615–2626.
60 Privitera E, Schiro R, Longoni D, Ronchi A, Rambaldi A, Bernasconi S, Ottolenghi S, Masera G, Biondi A: Constitutive expression of GATA-1, EPOR, alpha-globin, and gamma-globin genes in myeloid clonogenic cells from juvenile chronic myelocytic leukemia. Blood 1995;86:323–328.
61 Zipursky A, Brown EJ, Christensen H, Doyle J: Transient myeloproliferative disorder (transient leukemia) and hematologic manifestations of Down syndrome. Clin Lab Med 1999; 19:157–167.
62 Zipursky A, Brown E, Christensen H, Sutherland R, Doyle J: Leukemia and/or myeloproliferative syndrome in neonates with Down syndrome. Semin Perinatol 1997;21:97–101.
63 Homans AC, Verissimo AM, Vlacha V: Transient abnormal myelopoiesis of infancy associtred with trisomy 21. Am J Pediatat Hematol Oncol 1993;15:392–399.
64 Richards M, Welch J, Watmore A, Readett D, Vora AJ: Trisomy 21 associated transient neonatal myeloproliferation in the absence of Down's syndrome. Arch Dis Child (Fetal Neonatal Ed) 1998;79:F215–F217.
65 Kempsky HM, Craze JL, Chessells JM, Reeves BR: Cryptic deletions and inversions of chromosome 21 in a phenotypically normal infant with transient abnormal myelopoiesis: A molecular cytogenetic study. Br J Haematol 1998; 103:473–479.
66 Kurahashi H, Hara J, Yumura-Yagi K, Murayama N, Inoue M, Ishihara S, Tawa A, Okada S, Kawa-Ha K: Monoclonal nature of transient abnormal myelopoiesis in Down's syndrome. Blood 1991;77:1161–1163.
67 Barnett PLJ, Clark ACL, Garson OM: Acute nonlymphocytic leukemia after transient myeloproliferative disorder in a patient with Down syndrome. Med Pediatr Oncol 1990;18:347–353.
68 Kounami S, Aoyagi N, Tsuno H, Suzuki H, Kitano N, Koike M: Additional chromosome abnormalities in transient abnormal myelopoiesis in Down's syndrome patients. Acta Haematol 1997;98:109–112.
69 Zipursky A, Rose T, Skidmore M, Thorner P, Doyle J: Hydrops fetalis and neonatal leukemia in Down syndrome. Pediatr Hematol Oncol 1996;13:81–87.
70 Al-Kasim F, Doyle JJ, Massey GV, Weinstein HJ, Zipursky: Incidence and treatment of potentially lethal diseases in transient leukemia of Down syndrome: Pediatric Oncology Group Study. J Pediatr Hematol Oncol 2002;24:9–13.
71 Arai H, Ishida A, Nakajima W, Nishinomiya F, Yamazoe A, Takada G: Immunohistochemical study on transforming growth factor-beta1 expression in liver fibrosis of Down's syndrome with transient abnormal myelopoiesis. Hum Pathol 1999;30:474–476.
72 Hattori H, Matsuzaki A, Suminoe A, Ihara K, Nakayama H, Hara T: High expression of platelet-derived growth factor and transforming growth factor-beta 1 in blast cells from patients with Down syndrome suffering from transient myeloproliferative disorder and organ fibrosis. Br J Haematol 2001;115:472–475.
73 Tsao MS: Hepatic sinusoidal fibrosis in agnogenic myeloid metaplasia. Am J Clin Pathol 1989;91:302.
74 Miyauchi J, Ito Y, Kawano T, Tsunematsu Y, Shimizu K: Unusual diffuse liver fibrosis accompanying transient myeloproliferative disorder in Down's syndrome: A report of four autopsy cases and proposal of a hypothesis. Blood 1992;80:1521–1527.

75 Chessells JM: Myeloblastic syndromes; in Lilleyman J, Hann I, Blanchette V (eds): Pediatric Hematology. Churchill Livingstone, Edinburgh, 1999, pp 83–101.
76 Taub JW, Ravindranath Y: Down syndrome and the transient myeloproliferative disorder: Why is it transient? J Pediatr Hematol Oncol 2002;24:6–8.
77 Lin HP, Menaka H, Lim KH, Yong HS: Congenital leukemoid reaction followed by fatal leukemia: A case with Down's syndrome. Am J Dis Child 1980;134:939–941.
78 Zipursky A, Peeters M, Poon A: Megakaryoblastic leukemia and Down's syndrome – A review. Prog Clin Biol Res 1987;246:33–56.
79 Zipursky A, Poon A, Doyle J: Hematologic and oncologic disorders in Down syndrome in Lott, McCoy (eds): Down Syndrome: Advances in Medical Care. New York, Wiley-Liss, 1992, p 93–101.
80 Yamaguchi Y, Fujii H, Kazama H, Iinamu K, Shinomiya N, Aoki T: Acute myeloblastic leukemia associated with trisomy 8 and translocation t(8;21) in a child with Down syndrome. Cancer Genet Cytogenet 1997;97:32–34.
81 Kounami S, Aoyagi N, Tsuno H, Suzuki H, Kobayashi M, Koike M: Myelodysplastic syndrome after regression of transient abnormal myelopoiesis in a Down syndrome infant: Different clonal origin? Cancer Genet Cytogenet 1998;104:115–118.
82 Kempski HM, Chessells JM, Reeves BR: Deletions of chromosome 21 restricted to the leukemic cells of children with Down syndrome and leukemia. Leukemia 1997;11:1973–1977.
83 Abe K, Kajii T, Niikawwa N: Disomic homozygosity in 21-trisomic cells: A mechanism responsible for transient myeloproliferative syndrome. Hum Genet 1989;82:313–316.
84 Niikawa N, Deng HX, Abe K, Harada N, Okada T, Tsuchiya H, Akaboshi I, Matsuda I, Fukushima T, Kaneko Y, Kuwano A, Kajii T: Possible mapping of the gene for transient myeloproliferative syndrome at 21q11.2. Hum Genet 1991;87:561–566.
85 Shen JJ, Williams BJ, Zipursky A, Doyle J, Sherman SL, Jacobs PA, Shugar AL, Soukup SW, Hassold TJ: Cytogenetic and molecular studies of Down syndrome individuals with leukaemia. Am J Hum Genet 1995;56:915–925.
86 Cavani S, Perfumo C, Argusti A, Pierluigi M, Perroni L, Schmiegelow K, Petersen MB, Cotter FE, Strigini P, Dagna-Bricarelli F, Nizetic D: Cytogenetic and molecular study of 32 Down syndrome families: Potential leukaemia predisposing role of the most proximal segment of chromosome 21q. Br J Haematol 1998;103: 213–216.
87 Gamis AS, Hilden JM: Transient myeloproliferative disorder, a disorder with too few data and many unanswered questions: Does it contain an important piece of the puzzle to understanding hematopoiesis and acute myelogenous leukemia? J Pediatr Hematol Oncol 2002; 24:2–5.
88 Okuda T, van Deursen J, Hiebert SW, Grosveld G, Downing JR: AML1, the target of multiple chromosomal translocations in human leukemia, is essential for normal fetal liver hematopoiesis. Cell 1996;84:321–330.
89 Taub JW, Matherly LH, Stout ML, Buck SA, Gurney JG, Ravindranath Y: Enhanced metabolism of 1-beta-*D*-arabinofuranosylcytosine in Down syndrome cells: A contributing factor to the superior event free survival of Down syndrome children with acute myeloid leukemia. Blood 1996;87:3395–3403.
90 Holt SE, Brown EJ, Zipurksy A: Telomerase and the benign and malignant megakaryoblastic leukemias of Down syndrome. J Pediatr Hematol Oncol 2002;24:14–17.
91 Yoshioka K, Yamamoto S, Moriguchi N, Miyata H, Tsukiyama K, Isokawa S, Horiuchi F, Takemura T: Overexpression of bcl-2 in transient abnormal myelopoiesis associated with Down syndrome. Ann Hematol 2000;79: 319–321.
92 Le Bousse-Kerdiles MC, Martyre MC: Involvement of the fibrogenic cytokines. TGF-beta and bFGF, in the pathogenesis of idiopathic myelofibrosis. Pathol Biol (Paris) 2001;49: 153–157.
93 Arico M, Biondi A, Pui CH: Juvenile myelomonocytic leukemia. Blood 1997;90:479–488.

Georgina W. Hall, PhD, MRCP, MRCPath
Paediatric Haematology/Oncology Unit
John Radcliffe Hospital, Headley Way
Headington, Oxford OX3 9DU (UK)
E-Mail georgina.hall@ndcls.ox.ac.uk

Bain BJ (ed): Chronic Myeloproliferative Disorders.
Basel, Karger, 2003, pp 90–101 / Acta Haematol 2002;108:120–131

# Polycythaemia vera: Will New Markers Help Us Answer Old Questions?

Heike L. Pahl

Division of Experimental Anaesthesiology, Center for Clinical Research, University Hospital Freiburg, Germany

**Key Words**
Polycythaemia · Erythrocytosis · Polycythaemia vera · Myeloproliferative disorders

**Abstract**
The publication of the first molecular markers for polycythaemia vera (PV) has led to a renewed interest in finding the mutations leading to the development of this myeloproliferative disorder. A substantial amount of cell biological data on the PV stem cell, however, have been available for over two decades. This review will summarise, combine and integrate both fields of investigation. Based on this analysis, two alternative models for the development of myeloproliferative disorders are proposed.

## Introduction

In 1966, polycythaemia vera (PV) became one of the first diseases for which therapeutic decisions could be based on prospective, randomised clinical trials. That year saw the inception of the Polycythaemia Vera Study Group (PVSG) and the design of the first PVSG study protocol (PVSG-01), one of the first examples of evidence-based medicine. Ten years later, when the first results were published, coincidentally, substantial advances were also being made in understanding the cell biology of haematopoiesis. Culture techniques had become available with which haematopoietic precursor cells could be quantified and analysed. Using these new tools, many basic questions about PV stem cells were answered.

The power of cell biology, however, is limited. Ultimately, we need to understand diseases at the molecular level if we are to design specific and targeted therapies.

Thirty-six years after its institution, the PVSG is, unfortunately, no longer designing new studies. In parallel with its decline, basic science research on PV also diminished and few new insights were gained. Recently, however, there has been renewed interest in PV both on the scientific and on the clinical level. Study groups in Europe are conducting new randomised clinical trials. In addition, the first molecular markers for PV have been published in the past 3 years. Therefore, we may now be in the fortunate position to combine old and new knowledge in the field. Perhaps we are less than 10 years away from our ultimate goal: understanding the molecular aberrations that lead to the development of PV and designing specific therapies to reverse the consequences of these changes.

## PV: A Disease of the Pluripotent Stem Cell

Much of our early knowledge about the cellular alterations in PV comes from the study of 2 female patients, who are heterozygous for the X-chromosomally encoded gene glucose-6-phosphate dehydrogenase (G-6-PD). The two G-6-PD alleles in these patients could be distinguished by electrophoresis, a technique sensitive enough to allow the analysis of 50 erythrocytes or 20 granulocytes/macrophages as well as the classification of single haematopoietic colonies [1].

Using cells from these patients, Adamson et al. [2] demonstrated that peripheral erythrocytes, granulocytes, monocytes and platelets contained only a single G-6-PD isoform. In contrast, skin fibroblasts contained both isoforms in nearly equal amounts. These data demonstrated that PV arises from a single pluripotent haematopoietic stem cell.

### Old Truths about PV Stem Cells

Even though a stem cell is altered, the difference may become apparent only when it starts to differentiate. Thus, knowledge of the stem cell origin of PV immediately led to a subsequent question: at what developmental stage is the stem cell abnormality phenotypically expressed?

Using the $^{3}$H-thymidine suicide assay, Mladenovic and Adamson [3] reported that there is no significant difference between PV and healthy controls in the number of progenitor cells in S-phase at a given time. These data partially confirm but partially contradict an earlier report by Singer et al. [4], who demonstrated increased cycling of PV colony-forming units in culture (CFU-C) but no difference in proliferation between normal and PV burst-forming units-erythroid (BFU-E). In contrast, Fauser and Messner [5] report that an increased fraction of the circulating BFU-E are cycling in PV patients, while granulocyte-macrophage colony-forming units (GM-CFU) are similar to normal controls. Interpretation of these data is difficult, since Cashman et al. [6] reported that cell cycle activity of progenitor cells can be altered by storage at 4°C.

PV stem cells, however, clearly exhibit abnormal and unregulated proliferation in long-term bone marrow cultures [6]. Unlike normal marrow cells, which return to a quiescent state 5–7 days after a medium change, PV cells remain continuously in cycle [6]. PV cells apparently fail to respond to growth-inhibitory signals form the adherent cells in the culture system.

However, even though PV cells may divide more frequently, perhaps even continuously, there is no significant increase in the number of progenitor cells in PV patients [7]. Only a slight increase in the number of bone marrow CFU-E was noticed, the number of BFU-E and CFU-C were similar in both blood and bone marrow of PV patients and healthy controls [7]. This is in contrast to patients with chronic myeloid leukaemia (CML), a more aggressive myeloproliferative disorder, where a significant increase in all types of progenitor cells is seen, both in the bone marrow and in the periphery [7]. Thus, unlike leukaemic cells, PV stem cells do not appear to rapidly overgrow their normal counterparts. Rather, they must possess a more subtle growth advantage. This observation, of course, is consistent with the difference in the clinical course between CML and PV. While CML commences in a chronic phase but invariably accelerates to a blast crisis, PV progresses slowly and only rarely transforms to acute leukaemia.

### Endogenous Erythroid Colonies

Prchal and Axelrad [8] first reported an intriguing abnormality of PV cells. Unlike healthy progenitors, PV cells can form erythroid colonies in the absence of erythropoietin (EPO). These colonies have been termed 'endogenous erythroid colonies – EECs' and their culture has been used extensively to characterize the abnormal stem cell clone in PV patients.

Using the 2 previously mentioned G-6-PD heterozygous female PV patients, Adamson [9] established that all EECs arise from the neoplastic clone. Much more surprising was the finding that not all clonally derived PV cells are capable of giving rise to EECs. Rather, the malignant PV clone contains both EPO-independent and EPO-dependent cells [9–11]. Moreover, an individual BFU-E can give rise to both EPO-independent and EPO-dependent daughter cells [12]. Thus, the aberrant characteristic of EPO independence appears not to be genetically fixed within the mutant pluripotent stem cell.

If the ability to proliferate in the absence of EPO is not conferred to the altered PV stem cell by a mutation which is passed along to all its progeny, how does the PV clone gain its selective advantage?

Eaves and Eaves [13] first demonstrated that the percentage of cells capable of forming EECs increases with the stage of maturation. In a newly diagnosed PV patient, less than 5% of the primitive BFU-E were capable of forming EECs, while this was the case for 44% of the mature BFU-Es and 66% of the CFU-Es [13]. Similar results were shown for 4 additional patients. Two years later, the same authors confirmed their observation in a large cohort of 61 patients [7]. The proportion of progenitors capable of EEC formation increased 3–5 times during differentiation of BFU-E to CFU-E [7]. From these data it would appear that the PV clone gains a selective advantage as the cells mature along the erythroid lineage. This may occur through an increased proliferation of the abnormal BFU-E compared to the healthy cells. Alternatively, under conditions of low serum EPO, which have been demonstrated in PV patients [14], the normal precursors may not be able to divide or differentiate.

These data, however, are contradicted by two publications. Lacombe et al. [15] found a similar proportion of BFU-E- and CFU-E-derived EECs in 9 patients. These authors interpret these data to mean that the abnormal clone did not expand preferentially between the BFU-E and the CFU-E stage of differentiation. Fibach and Rachmilewitz [16] come to a similar conclusion using a two-step liquid culture. Peripheral blood precursor cells were either plated directly in methylcellulose with or without EPO and scored for BFU-E- and BFU-E-derived EECs, respectively. Alternatively, an aliquot of cells was expanded in liquid culture without EPO for 6 days and subsequently plated again with and without EPO for 7 days and scored for CFU-E- and CFU-E-derived EECs, respectively. The analysis of 6 patients revealed no difference in the percentage of BFU-E- or CFU-E-derived EEC colonies [16]. Nonetheless, the absolute number of colony-forming cells increased 4–10 times during liquid culture, demonstrating that the cells were actively proliferating [16]. Again, these data were interpreted to show that, even in the absence of EPO during liquid culture, the abnormal, EPO-independent clone has no growth advantage over its normal counterpart. Moreover, the PV stem cell defect appears to be expressed only at the CFU-E level.

Interpretation of all these studies is complicated by the above-described observation that the neoplastic PV clone contains both EPO-independent and EPO-dependent progenitors. Therefore, the percentage of EECs formed does not define the size of the abnormal clone [9]. Because of the presence of clonally derived, EPO-dependent cells, the size of the neoplastic clone is underestimated by the measurement of EECs. What is required is a cellular marker such as G-6-PD to unequivocally identify either the PV clone or the residual normal stem cells.

Adamson et al. [10] have elegantly demonstrated that in both G-6-PD heterozygous patients the percentage of normal BFU-E decreased over a 3-year period, as the disease progressed. At the time of last analysis, less than 10% of BFU-E were normal in these patients. Thus, it is possible that the patients analysed by Fibach and Rachmilewitz [16], after a disease duration of up to 12 years contained almost exclusively neoplastic cells. In this case, of course, no growth advantage would be observable.

Two observations argue against this explanation. First Lacombe et al. [15] apparently used newly diagnosed, untreated patients, although this is not unequivocally stated. It is reasonable to assume that these patients had at least some normal BFU-E. Nonetheless, no difference in BFU-E- and CFU-E-derived numbers of EECs was observed [15]. Second, Eaves and Eaves [13] report an increase in the proportion of EECs from CFU-E in 2 patients 12 and 20 years after diagnosis. Because G-6-PD analysis was not possible in these patients, it is not clear how many, if any, normal BFU-E remained in these patients. The discrepancy in the data must therefore be explained differently.

It is possible that the percentage of cells surviving in culture without EPO long enough to score as an EEC colony is simply lower in the BFU-E compartment under certain culture conditions. Cashman et al. [12] suggest that EPO independence 'can be more accurately viewed as a prolongation of the time that erythroid cells can survive and differentiate in the absence of EPO'. If this is the case, scoring mature BFU-E EECs after 14 days or primitive BFU-E-derived EECs after 21 days may detect only a small subset of the EPO-independent cells. Many neoplastic cells, while able to survive without EPO for a longer period of time than normal cells, may nevertheless not be able to remain viable and replicate for 14–21 days under these conditions. The apparent increase in CFU-E-derived EECs may thus more accurately reflect a failure to detect BFU-E-derived EECs in these cultures.

In this context it is important to note another technical detail. Eaves et al. [7, 13], who observe increased numbers of EECs from CFU-Es, used leucocyte-conditioned medium (LCM), which contains burst-promoting activity, in their methylcellulose assays. In contrast, both Lacombe et al. [15] and Fibach and Rachmilewitz [16], who saw no difference between CFU-E- and BFU-E-derived EECs, did not add LCM to either their plasma clot or their methylcellulose assays. Perhaps the culture conditions used by Eaves et al. [7, 13] allow a greater proportion of EECs to develop from CFU-E than from BFU-E, even if both precursor pools exclusively contain neoplastic cells. Thus, the observed increase in EEC formation between BFU-E and CFU-E cells may not, in fact, reflect a genuine growth advantage of the neoplastic clone during this stage of differentiation.

More recent data from Cashman et al. [6] support this interpretation. Bone marrow sampled from PV patients at the time of diagnosis was cultured for 7 weeks. On day 0, 6 of the 7 patients displayed a higher percentage of CFU-E-derived than BFU-E-derived EECs [6]. Again, G-6-PD analysis was not possible in these patients. However, normal BFU-E should have been present at least in some of these individuals since the disease was newly diagnosed. Nonetheless, over the entire culture period, the proportion of BFU-E giving rise to EECs remained constant. The authors concluded that under the long-term culture conditions employed, the PV clone does not have a positive selective advantage [9].

G-6-PD analysis of one patient by Adamson [9] and Adamson et al. [10], however, independently confirms the predominance of the neoplastic clone in the more mature CFU-E precursor compartment. While BFU-E-derived colonies from this patient contained both A and B isoforms of the enzyme, CFU-E-derived colonies displayed only the A isoform of the neoplastic clone. These data alone, irrespective of the studies detailed above, would argue that the neoplastic clone has a selective advantage in vivo.

The simplest explanation would be an increased proliferation of PV BFU-E, giving rise to large numbers of neoplastic CFU-E, which would therefore drastically outnumber the few normal CFU-E produced. If this model were correct, the CFU-E:BFU-E ratio of the abnormal PV clone should be much higher than that of the normal counterparts. However, this is not the case. Adamson et al. [10] report a CFU-E:BFU-E ratio of 4.8:1 for the abnormal clone, very close to that determined for 39 healthy controls. The remaining healthy precursors in the same PV patient, however, displayed a CFU-E:BFU-E ratio of 0.8:1, a dramatic decrease relative to the healthy marrow controls. Eaves et al. [7], while not distinguishing between the abnormal and the normal clone, find the overall CFU-E:BFU-E ratio in PV only very slightly increased relative to healthy controls. Increased BFU-E proliferation, then, cannot explain the vast predominance of the PV clone in the CFU-E compartment. Rather, it has been proposed that the PV clone must suppress proliferation and differentiation of normal stem cells [9, 10].

While this model is very attractive, there is only limited evidence to support it. The CFU-E:BFU-E ratios combined with G-6-PD analysis for the identification of residual normal cells are clearly the strongest data. However, it rests on the analysis of a single patient. A second series of experiments by Singer et al. [4], analysing both G-6-PD heterozygous patients, is intriguing as well. Bone marrow cells were exposed to $^{3}$H-thymidine, which kills cells actively synthesising DNA, i.e. proliferating cells. Following this treatment, only abnormal PV cells were killed, reflecting their increased cell cycle rate. However, surprisingly, the number of granulocytic colonies (called CFU-C in this publication) arising from normal stem cells increased. This suggests that PV CFU-C inhibit the growth of normal CFU-C in vivo and that this inhibition is lost when the PV cells are killed. This effect was only observed in the granulocytic lineage; abnormal BFU-E were not killed by $^{3}$H-thymidine treatment and consequently no increase in normal BFU-E was observed [4]. Therefore, erythroid and granulocytic colonies appear to differ in the method used to suppress their normal counterparts.

Unfortunately, these models are based on data obtained from very few, selected patients and must therefore be regarded as hypotheses. The recent publication of molecular markers, detailed below, may allow us to test these postulates on larger cohorts of patients.

## Pathophysiological and Diagnostic Markers in PV

Many abnormalities of PV cells have been described, for example increased Bcl-x expression [17], constitutive STAT-3 activation [18] and dysregulated kinase or phosphatase activity. However, these were reviewed recently [19] and have not been pursued further. Therefore, they will not be discussed here. This section will concentrate on cellular, genetic and molecular markers, which may lead either to the elucidation of the molecular cause of PV or, at least, to the development of more precise diagnostic tools.

### *Cellular Markers in PV*

#### *Endogenous Erythroid Colonies*

The first description of EEC growth by Prchal and Axelrad [8] already concluded with the suggestion that the presence of these abnormal colonies could serve as a diagnostic criterion to distinguish PV from secondary erythrocytosis. Since then, many researchers have supported this recommendation [14, 20, 21]. The usefulness and validity of EECs as an aid in the diagnosis of PV have been reviewed in two excellent articles [22, 23] and will therefore not be repeated here in detail. However, some aspects merit special consideration.

First, while many studies find peripheral blood EECs in all patients tested [15, 16, 24, 25], other publications state that some PV patients may display EECs in bone marrow cultures but not in the periphery [26, 27]. Therefore, the absence of peripheral blood EECs in one assay does not preclude a diagnosis of PV, but necessitates either several repetitions with larger number of cells or, if possible, the investigation of bone marrow. As Westwood and Pearson correctly emphasise, every assay has a lower limit of detection and EECs may be present in very small numbers in the peripheral blood. Hence, it is impossible to state that an individual has no EECs, rather he has, for example, less than 1 EEC in $10^5$ mononuclear cells (MNCs) [23].

Second, numerous studies have reported the presence of EECs in 50–85% of patients with essential thrombocythaemia (ET) [8, 28–32], [reviewed in ref. 23]. In a pivotal study, Shih and Lee [29] documented that the presence of EECs in patients with idiopathic thrombocytosis but normal haemoglobin levels predicts the onset of overt PV within several years. These observations were confirmed in part in a subsequent study [33]. However, in this latter cohort, many EEC-positive patients with ET had not progressed to PV at the time of publication. Clearly, the remarkably subtle growth advantage of the PV clone over normal stem cells could account for a prolonged normocythaemic phase of PV in selected patients. Women of childbearing age may be especially likely to present with EEC-positive ET rather than PV because their haematocrit may be lowered by menstruation.

EECs have also been reported in patients with idiopathic myelofibrosis (IMF) [11, 34] [reviewed in ref. 22]. Perhaps this finding represents the far end of PV progression. IMF commences in a hypercellular phase before pancytopenia develops [35]. It is possible that patients with EEC-positive IMF have progressed through an exceptionally short polycythaemic phase, which went clinically undetected, and are now in a state of post-polycythaemic myelofibrosis [36]. There is, of necessity, no clinical evidence for this hypothesis. Nonetheless, the occurrence of EECs in all three chronic myeloproliferative disorders (cMPDs) underscores the close pathophysiological relationship between these disorders.

*Endogenous Megakaryocytic Colonies*

The growth of endogenous megakaryocytic colonies (EMCs), i.e. growth in the absence of exogenously added growth factors, has received much less attention than the growth of EECs [reviewed in ref. 23]. This is perhaps due to the greater technical difficulties associated with these cultures. Unlike EECs, which are readily discernable by their red haemoglobinisation, EMCs need to be stained with an antibody against a megakaryocyte/platelet-specific protein for unequivocal identification. Nonetheless, the presence of EMCs is pathophysiologically interesting and its implication for the aetiology of cMPDs should be considered.

EMC growth has been detected in 60–100% of patients with ET [reviewed in ref. 23] and has been used successfully to discriminate ET from reactive thrombocytosis [31]. Like EECs, however, EMCs are also found in all cMPDs. A proportion of 50–80% of patients with PV display EMC growth [31, 32, 37] [reviewed in in ref. 23]. Likewise, 50–100% of IMF patients have detectable EMCs [38, 39]. Too few IMF patients have been analysed to determine at what frequency these individuals show isolated EEC or EMC growth or display both types of endogenous colonies simultaneously. A proportion of 50–80% of PV patients, who invariably grow EECs, show both EEC and EMC growth. In contrast, ET patients may display either EEC or EMC alone or grow both or neither types of colonies [32].

Because the elegant experiments performed by Cashman et al. [12], demonstrating the presence of both EPO-independent and EPO-dependent cells in the progeny of a single BFU-E, have not been performed on EMCs, it is not clear whether the ability to give rise to EMCs is conferred by mutation, or, as in EECs, remains pliable throughout the differentiation process. The presence of EMCs in all three cMPDs, however, strengthens the hypothesis that some of the pathophysiological changes giving rise to these disorders are shared among PV, ET and IMF.

*Growth Factor Independence and Hypersensitivity*

The discovery of EECs spurned a great controversy in the MPD field. Were EEC-forming erythroid precursors hypersensitive to minute amounts of EPO in the culture medium or were they completely independent of this growth factor? Elegant and meticulous experiments in a completely defined serum-free medium gave an unexpected answer: the EPO dose-response curve of PV and normal BFU-E is identical [40]. However, PV cells are 100–1,000 times more sensitive to insulin-like growth factor-1 (IGF-1) than healthy control cells. Thus, PV erythroid precursors are EPO independent and IGF-1 hypersensitive [40].

Using this improved serum-free culture medium, Axelrad et al. [41] have demonstrated that PV progenitor cells are exclusively hypersensitive to IGF-1 and not to other growth factors. The authors are careful to emphasise the importance of exact definitions. 'Hypersensitivity' denotes the fact that the half maximal number of colonies is obtained at a lower concentration of growth factor, when PV and normal cells are compared. In contrast, 'hyperreactivity' describes the observation that at a certain concentration of growth factor, more colonies are obtained from PV patients than from healthy controls. By these definitions, PV cells are hyperactive to interleukin-3 (IL-3) and granulocyte-macrophage colony-stimulating factor (GM-CSF), but exclusively hypersensitive to IGF-1 [40, 42, 43].

The precise mechanism conferring IGF-1 hypersensitivity in PV is not known. However, Mizra et al. [44] have shown that serum concentrations of the IGF-1-binding protein, IGF-binding protein-1 (IGFBP-1), are increased four-fold in patients with PV. Moreover, both basal and IGF-1-stimulated phosphorylation of the IGF-1 receptor is elevated in PV compared to healthy controls [45].

More recently, Axelrad et al. [46] have extended these findings by demonstrating that megakaryocytic precursors from ET patients are up to 1,000-fold more sensitive to thrombopoietin (TPO). This response is highly selective, as the cells show a normal response to 9 other cytokines. In yet unpublished experiments, the group has demonstrated that precursor cells from patients with IMF are hypersensitive to stem cell factor (SCF) and, again, not to other growth factors [47].

Having now studied all three cMPDs, Dr. Axelrad has formulated the hypothesis that precursor cells in each of the disorders acquire a lineage restricted hypersensitivity to a single growth factor, which is specific to the particular disease: IGF-1 in the erythroid lineage of PV, TPO in megakaryocytes of ET and SCF in stem cells of IMF. This could explain how alterations in the pluripotent stem cell, which underlie all three disorders, nonetheless lead to distinct clinical presentations with predominance of a particular lineage.

## *Cytogenetic Markers in PV*

Unlike CML, which exhibits the Philadelphia chromosome, there is no consistent cytogenetic abnormality detectable in PV. However, certain karyotypic changes are seen in a subset of patients and appear to be associated with disease development in these individuals. The evolution of PV, like that of most neoplasms, is likely to require several mutations accumulating in a single cell [48, 49]. Therefore, analysis of cytogenetic lesions evident in a subgroup of patients may identify genes whose activation or inactivation constitutes one of the multiple 'hits' necessary for the development of a PV stem cell.

### *Deletions of Chromosome 20q*

At diagnosis, around 10% of PV patients present with interstitial deletions of the long arm of chromosome 20 [50, 51]. While these deletions are not unique to cMPDs, as they also occur in 5% of patients with myelodysplastic syndromes (MDS) and in some patients with acute myeloid leukaemia (AML), they nonetheless represent the most common cytogenetic lesion in PV [50, 51]. In addition, they are detected in the pluripotent stem cell, suggesting that they contribute to its alteration. The laboratory of Dr. Green has compiled a detailed physical and transcriptional map of the common deleted region [52, 53]. The size of this common deleted region has now been reduced to 2.7 Mb in patients with MPDs and to 1.7 Mb if patients with MDS and AML are included [53]. This region contains 5 candidate genes expressed both in normal bone marrow and in purified CD34-positive cells [53]. Because loss of genetic material is associated with disease development, this region may harbour a tumour suppressor gene. Following the loss of one allele on the deleted chromosome 20, the second copy of the gene could be inactivated by one of several mechanisms ranging from point mutation to large deletions. Searching the remaining copy of chromosome 20 with DNA markers, Bench et al. [52] were unable to detect deletions in 28 patients. The hypothesised change on the remaining allele must therefore be smaller than the distance between the DNA markers used. At the time of writing, therefore, the gene on chromosome 20, which contributes to the development of cMPD has not been identified. The compilation of such detailed genetic information and the collection of a large panel of patients should, however, lead to its recognition shortly.

### *Loss of Heterozygosity on Chromosome 9*

The description of several families in which the propensity to develop cMPDs is inherited raises the possibility of discovering causative genetic lesions by classical genetic methods. Kralovics et al. [54] have reported loss of heterozygosity (LOH) in such families. A LOH denotes the absence of either the paternal or the maternal allele at a given chromosomal locus. This can occur through loss of an entire chromosome, through interstitial deletions or through mitotic recombination. Extending their findings from familial PV, Kralovics et al. conducted a genome-wide screen for LOH on 6 PV patients. LOHs were identified on chromosomes 9, 10 and 11 [55]. These were verified by further analysis of 20 PV and 6 ET patients. While the LOHs on chromosomes 10 and 11 were unique to individual patients, LOH of the short arm of chromosome

9 was seen in 6 of the 20 patients (33%) [55]. All three LOH were detected in CD34-positive cells, denoting their occurrence in pluripotent stem cells. Candidate genes from the affected region were analysed for mRNA expression. In 2 PV patients with 9p LOH, semi-quantitative reverse transcriptase-polymerase chain reaction (RT-PCR) revealed a surprising increase in the mRNA levels of the transcription factor NF-1B (also known as the CCAAT-binding factor in humans), both in granulocytes and in CD34-positive cells. Two PV patients without the LOH showed no such increase [55]. How mitotic recombination leading to an LOH might increase mRNA expression is completely unclear.

The overexpression of a transcription factor in PV CD34-positive cells is an intriguing finding, since such a change could evoke various downstream effects and easily alter cellular responses. However, data for only 2 patients have been published so far. Confirmation of this finding in a larger cohort of patients is therefore essential. Given that the mRNA increase was noted in peripheral granulocytes and that the LOH was present in one third of the patients studied, it should be possible to rapidly screen a sufficient number of patients to validate this observation.

## *Molecular Markers in PV*

The past 3 years have seen the publication of the first molecular markers for PV. For the first time, molecular tests for the diagnosis of PV have been proposed.

### *Reduced Expression of c-Mpl, the Thrombopoietin Receptor*

The first molecular marker described for PV was a reduced expression of the TPO receptor, c-Mpl. Moliterno et al. [36, 56] reported that c-Mpl expression was markedly reduced on platelets of 34/34 PV patients as well as 13/14 IMF patients but not in patients with CML or secondary erythrocytosis (SE). The expression of glycoprotein IIIa (gpIIIa) was unaffected [56]. Horikawa et al. [57] subsequently published that c-Mpl protein and mRNA expression was reduced in platelets of 17/17 ET patients. Again, gpIIb/IIIa protein and mRNA were normal. Harrison et al. [58] similarly found a 35–100% reduction in c-Mpl protein expression on platelets from 18 ET patients. In addition, this study confirmed the reduced c-Mpl expression in 4 PV and 2 IMF patients. Moreover, the two CML patients investigated here also displayed a reduction [58].

Moliterno and Spivak [59] have investigated the mechanism leading to the selective reduction of c-Mpl cell surface expression. These studies demonstrate that c-Mpl glycosylation and processing is defective in PV and that the extent of this deficiency correlates with disease duration and the degree of extramedullary haematopoiesis. Interestingly, the glycosylation and transport deficit is selective for the c-Mpl protein, since two other membrane proteins, gpIIb and multimerin, were processed normally [59]. The mechanism whereby a cell may selectively retain a specific protein within the endoplasmic reticulum/Golgi compartment remains elusive.

Recently, the same authors have demonstrated reduced c-Mpl expression and incomplete glycosylation in CD34-positive cells from PV patients [60]. This makes the c-Mpl-processing defect the first molecular marker described for PV stem cells. It remains to be determined whether the difference in c-Mpl expression detectable by FACS analysis is sufficient to unequivocally separate PV stem cells from their healthy counterparts. If this were feasible, the long-awaited possibility of working with a purified population of PV stem cells would become reality.

### *c-Mpl as a Marker for the Diagnosis of PV*

Moliterno et al. [61] have demonstrated the feasibility of using Western blotting to quantify c-Mpl protein for the diagnosis of PV. A Western blot system was established which used an antibody against gpIIb as an internal control and included reference positive and negative control samples on each membrane. In a cohort of 27 PV and 19 secondary erythrocytosis patients, this assay showed a sensitivity of 96% and a specificity of 95% for the distinction of PV from secondary erythrocytosis [61].

In contrast, Le Blanc et al. [62] compared c-Mpl expression in 15 PV patients versus 10 healthy controls using Western blots without an internal control antibody. Seven of the 15 PV patients tested displayed c-Mpl protein levels similar to those of the controls, while in 8 PV patients c-Mpl levels were markedly reduced. It is important to point out that Le Blanc et al. used a different c-Mpl antibody from that used by Moliterno et al. It is possible that the glycosylation defect underlying reduced c-Mpl detection in Western blots is not equally visible with every antibody but depends on the specific epitope recognised. Harrison et al. [58] using both radio-immunoassay and Western blotting for the detection of c-Mpl protein, report that, in a cohort of 18 ET and 8 patients with reactive thrombocytosis, these assays were insufficiently discriminatory for the diagnosis of ET. Technical differences may in large part account for the discrepancies. However, these studies emphasise the logistic difficulties involved in using Western blotting as a diagnostic tool.

*Quantification of PRV-1 mRNA for the Diagnosis of PV*

PRV-1 is a member of the uPAR/Ly6/CD59 family of receptors [63]. Flow cytometric analysis has recently determined that PRV-1 is identical to the well-known neutrophil antigen NB1/CD177, whose cDNA was only cloned last year [64]. The PRV-1 mRNA is 8- to 64-fold overexpressed in granulocytes from PV patients (n = 62) compared to healthy controls (n = 64) [63, 65, and Klippel and Pahl, unpubl. obs.]. Moreover, PRV-1 mRNA levels are normal in patients with secondary erythrocytosis, CML and AML [63, 65]. A quantitative RT-PCR assay has been developed for the measurement of PRV-1 mRNA levels [65]. Validation of this assay on 48 PV patients, 34 healthy controls and 8 patients with SE revealed a sensitivity and specificity of 100% in this cohort [65]. Thus, quantification of PRV-1 expression can be used as a sensitive and specific tool to distinguish between patients with PV and SE.

Interestingly, PRV-1 levels are also increased in 50% of patients with ET [65]. In a cohort of 35 ET patients, PRV-1 expression and the ability to form EECs was compared. All 20 ET patients who overexpressed PRV-1 gave rise to EECs while the PRV-1-negative patients developed no endogenous colonies [65]. Because Shih and Lee [29] have demonstrated that in ET patients the presence of EECs can predict the development of PV, we have proposed that EEC-positive/PRV-1-positive ET patients constitute a pathophysiologically distinct population from EEC-negative/PRV-1-negative patients. It is our hypothesis that these patients would, if left untreated, develop PV during the course of their disease. However, many ET patients require myelosuppressive treatment and this will most likely mask or prevent the development of overt erythrocytosis. Therefore, this hypothesis could be difficult to prove in an observational study.

Based on the published data, the German Society of Internal Medicine (Deutsche Gesellschaft für Innere Medizin; DGIM) has recommended the use of PRV-1 mRNA quantification for the diagnosis of PV. A German MPD Study Group (Deutsche MPE Initiative) has been formed, which will open a prospective pathophysiological study on newly diagnosed MPD patients. At diagnosis, the patients will be tested for PRV-1 expression and screened for novel molecular markers by cDNA array analysis. The aim of this trial is to determine whether the presence of PRV-1 or other molecular markers can predict the disease course. Patients will be followed for at least 10 years and all relevant clinical events will be recorded. The availability of both long-term clinical data and gene expression analysis should enable us to identify novel prognostic molecular markers in the cMPDs. Such a trial will also determine whether PRV-1-positive and PRV-1-negative ET patients differ in their rate of thromboembolic complications, their therapeutic requirements and their overall disease course. This study is required to determine the utility and validity of PRV-1 measurement in ET. For the distinction of PV from SE, and for the diagnosis of PV, however, this molecular marker appears very useful.

*Function of the PRV-1 Protein*

Because members of the uPAR/Ly6/CD59 family of receptors fulfil a diverse set of functions, the physiological role of PRV-1 cannot be deduced by its membership in this family. Moreover, unlike the PRV-1 mRNA, which is highly overexpressed in PV granulocytes, the PRV-1 protein is expressed equally on granulocytes from PV patients and healthy controls [66]. One could assume that the continued presence of PRV-1 mRNA in PV patients would lead to the production of increased amounts of protein. However, a regulatory mechanism may exist which limits the number of molecules of a given cell surface receptor produced or inserted into the cell membrane. PRV-1 protein produced in excess of that required for insertion into the cell membrane could be shed into the plasma, as has been described for uPAR [67]. We have developed a sensitive ELISA for the detection of PRV-1 and are currently investigating the PRV-1 plasma levels of PV patients and healthy controls.

As the function of the PRV-1 protein remains unknown, the role this marker may play in the pathophysiology of PV is not yet evident. If PRV-1 plasma levels are significantly increased in patients with PV, this protein may play a role in promoting the hyperproliferation of precursor cells in the bone marrow. If, on the other hand, PRV-1 plasma levels are similar in PV patients and healthy individuals, it is difficult to imagine how this protein could contribute to the pathophysiology of PV. In the latter scenario, PRV-1 mRNA overexpression simply provides a very consistent and routinely measurable abnormality to aid in the diagnosis of the disorder, without a concomitant role of its protein product.

However, the consistency with which PRV-1 mRNA overexpression is observed in all PV patients tested suggests that the cause of this aberrant transcription, perhaps the altered activity of a transcription factor, may well constitute the molecular alteration responsible for disease development. In this context it is interesting to note that we have recently characterized the PRV-1 promoter and noted the presence of a protein capable of binding the PRV-1 promoter in extracts from PV granulocytes, which was absent from healthy control granulocytes [68]. It is thus very possible that elucidation of the abnormality causing PRV-1 mRNA overexpression will contribute to our understanding of the molecular pathology of PV.

## Will New Markers Help Us Answer Old Questions?

*Clinical Classification*

In 1951, Dameshek [69] coined the term MPDs for four clinically related disorders. It later became apparent that one of the four, CML, has a distinct molecular etiology, identifiable by the *BCR/ABL* translocation [70]. For the remaining three disorders, the recent description of molecular markers raises the possibility of categorising them by molecular means as well.

All patients with PV display three consistent alterations: EEC formation, reduced c-Mpl expression and increased PRV-1 mRNA levels. Interestingly, a large proportion of patients with ET and IMF also show each of these changes. However, except for the correlation between PRV-1 expression and EEC formation in one cohort [65], the concordance or discordance of these phenomena has not been studied. The German MPD Initiative has postulated the existence of an 'EEC/c-Mpl/PRV-1' disease: we propose that the three molecular markers concur in most MPD patients [71]. Studies currently investigate this hypothesis.

Based on this postulate, MPD patients may be assigned to one of five diagnostic categories:

ET$^{\text{PRV-1-negative}}$ patients with ET (EEC-negative/c-Mpl-normal/PRV-1-negative)

ET$^{\text{PRV-1-positive}}$ patients with ET (EEC-positive/c-Mpl-low/PRV-1-positive)

PV patients with PV (EEC-positive/c-Mpl-low/PRV-1-positive)

IMF$^{\text{PRV-1-positive}}$ patients with IMF (EEC-positive/c-Mpl-low/PRV-1-positive)

IMF$^{\text{PRV-1-negative}}$ patients with IMF (EEC-negative/c-Mpl-normal/PRV-1-negative).

Creating new diagnostic categories is only sensible if the assignment provides clinically meaningful information. It is, therefore, up to the German MPD Initiative to prove in a prospective pathophysiological study whether these new categories are more meaningful and prognostic in terms of frequency of complications, therapeutic requirements and overall disease course than the present diagnostic entities.

*Molecular Etiology*

Molecular classification of clinically similar diseases, however, can be useful regardless of whether prognosis, rate of complications or therapeutic requirements differ significantly between the individual groups. Our ultimate goal in studying diseases is to understand the molecular changes causing disease development. Or rather, the penultimate goal. The final step will be accomplished when we have developed a targeted therapy that reverses the disease-inducing changes. If we find such an agent, we will need to know precisely which patients can benefit from it. This will be determined by the molecular alteration underlying disease development in each individual.

In order to determine the molecular etiology of MPDs, we need to develop testable hypotheses about the disease origin. I would like to put forward the following hypotheses on the molecular origin of myeloproliferative disorders (fig. 1):

(1) The diseases categorised above as ET$^{\text{PRV-1-positive}}$, PV and IMF$^{\text{PRV-1-positive}}$ arise from a common set of mutations in the pluripotent stem cell. These mutations confer on the progeny of this stem cell: (A) the ability to give rise to EECs (erythroid lineage); (B) a processing defect of the c-Mpl protein (stem cell and megakaryocytic lineage), and (C) overexpression of PRV-1 mRNA (myeloid lineage).

At this point, two models can be envisioned

(1a) Having acquired these mutations, the stem cell undergoes a further set of changes during differentiation. This second group of mutations is different in each of the three disease entities. During this stage, the MPD stem cell acquires the selective hypersensitivities described by Axelrad et al., which determine whether the disease develops into ET$^{\text{PRV-1-positive}}$, PV or IMF$^{\text{PRV-1-positive}}$. In this model, the disorders ET$^{\text{PRV-1-negative}}$ and IMF$^{\text{PRV-1-negative}}$ arise from an entirely different set of mutations and are in effect phenocopies of 'true ET' and 'true IMF', but molecularly entirely different disorders.

(1b) An alternative hypothesis is that the stem cell which has sustained the 'EEC/c-Mpl/PRV-1'-mutations always acquires IGF-1 hypersensitivity and gives rise to one molecular entity, which may present clinically either

(a) as an ET, which will ultimately develop into a PV (ET$^{\text{PRV-1-positive}}$);

(b) as PV or

(c) as IMF having quickly passed the hypercellular, polycythaemic stage (IMF$^{\text{PRV-1-positive}}$).

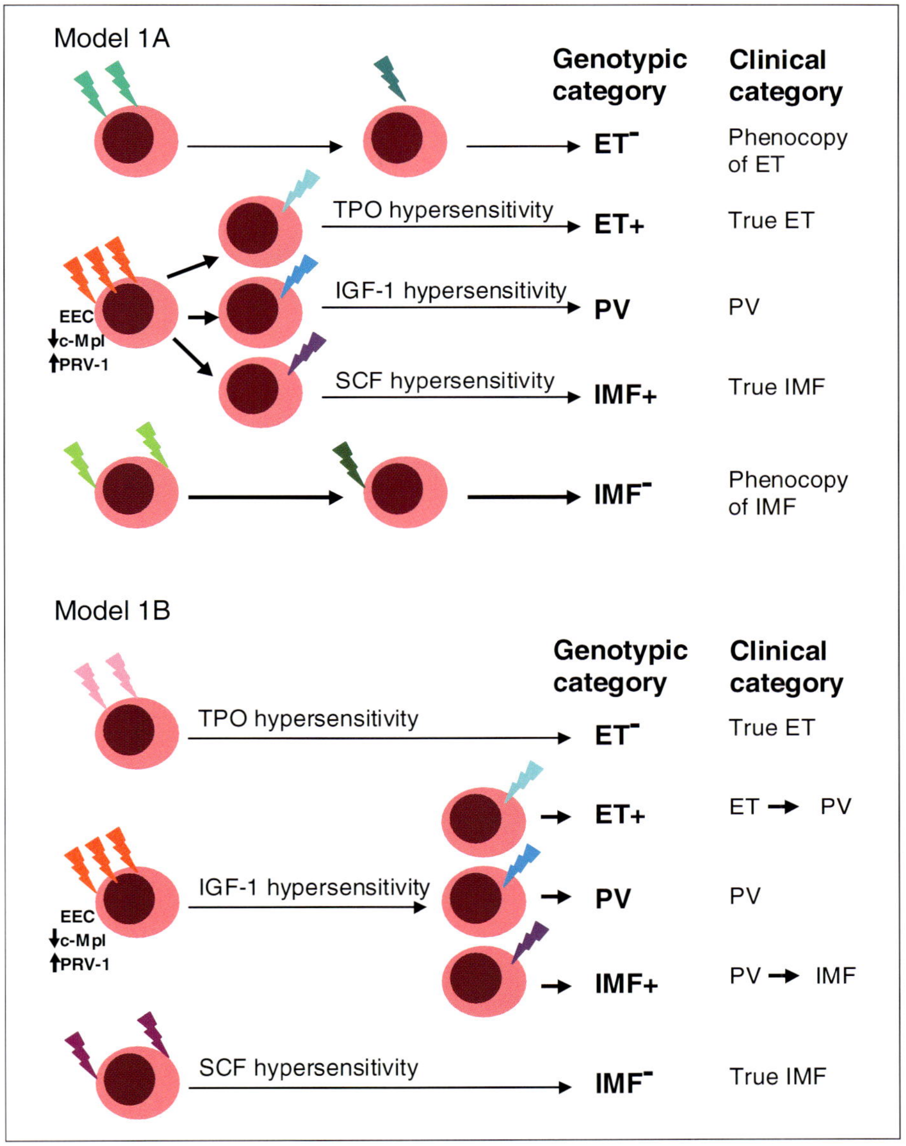


**Fig. 1.** Two alternative models for the molecular development of MPDs. Both models assume that a common set of mutations (red flashes) confer on a stem cell: (1) the ability to form EECs during erythroid development; (2) reduced c-Mpl expression; (3) PRV-1 overexpression after myeloid development. Model 1A: the 'EEC/c-Mpl/PRV-1' stem cell begins to differentiate and acquires different new mutation(s) (various blue and blue/purple flashes), which confer TPO, IGF-1 or SCF hypersensitivity. In this model, the 'EEC/c-Mpl/PRV-1'-negative forms of ET and IMF arise from an entirely different set of mutations (various green flashes). Model 1B: the 'EEC/c-Mpl/PRV-1' stem cell is always IGF-1 hypersensitive, and, as it differentiates, acquires new mutations (various blue and blue/purple flashes) which cause it to give rise to clinically different presentations (1) ET+: long phase of thrombocythaemia before conversion into PV; (2) PV or (3) PV presenting as IMF. 'True' ET and 'true' IMF arise from different mutations (various pink/purple flashes) in other stem cells that do not have the 'EEC/c-Mpl/PRV-1' mutation.

In this model, ET$^{PRV\text{-}1\text{-}positive}$ and IMF$^{PRV\text{-}1\text{-}positive}$ represent variant clinical presentations of PV. Here, ET$^{PRV\text{-}1\text{-}negative}$ and IMF$^{PRV\text{-}1\text{-}negative}$ represent the 'true' forms of these disorders and arise from entirely different sets of mutations. Hence, in model 1b, TPO and SCF hypersensitivity arise de novo, in stem cells which do not have the 'EEC/c-Mpl/PRV-1' mutation.

The models proposed here are easily testable. Whether TPO and SCF hypersensitivity occur in 'EEC-positive/c-Mpl-low/PRV-1positive' cells or are separately acquired is easily ascertained. Such experiments are currently in progress.

Whether the differentiation by molecular markers will prove clinically valuable depends both on the results of the German trial and on the development of a molecular therapy which specifically targets the 'EEC-positive/c-Mpl-low/PRV-1'-stem cell.

## Acknowledgements

The author gratefully acknowledges support by the Alfried-Krupp-Förderpreis für Junge Hochschullehrer, the SFB 364, Teilprojekt A-12, and the Else-Kröner-Fresenius-Stiftung.

## References


1 Prchal JF, Adamson JW, Steinmann L, Fialkow PJ: Human erythroid colony formation in vitro: Evidence for clonal origin. J Cell Physiol 1976;89:489–492.
2 Adamson JW, Fialkow PJ, Murphy S, Prchal JF, Steinmann BA: Polycythemia vera: Stem-cell and probable clonal origin of the disease. N Engl J Med 1976;295:913–916.
3 Mladenovic J, Adamson JW: Characteristics of circulating erythroid colony-forming cells in normal and polycythaemic man. Br J Haematol 1982;51:377–384.
4 Singer JW, Fialkow PJ, Adamson JW, Steinmann L, Ernst C, Murphy S, Kopecky KJ: Polycythemia vera. Increased expression of normal committed granulocytic stem cells in vitro after exposure of marrow to tritiated thymidine. J Clin Invest 1979;64:1320–1324.
5 Fauser AA, Messner HA: Pluripotent hemopoietic progenitors (CFU-GEMM) in polycythemia vera: Analysis of erythropoietin requirement and proliferative activity. Blood 1981;58:1224–1227.
6 Cashman JD, Eaves CJ, Eaves AC: Unregulated proliferation of primitive neoplastic progenitor cells in long-term polycythemia vera marrow cultures. J Clin Invest 1988;81:87–91.
7 Eaves AC, Henkelman DH, Eaves CJ: Abnormal erythropoiesis in the myeloproliferative disorders: An analysis of underlying cellular and humoral mechanisms. Exp Hematol 1980; 8:235–247.
8 Prchal JF, Axelrad AA: Bone-marrow responses in polycythemia vera. N Engl J Med 1974;290:1382.
9 Adamson JW: Analysis of myeloproliferative disorders using cell markers in culture. Hämatol Bluttransfus 1979;23:205–210.
10 Adamson JW, Singer JW, Catalano P, Murphy S, Lin N, Steinmann L, Ernst C, Fialkow PJ: Polycythemia vera. Further in vitro studies of hematopoietic regulation. J Clin Invest 1980; 66:1363–1368.
11 Lutton JD, Levere RD: Endogenous erythroid colony formation by peripheral blood mononuclear cells from patients with myelofibrosis and polycythemia vera. Acta Haematol 1979;62: 94–99.
12 Cashman J, Henkelman D, Humphries K, Eaves CJ, Eaves AC: Individual BFU-E in polycythemia vera produce both erythropoietin dependent and independent progeny. Blood 1983;61:876–884.
13 Eaves CJ, Eaves AC: Erythropoietin (Ep) dose-response curves for three classes of erythroid progenitors in normal human marrow and in patients with polycythemia vera. Blood 1978; 52:1196–1210.
14 Shih LY, Lee CT, See LC, Ou YC, Dunn P, Wang PN, Kuo MC, Wu JH: In vitro culture growth of erythroid progenitors and serum erythropoietin assay in the differential diagnosis of polycythaemia. Eur J Clin Invest 1998; 28:569–576.
15 Lacombe C, Casadevall N, Varet B: Polycythaemia vera: In vitro studies of circulating erythroid progenitors. Br J Haematol 1980;44: 189–199.
16 Fibach E, Rachmilewitz EA: Proliferation and differentiation of erythroid progenitors in liquid culture: Analysis of progenitors derived from patients with polycythemia vera. Am J Hematol 1990;35:151–156.
17 Silva M, Richard C, Benito A, Sanz C, Olalla I, Fernandez-Luna JL: Expression of Bcl-x in erythroid precursors from patients with polycythemia vera. N Engl J Med 1998;338:564–571.
18 Roder S, Steimle C, Meinhardt G, Pahl HL: STAT3 is constitutively active in some patients with polycythemia rubra vera. Exp Hematol 2001;29:694–702.
19 Pahl HL: Towards a molecular understanding of polycythemia rubra vera. Eur J Biochem 2000;267:3395–3401.
20 Lemoine F, Najman A, Baillou C, Stachowiak J, Boffa G, Aegerter P, Douay L, Laporte JP, Gorin NC, Duhamel G: A prospective study of the value of bone marrow erythroid progenitor cultures in polycythemia. Blood 1986;68:996–1002.
21 Shih LY, Lee CT, Ou YC: Prediction of clinical course in patients with idiopathic erythrocytosis by endogenous erythroid colony assay but not by serum erythropoietin levels. Exp Hematol 1997;25:288–292.
22 Weinberg RS: In vitro erythropoiesis in polycythemia vera and other myeloproliferative disorders. Semin Hematol 1997;34:64–69.
23 Westwood NB, Pearson TC: Diagnostic applications of haemopoietic progenitor culture techniques in polycythaemias and thrombocythaemias. Leuk Lymphoma 1996;22:95–103.
24 Biljanovic-Paunovic L, Ruvidic R, Pavlovic-Kentera V: Endogenous BFU-E in peripheral blood in diagnosis of polycythemia vera. Eur J Haematol 1990;45:262–266.
25 Beckman BS, Anderson WF, Beltran GS, Beckman EN, Fisher JW: Diagnostic use of CFU-E formation from peripheral blood in polycythemia vera. Am J Clin Pathol 1983;79:496–499.
26 Weinberg RS, Worsley A, Gilbert HS, Cuttner J, Berk PD, Alter BP: Comparison of erythroid progenitor cell growth in vitro in polycythemia vera and chronic myelogenous leukemia: Only polycythemia vera has endogenous colonies. Leuk Res 1989;13:331–338.
27 Masters GS, Baines P, Jacobs A: Erythroid colony growth from peripheral blood and bone marrow in polycythaemia. J Clin Pathol 1990; 43:937–941.
28 Turhan AG, Cashman JD, Eaves CJ, Humphries RK, Eaves AC: Variable expression of features of normal and neoplastic stem cells in patients with thrombocytosis. Br J Haematol 1992;82:50–57.
29 Shih L-Y, Lee C-T: Identification of masked polycythemia vera from patients with idiopathic marked thrombocytosis by endogenous erythroid colony assay. Blood 1994;83:744–748.
30 Florensa L, Besses C, Woessner S, Sole F, Acin P, Pedro C, Sans-Sabrafen J: Endogenous megakaryocyte and erythroid colony formation from blood in essential thrombocythaemia. Leukemia 1995;9:271–273.
31 Rolovic Z, Basara N, Gotic M, Sefer D, Bogdanovic A: The determination of spontaneous megakaryocyte colony formation is an unequivocal test for discrimination between essential thrombocythaemia and reactive thrombocytosis. Br J Haematol 1995;90:326–331.
32 Juvonen E, Ikkala E, Oksanen K, Ruutu T: Megakaryocyte and erythroid colony formation in essential thrombocythaemia and reactive thrombocytosis: Diagnostic value and correlation to complications. Br J Haematol 1993; 83:192–197.
33 Jantunen R, Juvonen E, Ikkala E, Oksanen K, Anttila P, Ruutu T: Development of erythrocytosis in the course of essential thrombocythemia. Ann Hematol 1999;78:219–222.
34 Partanen S, Ruutu T, Vuopio P: Circulating haematopoietic progenitors in myelofibrosis. Scand J Haematol 1982;29:325–330.
35 Smith BD, Moliterno AR: Biology and management of idiopathic myelofibrosis. Curr Opin Oncol 2001;13:91–94.
36 Moliterno AR, Siebel KE, Sun AY, Hankins WD, Spivak JL: A novel thrombopoietin signaling defect in polycythemia vera platelets. Stem Cells 1998;16:185–192.
37 Battegay EJ, Thomssen C, Nissen C, Gudat F, Speck B: Endogenous megakaryocyte colonies from peripheral blood in precursor cell cultures of patients with myeloproliferative disorders. Eur J Haematol 1989;42:321–326.
38 Croizat H, Amato D, McLeod DL, Eskinazi D, Axelrad AA: Differences among myeloproliferative disorders in the behavior of their restricted progenitor cells in culture. Blood 1983; 62:578–584.
39 Taksin AL, Couedic JP, Dusanter-Fourt I, Masse A, Giraudier S, Katz A, Wendling F, Vainchenker W, Casadevall N, Debili N: Autonomous megakaryocyte growth in essential thrombocythemia and idiopathic myelofibrosis is not related to a c-mpl mutation or to an autocrine stimulation by Mpl-L. Blood 1999;93:125–139.
40 Correa PN, Eskinazi D, Axelrad AA: Circulating erythroid progenitors in polycythemia vera are hypersensitive to insulin-like growth factor-1 in vitro: Studies in an improved serum-free medium. Blood 1994;83:99–112.
41 Kahn CR: Insulin resistance, insulin insensitivity, and insulin unresponsiveness: A necessary distinction. Metab Clin Exp 1978;27:1893–1902.
42 Dai C-H, Krantz SB, Means RT, Horn ST, Gilbert HS: Polycythemia vera blood burst-forming units-erythroid are hypersensitive to interleukin-3. J Clin Invest 1991;87:391–396.

43 Dai CH, Krantz SB, Dessypris EN: Polycythemia vera II: Hypersensitivity of bone marrow erythroid, granulocyte-macrophage, and megakaryocyte progenitor cells to interleukin 3 and granulocyte-macrophage colony-stimulating factor. Blood 1992;80:891–899.
44 Mirza AM, Ezzat S, Axelrad AA: Insulin-like growth factor binding protein-1 is elevated in patients with polycythemia vera and stimulates erythroid burst formation in vitro. Blood 1997; 89:1862–1869.
45 Mirza AM, Correa PN, Axelrad AA: Increased basal and induced tyrosine phosphorylation of the insulin-like growth factor 1 receptor b subunit in circulating mononuclear cells of patients with polycythemia vera. Blood 1995;86: 877–882.
46 Axelrad AA, Eskinazi D, Correa PN, Amato D: Hypersensitivity of circulating progenitor cells to megakaryocyte growth and development factor (PEG-rHu MGDF) in essential thrombocythemia. Blood 2000;96:3310–3321.
47 Axelrad AA: Megakaryocytic Hypersensitivity to thrombopoietin in essential thrombocythemia. Millennium Meeting on Myeloproliferative Diseases and Myelodysplastic Syndromes, New York, Oct 2000.
48 Kennedy EP, Hamilton SR: Genetics of colorectal cancer. Semin Surg Oncol 1998;15:126–130.
49 Ilyas M, Straub J, Tomlinson IP, Bodmer WF: Genetic pathways in colorectal and other cancers. Eur J Cancer 1999;35:335–351.
50 Rege-Cambrin G, Mecucci C, Tricot G, Michaux JL, Louwagie A, Van Hove W, Francart H, Van den Berghe H: A chromosomal profile of polycythemia vera. Cancer Genet Cytogenet 1987;25:233–245.
51 Hinshelwood S, Bench AJ, Green AR: Pathogenesis of polycythaemia vera. Blood Rev 1997;11:224–232.
52 Bench AJ, Aldred MA, Humphray SJ, Champion KM, Gilbert JG, Asimakopoulos FA, Deloukas P, Gwilliam R, Bentley DR, Green AR: A detailed physical and transcriptional map of the region of chromosome 20 that is deleted in myeloproliferative disorders and refinement of the common deleted region. Genomics 1998; 49:351–362.
53 Bench AJ, Nacheva EP, Hood TL, Holden JL, French L, Swanton S, Champion KM, Li J, Whittaker P, Stavrides G, Hunt AR, Huntly BJ, Campbell LJ, Bentley DR, Deloukas P, Green AR: Chromosome 20 deletions in myeloid malignancies: Reduction of the common deleted region, generation of a PAC/BAC contig and identification of candidate genes. UK Cancer Cytogenetics Group (UKCCG). Oncogene 2000;19:3902–3913.
54 Kralovics R, Stockton DW, Prchal JT: Genome-wide search for the genetic defect in polycythemia vera. Blood 2000;96:741a.
55 Kralovics R, Guan YL, Prchal JT: Acquired uniparental disomy of chromosome 9p is a frequent stem cell defect in polycythemia vera. Exp Hematol 2002;30:229–236.
56 Moliterno AR, Hankins D, Spivak JL: Impaired expression of the thrombopoietin receptor by platelets from patients with polycythemia vera. N Engl J Med 1998;338:572–580.
57 Horikawa Y, Matsumura I, Hashimoto K, Shiraga M, Kosugi S, Tadokoro S, Kato T, Miyazaki H, Tomiyama Y, Kurata Y, Matsuzawa Y, Kanakura Y: Markedly reduced expression of platelet c-mpl receptor in essential thrombocythemia. Blood 1997;90:4031–4038.
58 Harrison CN, Gale RE, Pezella F, Mire-Sluis A, Machin SJ, Linch DC: Platelet c-mpl expression is dysregulated in patients with essential thrombocythaemia but this is not of diagnostic value. Br J Haematol 1999;107:139–147.
59 Moliterno AR, Spivak JL: Posttranslational processing of the thrombopoietin receptor is impaired in polycythemia vera. Blood 1999;94: 2555–2561.
60 Moliterno AR, Williams DM, Spivak JL: Enhanced in vitro viability of polycythemia vera CD34+ cells is associated with abnormal thrombopoietin receptor expression and aberrant signal transduction. Blood 2001;98:469a.
61 Moliterno AR, Silver RT, Spivak JL: A diagnostic assay for polycythemia vera. Blood 2000;96:512a.
62 Le Blanc K, Andersson P, Samuelsson J: Marked heterogeneity in protein levels and functional integrity of the thrombopoietin receptor c-mpl in polycythaemia vera. Br J Haematol 2000;108:80–85.
63 Temerinac S, Klippel S, Strunck E, Röder S, Lübbert M, Lange W, Azemar M, Meinhardt G, Schaefer HE, Pahl HL: Cloning of PRV-1, a novel member of the uPAR receptor superfamily, which is overexpressed in polycythemia rubra vera. Blood 2000;95:2569–2576.
64 Kissel K, Santoso S, Hofmann C, Stroncek D, Bux J: Molecular basis of the neutrophil glycoprotein NB1 (CD177) involved in the pathogenesis of immune neutropenias and transfusion reactions. Eur J Immunol 2001;31:1301–1309.
65 Klippel S, Strunck E, Temerinac S, Kralovics R, Prchal JT, Griesshammer M, Heimpel H, Pahl HL: Quantification of PRV-1 expression, a molecular marker for the diagnosis of polycythemia vera. Blood 2001;98:470a.
66 Klippel S, Strunck E, Busse CE, Behringer D, Pahl HL: Biochemical Characterization of PRV-1, a novel hematopoietic cell surface receptor, which is overexpressed in polycythemia rubra vera. Blood, in press.
67 Ronne E, Pappot H, Grondahl-Hansen J, Hoyer-Hansen G, Plesner T, Hansen NE, Dano K: The receptor for urokinase plasminogen activator is present in plasma from healthy donors and elevated in patients with paroxysmal nocturnal haemoglobinuria. Br J Haematol 1995;89:576–581.
68 Strunck E, Bliestle H, Temerinac S, Pahl HL: Characterization of the promoter for PRV-1, a gene overexpressed in polycythemia vera. Blood 2000;96:511a.
69 Dameshek W: Some speculations on the myeloproliferative syndromes. Blood 1951;6:372–375.
70 Bentz M, Cabot G, Moos M, Speicher MR, Ganser A, Lichter P, Dohner H: Detection of chimeric BCR-ABL genes on bone marrow samples and blood smears in chronic myeloid and acute lymphoblastic leukemia by in situ hybridization. Blood 1994;83:1922–1928.
71 The German MPD Initiative. First Falkau Conference, June 2001.

Prof. Heike L. Pahl
Division of Experimental Anaesthesiology
Centre for Clinical Research
University Hospital Freiburg
Breisacher Strasse 66
D–79106 Freiburg (Germany)
E-Mail pahl@uni-freiburg.de

Bain BJ (ed) Chronic Myeloproliferative Disorders.
Basel, Karger, 2003, pp 102–108 / Acta Haematol 2002;108:113–119

# Cytogenetic and Molecular Genetic Aspects of Idiopathic Myelofibrosis

John T. Reilly

Academic Unit of Haematology, Division of Genomic Medicine, Royal Hallamshire Hospital, Sheffield, UK

**Key Words**
Cytogenetics · Idiopathic myelofibrosis · Prognosis · 13q · 20q · Trisomy 1q

**Abstract**
Idiopathic myelofibrosis is a chronic myeloproliferative disorder in which the characteristic fibroblast proliferation is thought to be a secondary phenomenon resulting from the inappropriate release of megakaryocyte- and/or monocyte-derived growth factors, including PDGF, TGF-β, bFGF and calmodulin. In contrast, the haematopoietic cells are clonal, although the underlying pathogenetic mechanisms remain essentially unknown. Cytogenetic studies have highlighted that 13q–, 20q–, +8 and abnormalities of chromosomes 1, 7 and 9 constitute more than 80% of the chromosomal changes. A third of idiopathic myelofibrosis cases have abnormal karyotypes at diagnosis, a figure that increases if follow-up analyses are performed. Evolution to more complex karyotypes may accompany clinical progression, with abnormalities increasing to around 90% following acute leukaemic transformation. Cytogenetic abnormalities have been associated with prognosis and to a lack of treatment response to androgens. Oncogene mutations are rare and include point mutations in *N-RAS*, *c-KIT* and *TP53*.

## Introduction

Idiopathic (primary) myelofibrosis, or agnogenic myeloid metaplasia, is a chronic myeloproliferative disorder characterised by splenomegaly, varying degrees of bone marrow fibrosis, extramedullary haematopoiesis and teardrop poikilocytosis [1]. Although the disorder was first described in 1879 by Heuck [2], it was not until 1951, following Damshek's [3] seminal paper, that idiopathic myelofibrosis was regarded as one of the chronic myeloproliferative disorders. It is an uncommon disease, with a reported annual incidence of around 0.4 per 100,000 population [4, 5], with the highest rates being documented among Ashkenazi Jews in northern Israel [6]. The disease's aetiology remains largely unknown, although environmental factors may be relevant in a minority of patients, in view of the reported links with radiation [7, 8], as well as benzene and toluene exposure [9–11]. With reported median survivals ranging from 3 to 5 years, idiopathic myelofibrosis has the worst outcome among the chronic myeloproliferative disorders [12].

## Haematopoietic Clonality

Since the late 1970s a number of studies have suggested that haematopoiesis in idiopathic myelofibrosis is clonal and results from the malignant proliferation of a pluripotential stem cell. Such a conclusion was first reached by the analysis of glucose-6-phosphate-dehydrogenase alloenzymes in a black female patient who was heterozygous for this X-linked gene [13]. A number of groups have confirmed this finding by probing for X-linked restriction fragment length polymorphisms, taking advantage of the nonrandom inactivation of two X-linked genes, hypoxanthine phosphoribosyl transferase and phosphoglycerate kinase. Monoclonal haematopoiesis was detected in all bone marrow cells and at every stage of the disease, from the early cellular phase to advanced myelofibrosis [14–17]. However, these initial studies lacked appropriate controls to exclude the possibility of extreme lyonization. The most appropriate control is DNA from T cells [18] and the demonstration in female patients of clonal granulocytes, in the presence of polyclonal T cells, is now accepted as evidence for myeloid clonality. Two recent studies, using T cells as controls, have confirmed the clonal nature of idiopathic myelofibrosis, although the number of cases studied remains small [19, 20]. Idiopathic myelofibrosis is also characterised by an increased number of circulating pluripotent (CFU-GEMM) and lineage-restricted precursor cells (BFU-E, CFU-GM and CFU-MK), with mean levels ranging from 8- to 167-fold higher than those in control subjects [21–24]. Interestingly, Barosi et al. [25] have shown that circulating stem cell quantification, as defined by CD34 positivity, not only reliably differentiates myelofibrosis from other Ph-negative chronic myeloproliferative disorders, but can also provide important prognostic information.

In contrast to haematopoietic cells, idiopathic myelofibrosis bone marrow fibroblasts are polyclonal, indicating a reactive, or non-neoplastic origin. This view is supported by the physical characteristics of myeloproferative bone marrow fibroblasts, which appear identical to control fibroblasts [26], and by the fact that successful allogeneic bone marrow transplantation is accompanied by a reversal of marrow fibrosis [27]. It is now accepted that the fibroblast proliferation results from the inappropriate release of megakaryocyte/platelet- and/or monocytic-derived growth factors, including platelet-derived growth factor, transforming growth factor-β, basic fibroblast growth factor and calmodulin [28, 29].

## Cytogenetic Analysis

Cytogenetic studies have provided additional evidence for the current pathogenetic models of stromal proliferation and extramedullary haematopoieisis, features that characterise idiopathic myelofibrosis. Wang et al. [30], for example, noted that bone marrow fibroblast karyotypes from individual patients with idiopathic myelofibrosis differed from those found in the patient's bone marrow or peripheral blood cells. These findings are in line with the widely held view that the fibroblast proliferation is polyclonal, or reactive, and is not part of the underlying neoplastic process. Similarly, the current view is that splenic extramedullary haematopoiesis results from the sequestration, accumulation and proliferation of circulating myeloid progenitor cells in splenic cords [31]. Recently, Mesa et al. [32] have strengthened this hypothesis by demonstrating excellent concordance between bone marrow and splenic cytogenetic clones.

Standard cytogenetic analysis has been a pivotal step in the elucidation of pathogenetically relevant oncogenes in many haematological malignancies, although, until recently, the karyotypic data for idiopathic myelofibrosis have been both sparse and confusing. The publication of three large cytogenetic studies, however, involving a total of 265 well-characterised patients, has helped to clarify the situation [33–35]. All three studies, and a literature review of 157 abnormal cases [36], reveal that 13q–, 20q–, +8 and abnormalities of chromosome 1, 7 and 9 constitute more than 80% of the chromosomal changes. Interestingly, a similar spectrum of abnormalities characterise both post-polycythaemic and post-thrombocythaemic myelofibrosis [35]. It has been suggested that abnormalities of chromosomes 1, 5, 7 and 9 represent cytogenetic subclones indicative of genomic instability, since they are frequently associated with additional co-existing lesions. A third of cases with idiopathic myelofibrosis have abnormal karyotypes at diagnosis, a figure that increases if follow-up analyses are performed [34, 37, 38]. In general, age, duration of disease, or exposure to chemotherapy, does not appear to alter the pattern of chromosomal abnormalities, although the likelihood of detecting a cytogenetic clone increases with time. Evolution to more complex karyotypic patterns may accompany clinical progression [39, 40], a finding consistent with the multi-step process of leukaemogenesis. Indeed, cytogenetic abnormalities, typically complex changes, increase to approximately 90% following acute transformation [37].

### *13q Deletions*

Deletions involving chromosome 13 are the most common cytogenetic abnormality [34, 35, 38], occurring in approximately 25% of cases with an abnormal karyotype. Commonly, the deletion has been interpreted as interstitial with proximal and distal breakpoints in 13q12 and

13q14–q22, respectively [34, 41]. Since the retinoblastoma susceptibility gene (RB-1) is located in this region, it has been postulated that its inactivation may be pathologically important [42]. Pastore et al. [43], however, although reporting loss of heterozygosity at the RB-1 locus in 5/8 cases of idiopathic myelofibrosis, found no structural alteration of the remaining allele. Interestingly, routine cytogenetic analysis of these 5 cases failed to demonstrate a 13q abnormality suggesting that cryptic deletions may be common [Gaidano, pers. commun]. In contrast, Juneau et al. [44], using fluorescence in situ hybridization (FISH) to detect RB-1 deletion in a series of 24 patients, reported hemizygous RB-1 deletions in 6 cases with cytogenetic 13q deletions, but found no evidence of RB-1 deletion in 16 patients without del(13)(q14). Similarly, Tanaka et al. [45], reported allelic loss of RB-1, as well as of the microsatellite loci D13S319 and D13S25, in 9 out of 10 cases of myeloid malignancy, including 1 case of transformed myelofibrosis with del(13)(q12q14). Two further studies have confirmed that the genetic loss associated with 13q14 deletions is large and involves the gene-rich region around RB-1, D13S319 and D13S25 [46, 47]. Similar findings have been reported for chromosome 20q deletions and it has been proposed that cryptic deletions are, in fact, rare [20]. However, it should be stressed that these studies do not exclude the existence of very small deletions outside the study area as well as those that would be beyond the resolution of current FISH technology.

Translocations involving 13q14 have been reported [44, 46, 48], and it was expected that detailed analysis of these unusual cases would reduce the critically deleted region and so facilitate the identification of pathogenetically relevant genes. Unfortunately, a number of groups have shown that large submicroscopic deletions, involving *RB-1*, D13S319 and D13S25, are characteristic of these apparently balanced translocations [46, 47, 49, 50]. It would seem, therefore, that most translocations involving 13q14 serve the same function as cytogenetic deletions and that their study may not help in identifying pathogenetically important genes. It is possible that more than 1 gene is involved on chromosome 13. Macdonald et al. [51], for example, reported a case of myelofibrosis with a t(4;13) (q25;q12) and provided evidence for involvement of a novel gene located at 13q12.

*20q Deletion*

Deletion of chromosome 20q is the second most common finding and has been reported by most groups [33–35, 37, 52]. Deletion-mapping studies using conventional cytogenetics [53], FISH [54] and microsatellite PCR/Southern blot analysis [20] have delineated a commonly deleted region within the proximal two-thirds of band 20q12. Wang et al. [55] narrowed the segment to an area of 8 Mb flanked by a proximal boundary between markers D20S206 and D20S107 and a distal boundary between D20S119 and D20S424. Recently, physical mapping of translocation breakpoints in a variety of myeloid leukemia cell lines have been reported [56]. Six candidate genes were identified, including a human homologue of a Drosophila tumour suppressor gene, h-l(3)mbt, that may be pathogenetically relevant to myeloid malignancies.

*Trisomy 1q*

Structural changes leading to partial duplication of the long arm of chromosome 1 are the third commonest abnormality in idiopathic myelofibrosis [34, 35]. These rearrangements follow a definite non-random pattern and, although the whole of 1q may be involved, always appear to include the specific segment, 1q23–1q32 [57–59]. Duplicated 1q has been found at all stages of disease, including at diagnosis [34]. The inability to identify common breakpoints, or a preferential translocation site, suggests that an increase in copy number of gene(s) located on 1q may be more important than the position effect due to the juxtaposition of specific DNA sequences after the duplication events. In support of this concept, Zanke et al. [60] have demonstrated an associated amplification and overexpression of a haematopoietic protein tyrosine phosphatase in patients with partial trisomy 1q. Unbalanced translocations that result in 1q duplications have also been reported in idiopathic myelofibrosis. For example, 3 cases with a der(1)t(1;7)(7pter→7p11:1p1?→1qter), resulting in both monosomy 7q and trisomy 1q, were reported by Geraedts et al. [61]. The pathogenetic event is unclear, but could involve monosomy 7q, duplication 1q and/or the precise location of the translocation breakpoint. Furthermore, a der(1)t(1;9), that results in both a trisomy 1q and trisomy 9p, has also been documented in a small number of cases [34, 62].

*Trisomy 8*

Trisomy 8 is a well-recognised abnormality in idiopathic myelofibrosis [36] and has recently been associated with a poor prognosis [35]. The genetic consequences of chromosome 8 duplication remain unclear, and no molecular mapping has been reported.

### Cytogenetics and Prognosis

The clinical course and prognosis of idiopathic myelofibrosis is extremely variable, with survival ranging from months to decades. As a result, a number of groups have attempted to identify clinical and/or laboratory features that predict survival, with conflicting results [29]. Chromosomal abnormalities, for example, have been associated with a shorter survival in several recent studies [33, 34, 38], in contrast to earlier reports [63, 64]. Anaemia, however, has been consistently associated with a shorter survival [34, 38, 65, 66] and forms the basis of several useful prognostic schemes [34, 38]. In one prognostic model the median survival of patients below the age of 69 years who did not have either anaemia (haemoglobin <10 g/dl) or an abnormal karyotype was 180 months, compared to 22 months for the age-matched cohort with both anaemia and abnormal cytogenetics [34]. A recent study, however, suggested that it is the presence of specific clonal abnormalities, namely trisomy 8 and deletion of 12p, rather than an abnormal karyotype per se, that is prognostically important in idiopathic myelofibrosis [35]. This conclusion for trisomy 8 in idiopathic myelofibrosis is consistent with previous findings for trisomy 8 in other chronic myeloid disorders [67, 68]. Cytogenetic abnormalities have also been linked to treatment response, since anaemia appears to responds less well to androgen therapy in patients with karyotypic abnormalities [69].

A number of rare karyotypic abnormalities, unrelated to therapy, have been associated with a poor outcome. Trisomy 13, a non-random aberration in myelofibrosis [70], confers a poor prognosis due to early blast transformation [71, 72], as appears to be the case for der(1)t(1;9) [62] and t(6;10) (q27;q11) [73]. Patton et al. [74] have reported an association between a lack of committed erythroid progenitor cells in idiopathic myelofibrosis and chromosome 11 abnormalities and suggested the existence of genes on this chromosome which play a key role during erythropoiesis.

Interestingly, the incidence of chromosomal abnormalities is significantly less in younger patients [75, 76], a fact that might explain their better prognosis. Indeed, normal cytogenetic findings are characteristic of paediatric cases which, coupled with their long-term survival, suggests that they may have a different pathogenesis and require a more conservative management [77].

Finally, it should be pointed out that the prognostic usefulness of bone marrow cytogenetic information in idiopathic myelofibrosis can be limited by the difficulty of obtaining an adequate number of analysable metaphases. Nevertheless, it is usually possible to obtain satisfactory results following a standard 24- to 72-hour culture of peripheral blood without stimulation [34, 46], a fact that may be explained by the expanded pool of myeloid progenitors in the peripheral blood [21]. Tefferi et al. [78] have attempted to improve the detection rate by undertaking interphase cytogenetics, incorporating a panel of fluorescence in situ probes capable of detecting most of the recurrent cytogenetic lesions known in idiopathic myelofibrosis. However, although the technique was found to be complementary to standard cytogenetics, its additional cost and the fact that it did not always provide the necessary karyotypic information, limited its usefulness.

### Genetic Studies

Reports of oncogene mutational analysis in IMF are scarce and add little to our pathogenetic understanding. *RAS* mutations, predominantly affecting codon 12 of *N-RAS*, have been described by a number of groups [37, 79, 80], although they appear uncommon, occurring in only 6% of patients in chronic phase [37]. Interestingly, in the case reported by Buschle et al. [80], the mutation was identified in granulocytes, monocytes, B lymphocytes and T lymphocytes, indicating a pluripotential stem cell origin of the disease. *TP53* inactivation has been associated with transformation in a variety of *BCR-ABL*-negative chronic myeloproliferative disorders, although the number of patients studied with idiopathic myelofibrosis was small [81].

A number of studies have investigated the potential role of *c-KIT* in the pathogenesis of idiopathic myelofibrosis. *c-KIT* expression, for example, is increased in chronic myeloproliferative disorders, including myelofibrosis, and the receptor may be expressed for longer than normal during progenitor cell development [82, 83]. Interestingly, Nakata et al. [84] reported *c-KIT* mutation (Asp52Asn) in 2 out of 6 patients with idiopathic myelofibrosis and suggested that this acquired abnormality of the extracellular domain resulted in enhanced sensitivity of the patients' progenitor cells to stem cell factor [85]. The present author, however, has not been able to confirm the presence of the Asp52Asn mutation in a large cohort of well-characterised patients, suggesting that *c-KIT* mutations are rare in idiopathic myelofibrosis [unpubl. obs.]. Nevertheless, a significant increase in circulating c-KIT+, CD34+ stem cells is characteristic of IMF and it is possible that c-KIT may have a pathogenetic role involving paracrine or autocrine mechanisms. Moliterno et al. [86] have reported that the thrombopoietin receptor (c-mpl) is markedly reduced or absent in idiopathic myelofibrosis megakaryocytes, suggesting that their proliferation and survival are independent of thrombopoietin [86]. However, the underlying mechanism for reduced receptor expression remains unclear as no mutations, or gene rearrangements, have been reported [87]. Finally, Wang et al. [88] have studied the hypermethylation of *CDKN2B* (p15INK4b) and *CDKN2A* (p16INK4a) and have shown a relationship with leukaemic transformation. It is possible, therefore, that hypomethylation agents may be a novel approach for future therapy.

### Future Approaches

The recent clarification of cytogenetic abnormalities in myelofibrosis has failed to enhance significantly our understanding of the pathogenesis of the disease, since the identification of non-random chromosomal abnormalities has not yet resulted in the identification of involved genes. Novel strategies are required, and one possible approach is the use of oligonucleotide microarrays to facilitate the analysis of gene expression. Jones et al. [89], for example, have reported preliminary results using this technique on purified CD34+ cells from a small cohort of patients with idiopathic myelofibrosis. Genes encoding a number of transcription factors were noted to be upregulated, including those encoding AP-1 proteins JunB and JunD and the haematopoietic transcription factor GATA2. In addition, regulation of G-protein signalling, including the guanine nucleotide exchange factor NET1, arrestin 2, the G0/G1 switch regulatory gene were all overexpressed, while key cell cycle proteins were downregulated. It was hypothesised that dysregulation of these intracellular signalling pathways and/or decreased expression of important cell cycle proteins may play a pathogenetic role.

### References


1 Ward HP, Block MH: The natural history of agnogenic myeloid metaplasia (AMM) and a critical evaluation of its relationship with the myeloproliferative syndrome Medicine (Baltimore) 1971;50:357–420.

2 Heuck G: Zwei Fälle von Leukämie mit eigentümlichem Blut- resp. Knockenmarksbefund. Virchows Archiv 1879;78:475–496.

3 Dameshek W: Some speculations on the myeloproliferative syndromes. Blood 1951;6:372–375.

4 Mesa RA, Li CY, Schroeder G, Tefferi A: Clinical correlates of splenic histopathology and splenic karyotype in myelofibrosis with myeloid metaplasia. Blood 2001;97:3665–3667.

5 Kutti J, Ridell B: Epidemiology of the myeloproliferative disorders: Essential thrombocythaemia, polycythaemia vera and idiopathic myelofibrosis. Pathol Biol 2001;49:164–166.

6 Chaiter Y, Brenner B, Aghai E, Tatarsky I: High incidence of myeloproliferative disorders in Ashkenazi Jews in northern Israel. Leuk Lymph 1992;7:251–255.

7 Andersen RE, Hoshino T, Yamamoto T: Myelofibrosis with myeloid metaplasia in survivors of the atomic bomb in Hiroshima. Ann Intern Med 1964;60:1–18.

8 Johnson SA, Bateman CJ, Beard ME, Whitehouse JM, Water AH: Long-term haematological complications of thorotrast. Q J Med 1977;182:259–271.

9 Rawson R, Parker F, Jackson H: Industrial solvents as possible etiologic agents in myeloid metaplasia. Science 1941;93:541–542.

10 Hu H: Benzene-associated myelofibrosis. Ann Intern Med 1987;106:171–173.

11 Bosch X, Campistol JM, Montoliu J, Cervantes F, Revert, L: Toluene-associated myelofibrosis. Blut 1989;58:219–220.

12 Rozman C, Giralt M, Feliu E, Rubio D, Cortes MT: Life expectancy of patients with chronic nonleukemic myeloproliferative disorders. Cancer 1991;67:2658–2663.

13 Jacobson RJ, Salo A, Fialkow PJ: Agnogenic myeloid metaplasia: A clonal proliferation of hematopoietic stem cells with secondary myelofibrosis. Blood 1978;51:189–194.

14 Lucas GS, Padua RA, Masters GS, Oscier D, Jacobs A: The application of X chromosome gene probes to the diagnosis of myeloproliferative disease. Br J Haematol 1989;72:530–533.

15 Kreipe H, Jaquet K, Felgner J, Radzun H-J, Parwaresch MR: Clonal granulocytes and marrow cells in the cellular phase of agnogenic myeloid metaplasia. Blood 1991;78:1814–1817.

16 Anger B, Janssen JW, Schrezenmeier H, Hehlmann R, Heimpel H, Bartram CR: Clonal analysis of chronic myeloproliferative disorders using X-linked DNA polymorphisms. Leukemia 1990;4:258–261.

17 Taylor KM, Shetta M, Talpaz M, Kantarjian HM, Hardikar S, Chinault AC, McCredie KB, Spitzer G: Myeloproliferative disorders: Usefulness of X-linked probes in diagnosis. Leukemia 1989;3:419–422.

18 Gale RE, Wheadon H, Boulos P, Linch DC: Tissue specificity of X-chromosome inactivation patterns. Blood 1994;83:2899–2905.

19 Tsukamoto N, Morita K, Maehara T, Okamoto K, Sakai H, Karasawa M, Naruse T, Omine M: Clonality in chronic myeloproliferative disorders defined by X-chromosome linked probes: Demonstration of heterogeneity in lineage involvement. Br J Haematol 1994;86:253–258.

20 Asimakopoulos FA, Gilbert JG, Aldred MA, Pearson TC, Green AR: Interstitial deletion constitutes the major mechanism for loss of heterozygosity on chromosome 20q in polycythemia vera. Blood 1996;88:2690–2698.

21 Hibbin JA, Njoku OS, Matutes E, Lewis SM, Goldman JM: Myeloid progenitor cells in the circulation of patients with myelofibrosis and other myeloproliferative disorders. Br J Haematol 1984;57:495–503.

22 Han ZC, Briere J, Nedellec G, Abgrall JF, Senzebe L, Parent D, Guern G: Characteristics of circulating megakaryocyte progenitors (CFU-MK) in patients with myelofibrosis. Eur J Haematol 1988;40:130–135.

23 Carlo-Stella C, Cazzola M, Gasner A, Barosi G, Dezza L, Meloni F, Pedrazzoli P, Hoelzer D, Ascari E: Effects of recombinant alpha and gamma interferons on the in vitro growth of circulating hematopoietic progenitor cells (CFU-GEMM, CFU-Mk, BFU-E and CFU-GM) from patients with myelofibrosis and myeloid metaplasia. Blood 1987;70:1014–1019.

24 Colovic MD, Wiernik PH, Jankovic GM, Vidovic AD, Janosevic S, Basara NM: Circulating haematopoietic progenitor cells in primary and secondary myelofibrosis: Relation to collagen and reticulin fibrosis. Eur J Haematol 1999;62:155–159.

25 Barosi G, Viarengo G, Pecci A, Rosti V, Piaggio G, Marchetti M, Frassoni F: Diagnostic and clinical relevance of the number of circulating CD34+ cells in myelofibrosis with myeloid metaplasia. Blood 2001;98:3249–3255
26 Castro-Malaspina H, Gay RE, Thanwar SL: Characteristics of bone marrow fibroblast colony forming cells and their progeny in patients with myeloproliferative disorders. Blood 1982; 59:1046–1056.
27 Guardiola P, Esperou H, Cazals-Hatem D, Ifrah N, Jouet J-P, Buzyn A, Sutton L, Gratecos N, Tilly H, Libure B, Gluckman E: Allogeneic bone marrow transplantation for agnogenic myeloid metaplasia. Br J Haematol 1997;98: 1004–1009.
28 Rameshwar P, Denny TN, Stein D, Gascon P: Monocytic adhesion in patients with bone marrow fibrosis is required for the production of fibrogenic cytokines. Potential role for interleukin-1 and TGF-beta. J Immunol 1994;15: 2819–2830.
29 Reilly JT: Idiopathic myelofibrosis: Pathogenesis, natural history and management. Blood Rev 1997;11:233–242.
30 Wang JC, Lang HD, Lichter S, Weinstein M, Benn P: Cytogenetic studies of bone marrow fibroblasts cultured from patients with myelofibrosis and myeloid metaplasia. Br J Haematol 1992;80:184–188.
31 Zhang B, Lewis SM: The splenomegaly of myeloproliferative and lymphoproliferative disorders: Splenic cellularity and vascularity. Eur J Haematol 1989;43:63–66.
32 Mesa RA, Li CY, Schroeder G, Tefferi A: Clinical correlates of splenic histopathology and splenic karyotype in myelofibrosis with myeloid metaplasia. Blood 2001;97:3665–3667.
33 Demory JL, Dupriez B, Fenaux P, Lai JL, Beuscart R, Jouet JP, Derminatti M, Bauters F: Cytogenetic studies and their prognostic significance in agnogeneic myeloid metaplasia: A report on 47 cases. Blood 1988;72:855–859.
34 Reilly JT, Snowden JA, Spearing RL, Fitzgerald PM, Jones N, Watemore A, Potter A: Cytogenetic abnormalities and their prognostic significance in idiopathic myelofibrosis: A study of 106 cases. Br J Haematol 1997;98:96–102.
35 Tefferi A, Mesa RA, Schroeder G, Hanson CA, Li C-Y, Dewald GW: Cytogenetic findings and their clinical relevance in myelofibrosis with myeloid metaplasia. Br J Haematol 2001;113: 763–771.
36 Bench AJ, Nacheva EP, Champion KM, Green AR: Molecular genetics and cytogenetics of myeloproliferative disorders. Baillières Clin Haematol 1998;11:819–848.
37 Reilly JT, Wilson J, Barnett D, Watmore A, Potter, A: Karyotypic and ras gene mutation analysis in idiopathic myelofibrosis. Br J Haematol 1994;88:575–581.
38 Dupriez B, Morel P, Demory J-L, Lai JL, Simon M, Plantier I, Bauters F: Prognostic factors in agnogenic myeloid metaplasia: a report on 195 cases with a new scoring system. Blood 1996;88:1013–1018.
39 Trautmann U, Rubbert A, Gramatzki M, Henschke F, Gebhart E: Multiple chromosomal changes and karyotypic evolution in a patient with myelofibrosis. Cancer Genet Cytogenet 1992;61:6–10.
40 Kerim S, Rege-Cambrin G, Scaravaglio P, Godio L, Saglio G, Aglietta M: Trisomy 8 and an unbalanced t(5;17)(q11;p11) characterize two karyotypically independent clones in a case of idiopathic myelofibrosis evolving to acute nonlymphoid leukemia. Cancer Genet Cytogenet 1991;52:63–69.
41 Morris CM, Cochrane JM, Benjes SM, Crossen PE, Fitzgerald PH: Molecular definition of interstitial deletions of chromsome 13 in leukaemia cells. Genes Chromosomes Cancer 1991;3:455–460.
42 Lebowitz P, Papac R, Ghosh PK: Impaired retinoblastoma suscepibility (Rb) gene expression in agnogenic myeloid metaplasia. Blood 1990;76(suppl 1): 236a.
43 Pastore C, Nomdedeu J, Volpe G, Guerrasio A, Cambrin GR, Parvis G, Pautasso M, Daglio C, Mazza U, Saglio G, Gaidano G: Genetic analysis of chromosome 13 deletions in bcr/abl negative chronic myeloproliferative disorders. Genes Chromosomes Cancer 1995;14:106–111.
44 Juneau AL, Kaehler M, Christensen ER, Schad CR, Zinsmeister AR, Lust J, Hanson C, Dewald GW: Detection of RB1 deletions by fluorescence in situ hybridization in malignant hematologic disorders. Cancer Genet Cytogenet 1998;103:117–123.
45 Tanaka K, Arif M, Eguchi M, Guo SX, Hayashi Y, Asaoku H, Kyo T, Dohy H, Kamada N: Frequent allelic loss of the RB, D13S319 and D13S25 locus in myeloid malignancies with deletion/translocation at 13q14 of chromosome 13, but not in lymphoid malignancies. Leukemia 1999;13:1367–1373.
46 Sinclair EJ, Forrest EC, Reilly JT, Watemore AE, Potter AM: Fluorescence in situ hybridization analysis of 25 cases of idiopathic myelofibrosis and two cases of secondary myelofibrosis: Monoallelic loss of RB1, D13S319 and D13S25 loci associated with cytogenetic deletion and translocation involving 13q14. Br J Haematology2001;113:365–368.
47 La Starza R, Wlodarska I, Aventin A, Falzetti B, Crescenzi B, Martelli MF, van de Berghe H, Mecucci C: Molecular delineation of 13q deletion boundaries in 20 patients with myeloid malignancies. Blood 1998;91:231–237.
48 Dewald GW, Wright PI: Chromosome abnormalities in the myeloproliferative disorders. Semin Oncol 1995;22:341–354.
49 Gardiner AC, Corcoran MM, Oscier DG: Cytogenetic, fluorescence in situ hybridisation and clinical evaluation of translocations with concomitant deletion at 13q14 in chronic lymphatic leukemia. Genes Chromosomes Cancer 1997;20:73–81.
50 Hawthorn LA, Chapman R, Oscier D, Cowell JK: The consistent 13q14 translocation breakpoint in chronic B-cell leukemia (BCLL) involves deletion of the D13S25 locus which lies distal to the retinoblastoma predisposition gene. Oncogene 1993;8:1415–1419.
51 Macdonald DHC, Lahiri D, Chase A, Sohal J, Goldman JM, Cross NCP: A case of myelofibrosis with a t(4;13)(q25;q12): evidence for involvement of a second 13q12 locus in chronic myeloproliferative disorders. Br J Haematol 1999;105:771–774.
52 Mertens F, Johansson B, Heim S, Kristoffersson U, Mitelman F: Karyotypic patterns in chronic myeloproliferative disorders: Report on 74 cases and review of the literature. Leukemia 1991;5:214–220.
53 Nacheva E, Holloway T, Carter N, Grace C, White N, Green AR: Characterization of 20q deletions in patients with myeloproliferative disorders and myelodysplastic syndromes. Cancer Genet Cytogenet 1995;80:87–94.
54 Roulston D, Espinosa R, Stoffel M, Bell GI, Le Beau MM: Molecular genetics of myeloid leukemia: Identification of the commonly deleted segment of chromosome 20. Blood 1993;82: 3424–3429.
55 Wang PW, Eisenbart JD, Espinosa R, Davis EM, Larson RA, Le Beau MM: Refinement of the smallest commonly deleted segment of chromosome 20 in malignant myeloid diseases and development of a PAC-based physical and transcription map. Genomics 2000;67:28–39.
56 MacGrogan D, Alvarez S, DeBlasio T, Jhanwar SC, Nimer SD: Identification of candidate genes on chromosome band 20q12 by physical mapping of translocation breakpoints found in myeloid leukemia cell lines. Oncogene 2001; 20:4150–4160.
57 Rowley JD: Abnormalities of chromosome no 1 in haematological malignancies. Lancet 1978;i:554–555.
58 Gahrton G, Friburg K, Lindsten J, Zech L: Duplication of part of the long arm of chromosome 1 in myelofibrosis terminating in acute myeloblastic leukemia. Hereditas 1978;88:1–5.
59 Donti E, Tabilio A, Bocchini F, Falzetti F, Martelli MF, Grignani F, Donti GV: Partial trisomy 1q in idiopathic myelofibrosis. Leuk Res 1990;14:1035–1040.
60 Zanke B, Squire J, Griesser H, Henry M, Suzuki H, Patterson B, Minden M, Mak TW: A hematopoietic protein tyrosine phosphatase (hePTP) gene that is amplified and overexpressed in myeloid malignancies maps to chromosome 1q32.1. Leukemia 1994;8:236–244.
61 Geraedts JPM, den Ottolander GJ, Ploem JE, Muntinghe OG: An identical translocation between chromosome 1 and 7 in three patients with myelofibrosis and myeloid metaplasia. Br J Haematol 1980;44:569–575.
62 Rege-Cambrin G, Speleman F, Kerim S, Scaravaglio P, Carozzi F, Dal Cin P, Michaux JL, Offner F, Saglio G, Van den Berghe H: Extra translocation +der(1q9p) is a prognostic indicator in myeloproliferative disorders. Leukemia 1991;5:1059–1063.
63 Whang-Peng J, Lee E, Knutsen T, et al: Cytogenetic studies in patients with myelofibrosis and myeloid metaplasia. Leuk Res 1978;2:41–56.

64 Miller JB, Testa JR, Lindgren V, Rowley JD: The pattern and clinical significance of karyotypic abnormalities in patients with idiopathic and postpolycythemic myelofibrosis. Cancer 1985;55:582–591.
65 Varki A, Lottenberg R, Griffith R, Reinhard E: The syndrome of idiopathic myelofibrosis: A clinicopathologic review with emphasis on the prognostic variables predicting survival. Medicine 1983;62:353–371.
66 Barosi G, Berzuini C, Liberato LN, Costa A, Polino G, Ascari E: A prognostic classification of myelofibrosis with myeloid metaplasia. Br J Haematol 1988;70:397–401.
67 Byrd JC, Lawrence D, Arthur DC, Pettenati MJ, Tantravahi R, Qumsiyeh M, Stamberg J, Davey FR, Schiffer CA, Bloomfield CD: Patients with trisomy 8 in acute myeloid leukemia are not cured with cytarabine-based chemotherapy: Results from Cancer and Leukemia Group B 8461. Clin Cancer Res 1998;4:1235–1241.
68 Sole E, Espinet B, Sanz GF, Cervera J, Calasanz MJ, Luno E, Prieto F, Granada I, Hernandez JM, Cigudosa JC, Diez JL, Bureo E, Marques ML, Arranz, E, Rios R, Martinez Climent JA, Vallespi T, Florensa L, Woessner S: Incidence, characterization and prognostic significance of chromosomal abnormalities in 640 patients with primary myelodysplastic syndromes. Grupo Cooperativo Espanol de Citogenetica Hematologica. Br J Haematol 2000; 108:346–356.
69 Besa EC, Nowell PC, Geller NL, Gardner FH: Analysis of the androgen response of 23 patients with agnogenic myeloid metaplasia: The value of chromosomal studies predicting response and survival. Cancer 1982;49:308–313.
70 Hsu LY, Greenberg ML, Kohan S, Wittman R: Trisomy 13 in bone marrow cells in acute myelocytic leukemia and myelofibrosis. Clin Genet 1979;15:327–331.
71 Zojer N, Meran JG, Vesely M, Gruner H, Ackermann J, Dellinger C, Zimmer-Roth I, Heinz R, Drach J, Ludwig H: Trisomy 13 is associated with poor prognosis in idiopathic myelofibrosis with myeloid metaplasia. Leuk Lymph 1999;35:415–421.
72 Ferrara F, Vicari L, Festa B, Di Noto R, Pane F, Sebastio L, Cimino R: Trisomy 13 in a case of myelofibrosis with myeloid metaplasia with early blastic transformation. Haematologica 1995;80:434–436.
73 Cox MC, Panetta P, Venditti A, Abruzzese E, Del Poeta G, Cantone M, Amadori S: New reciprocal translocation t(6;10)(q27;q11) associated with idiopathic myelofibrosis. Leuk Res 2001;25:349–351.
74 Patton WN, Bunce CM, Larkins S, Brown G: Defective erythropoiesis in primary myelofibrosis associated with a chromosome 11 abnormality. Br J Cancer 1991;64:128–131.
75 Cervantes F, Barosi G, Demory J-L, Reilly J, Guarnone R, Dupriez B, Pereira A, Montserrat E: Myelofibrosis with myeloid metaplasia in young individuals: Disease characteristics, prognostic factors and identification of risk groups. Br J Haematol 1998;102:684–690.
76 Cervantes F, Barosi G, Hernández-Boluda J-C, Marchetti M, Montserrat E: Myelofibrosis with myeloid metaplasia in adult individuals 30 years old or younger: Presenting features, evolution and survival. Eur J Haematol 2001;66: 324–327.
77 Altura RA, Head DR, Wang WC: Long-term survival of infants with idiopathic myelofibrosis. Br J Haematol 2000;109:459–462.
78 Tefferi A, Meyer RG, Wyatt WA, Dewald GW: Comparison of peripheral blood interphase cytogenetics with bone marrow karyotype analysis in myelofibrosis with myeloid metaplasia. Br J Haematol 2001;115:316–319.
79 Janssen JWG, Steenvoorden ACM, Lyons J, Anger B, Bohlke JU, Bos JL, Seliger H, Bartram CR: Ras gene mutations in acute and chronic myelocytic leukemias, chronic myeloproliferative disorders, and myelodysplastic syndromes. Proc Natl Acad Sci USA 1987;84: 9228–9232.
80 Buschle M, Janssen JW, Drexler H, Lyons J, Anger B, Bartram CR: Evidence for pluripotent stem cell origin of idiopathic myelofibrosis: Clonal analysis of a case characterised by a N-ras gene mutation. Leukemia 1988;2:658–660.
81 Gaidano G, Guerrasio A, Serra A, Carozzi F, Cambrin GR, Petroni D: Mutations in the p53 and ras family genes are associated with tumour progression of bcr/abl negative chronic myeloproliferative disorders. Leukemia 1993; 7:946–953.
82 Siitonen T, Savolainen ER, Koistinen P: Expression of the c-kit proto-oncogene in myeloproliferative disorders and myelodysplastic syndromes. Leukemia 1994;8:631–637.
83 Siitonen T, Zheng A, Savolainen E-R, Koistinen P: Spontaneous granulocyte-macrophage colony growth by peripheral blood mononuclear cells in myeloproliferative disorders. Leuk Res 1996;20:187–195.
84 Nakata Y, Kimura A, Katoh O, Kawaishi K, Hyodo H, Abe K, Kuramoto A, Satow Y: c-kit point mutation of extracellular domain in patients with myeloproliferative disorders. Br J Haematol 1995;91:661–663.
85 Kimura A, Nakata Y, Katoh O, Hyodo H: c-kit mutations in patients with myeloproliferative disorders. Leuk Lymph 1997;25:281–287.
86 Moliterno AR, Siebel KE, Sun AY, Hankins WD, Spivak JL: A novel thrombopoietin signaling defect in polycythemia vera platelets. Stem Cells 1998;16(suppl 2): 185–192.
87 Taskin AL, Cohen-Sokal K, Le Couedic JP, Abina MA, Villeval JL, Debili N, Casadevall N, Vainchenker W, Wendling F: Myelofibrosis: Experimental models and human studies. Stem Cells 1998;16(suppl 2):155–164.
88 Wang JC, Chen W, Nallusamy S, Chen C, Novetsky AD: Hypermethylation of the P15INK4b and P16INK4a in agnogenic myeloid metaplasia (AMM) and AMM in leukemic transformation. Br J Haematol 2002;116:582–586.
89 Jones LC, Tefferi A, Wachsman W, Hofmann WK, Koeffler HP: Detection of aberrant signalling pathways in CD34+ cells from patients with myelofibrosis with myeloid metaplasia using oligonucleotide microarrays. Blood 2001; 98(suppl):626a.

Dr. John T. Reilly
Academic Unit of Haematology
Division of Genomic Medicine
Royal Hallamshire Hospital
Sheffield S10 2JF (UK)
E-Mail j.t.reilly@sheffield.ac.uk

Bain BJ (ed): Chronic Myeloproliferative Disorders.
Basel, Karger, 2003, pp 109–119 / Acta Haematol 2002;108:55–65

# Cytogenetic and Molecular Genetic Aspects of Essential Thrombocythemia

David P. Steensma[a,b] Ayalew Tefferi[a]

[a]Division of Hematology and Internal Medicine, and [b]Department of Oncology, Mayo Clinic, Rochester, Minn., USA

**Key Words**
Essential thrombocythemia · Chronic myeloproliferative disorders · Thrombocytosis · Leukemic transformation

**Abstract**
Essential thrombocythemia (ET) is a chronic myeloid disorder that is characterized by thrombocytosis, thrombohemorrhagic and vasomotor symptoms, a long median survival, and a low risk of transformation to leukemia. ET can be difficult to distinguish from secondary (reactive) thrombocytosis, and the diagnosis of ET can only be made after the exclusion of other marrow disorders with similar features. Although ET has been assumed to be a clonal process, recent studies have suggested that a substantial number of cases classified as ET may actually not be clonal, and nonclonality may be associated with a lower risk of thrombosis. The lack of a characteristic cytogenetic marker for ET confounds analyses of clonality and offers no insight into disease pathogenesis. There is controversy over the proper classification of thrombocytosis associated with the pathological *BCR-ABL* gene rearrangement; such cases are not clearly distinguishable from chronic myelogenous leukemia (CML) and should be provisionally classified as CML. New insights are emerging into the role of the megakaryocytopoiesis regulator thrombopoietin (TPO) and its receptor, c-Mpl, in ET and related disorders, but TPO-Mpl dynamics appear to be complex. In some familial thrombocythemic syndromes, mutations in the 5′ untranslated region of TPO have recently been described, but these have not yet been observed in sporadic ET. In the future, global analysis of gene expression patterns may help overcome diagnostic dilemmas, refine disease classification, and lead to an improved understanding of the pathogenesis of ET.

## Introduction

Essential thrombocythemia (ET) is a bone marrow disorder characterized by excessive production of platelets and a significant risk of thrombohemorrhagic complications and vasomotor symptoms. ET usually follows an indolent clinical course, with a median survival of more than 20 years and a relatively low risk (<10%) of leukemic transformation [1, 2].

ET is currently classified with the chronic myeloproliferative disorders (CMPDs) [3]. Although ET is the most common cause of primary thrombocytosis, most thrombocytosis is actually not primary, even when the degree of thrombocytosis is extreme (i.e. a peripheral blood platelet count $>1{,}000 \times 10^9$/l) [4, 5]. More than 85% of the time, an elevated circulating platelet count is not due to a primary bone marrow problem but is a consequence of a marrow reaction to another pathological condition [4]. A wide variety of conditions can result in so-called reactive or secondary thrombocytosis; most prominent among these are the broad spectrum of inflammatory states and iron deficiency [6]. In some cases, distinguishing idiopathic ET from these secondary causes of thrombocytosis can be very difficult. This distinction remains the chief diagnostic challenge with respect to this class of conditions [7–9].

Even when a patient exhibits no clear evidence for a secondary cause of thrombocytosis, the diagnosis of ET often remains tenuous. It may be difficult to clearly distinguish ET from the other myeloid disorders such as the CMPDs, the myelodysplastic syndromes (especially the so-called 5q– syndrome, which can be associated with striking thrombocytosis and minimal dysplasia), and a variety of poorly characterized atypical chronic myeloid disorders. Among the CMPDs – a group that, in addition to ET, also includes polycythemia vera, myelofibrosis with myeloid metaplasia, and chronic myelogenous leukemia (CML) – only CML has a specific diagnostic hallmark: the Philadelphia chromosome, t(9;22)(q34;q11), or its molecular equivalent, the pathological *BCR-ABL* translocation [3]. At present, ET remains a diagnosis of exclusion; some have even proposed that ET should be considered a 'working diagnosis' rather than a specific disease [10, 11].

### The Challenge of Determining Clonality

It has generally been assumed that the CMPDs are clonal disease states [12]. However, they were not originally defined this way. William Dameshek [13] proposed the classification category 'myeloproliferative syndromes' (including ET) in 1951 on the basis of the overlapping clinical and pathological features of the disorders, long before there was any clear understanding of the concept of clonality or of the genetics of hematological malignancies. In fact, Dameshek's intuitive and clinically useful classification predated such fundamental genetic knowledge as the structure of DNA (1953) and the true number of human chromosomes (1956) [14, 15].

Three decades after Dameshek's contribution, Fialkow et al. [16], in 1981, provided the first convincing evidence that ET could be a clonal disorder. In 1977, the same investigators had demonstrated the clonal nature of CML [17], and the other myeloproliferative syndromes would soon follow. These investigators used the X-linked glucose-6-phosphate dehydrogenase (G6PD) locus as a cell marker in order to analyze clonality of the blood cells from 3 patients who were considered to have either probable or definite ET on clinical grounds. Both the A and B types of G6PD were found in nonhematopoietic tissues in each of these cases, but only a single enzyme type was found in the granulocytes, red blood cells, and platelets from each patient. This strongly suggested that the blood cells originated from genetically identical (clonal) multipotential stem cells [16]. Techniques such as G6PD analysis take advantage of the fact that in any single cell of a female, one of the two X-inherited chromosomes was randomly inactivated at an early stage of embryonic development (the Lyon hypothesis), so that mosaicism of X-linked markers such as G6PD can be used as a surrogate marker for clonality in women who are heterozygous for that marker.

Other X-linked markers, including the hypoxanthine phosphoribosyltransferase (HPRT) and phosphoglycerate kinase (PGK) genes, have also been used to study clonality. Unlike the original studies of G6PD, which measured the activity or structure of a protein, HPRT and PGK and several other X-chromosome inactivation assays are based on the well-recognized association between methylation patterns and gene activation status. (The inactive X chromosome is heavily methylated during lyonization, and, more generally, gene expression is inversely correlated with the degree of methylation.) Differences in methylation status can be measured by using restriction enzymes in order to detect methylation-dependent restriction fragment-length polymorphisms [18].

Yet another useful tool for determining clonality which has been used in studying ET and related disorders involves the measurement of variations in mRNA transcript expression from X-associated polymorphic genes using the reverse-transcriptase polymerase chain reaction. Three genes that have been widely used for this sort of analysis are iduronate-2-sulphatase *(IDS)*, the palmitoylated membrane protein p55 *(p55)*, and *G6PD* (measured at the gene level instead of the protein level) [19].

Recently, however, it has become clear that although the study of X-linked DNA and gene products provides powerful tools for analyzing clonality, such techniques have significant limitations [20]. For example, G6PD-based analysis of clonality is of no value if a woman has inherited two identical G6PD alleles, which is not an uncommon scenario. And, of course, X-linked gene analysis is of no value for studying clonality in men, who must make do with only one X chromosome. Making matters worse, the phenomenon of 'skewing' in normal persons makes interpretation of X chromosomal inactivation patterns problematical [21]. X-linked gene expression is customarily said to be 'skewed' if 75% or more of cells in a particular population express a single allele (i.e. an allele ratio of ≥3:1) [22]. Of course, this number is completely arbitrary, and other investigators have used a cutoff of 90% to define skewing, sacrificing sensitivity for the sake of specificity [18].

Using either cutoff, significant skewing appears to occur in a proportion of apparently hematologically normal females [23, 24]. It has been speculated that such 'normal skewing' is due to the small size of the stem cell pool during early embryonic development, at the time lyonization occurs [25]. But extreme skewing can also be acquired with increasing age, and normal elderly women often have evidence of 'clonal' hematopoiesis [22, 26]. The incidence of this finding is controversial, and estimates have ranged from less than 5% to more than 30% [24].

When the clonality of myeloid cells is in question, X-chromosome inactivation patterns in T cells can be used as a control to correct for normal skewing [27]. However, in one recent analysis, a clonal granulocyte inactivation pattern was detected in 23% of normal elderly women who had polyclonal T cells [26]. In the CMPDs, T cells may still be of value as a control assay, as they are generally (but not always) polyclonal [28]. Using such techniques, clonal X inactivation patterns in myeloid cells can be detected in 63–82% of patients with CMPDs [22, 26].

One relatively new X-associated technique takes advantage of the human androgen receptor gene *(HUMARA)*, localized to Xq11–12 and characterized by a highly polymorphic trinucleotide CAG repeat in the first of the eight exons of the gene [29]. Heterozygosity at this locus has been reported in more than 90% of white women and in 67% of Japanese women [30]. *HUMARA* analysis exploits the close proximity of this highly polymorphic repeat to several methylation sites, which has formed the basis for a polymerase chain reaction (PCR)-based analysis of clonality [31, 32]. Early versions of this PCR assay had some technical limitations, and DNA that was scanty or damaged yielded biased and inaccurate results [33]. A newer *HUMARA* assay, which analyzes the methylation status of the gene by bisulfite modification instead of via a methylation-sensitive restriction enzyme, may offer improved test characteristics [34, 35].

Autosomal chromosomal mutations offer a way to escape the limitations of X chromosome inactivation analysis, and, when present, such mutations offer an unambiguous marker of clonality that can be detected by conventional karyotyping and/or fluorescence in situ hybridization techniques [36]. Unfortunately, autosomal chromosomal abnormalities such as the chromosomal translocations and deletions characteristic of other myeloid disorders are actually quite rare in ET, occurring in less than 10% of cases [37]. For this reason, X-associated gene and protein techniques have been the primary tools used in ET in an attempt to better understand clonality patterns, despite the limitations described above.

Studies using X-linked gene techniques such as *HUMARA* polymorphism assays are conflicting and demonstrate a much more heterogeneous picture of ET than was previously recognized. Using *HUMARA*, a clonal pattern of hematopoiesis was detected in unfractionated nucleated blood cells from 22 of 32 ET patients (69%) in one analysis. In the same study, G6PD transcript analysis of purified platelet preparations showed clonal megakaryocytopoiesis in 2 of the 10 patients who had a polyclonal pattern in unfractionated cells [31]. The same group of investigators recently extended their observations to 53 ET patients, and a high proportion of monoclonality continued to be found, with several more patients detected in whom monoclonality was restricted to the megakaryocyte lineage [38]. With rare exceptions, single-lineage restriction of clonality has not been observed in other myeloid disorders [19].

In contrast to the above analyses, in a high-profile study of 46 patients classified as having ET on the basis of the Polycythemia Vera Study Group diagnostic criteria [39], *HUMARA* analysis and other X-chromosomal inactivation assays (i.e. *G6PD*, *PGK*, *p55*, and *IDS*) revealed an uninterpretable pattern in 23 patients (50%), whereas clonal hematopoiesis was demonstrable in only 10 patients (22%) [19]. The patients with a polyclonal pattern had less thrombotic risk, which is important to recognize because thrombohemorrhagic complications are one of the main clinical problems associated with ET and the possibility of their occurring is a chief reason for initiating drug therapy [1, 40–44]. Others have confirmed this clinical correlation between clonality and thrombotic risk [36]. These studies have raised serious doubts on the previous understanding of ET as a universally clonal disorder (fig. 1). The considerable variation in clinical behavior of patients with ET may eventually be explained as a better understanding of the heterogeneous biological characteristics of the disorder emerges [19, 45, 46].

### *BCR-ABL*-Positive ET: Does It Exist?

The presence of the Philadelphia chromosome (a reciprocal translocation between the long arms of chromosomes 9 and 22) or its molecular equivalent, the *BCR-ABL* fusion gene, is considered the diagnostic hallmark of

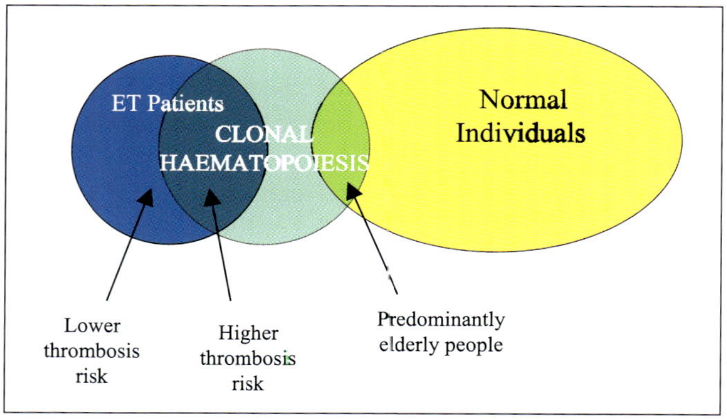


**Fig. 1.** Clonality and disease are not synonymous. Apparently, clonal hematopoiesis may be observed in healthy individuals, especially among the elderly, whereas new data suggest that not all cases of myeloproliferative disorders such as essential thrombocythemia are clonal.

CML. Although the *BCR-ABL* fusion is also frequently observed in cases of acute leukemia, the presence of *BCR-ABL* in a patient with a more indolent myeloid disorder would prompt most pathologists and clinicians to assign the diagnosis of CML, regardless of atypical clinical features. There is not yet universal agreement on this point, however. Some investigators maintain that, in the right clinical setting, the diagnosis of ET or another non-CML myeloid disorder must be considered despite the presence of the *BCR-ABL* fusion [47, 48].

Less than 30 well-characterized cases of 'Philadelphia-positive ET' have been reported [47–55]. When the literature on this putative entity was reviewed in 1996, 10 of 16 cases previously described were noted to have terminated in a leukemic phase similar to CML blast crisis [47]. This is a much higher rate of leukemic transformation than would be expected with ET in general, but it is consistent with the usual clinical course of CML [37]. Cases of 'Philadelphia-positive ET' have been considered by some to be distinct from CML on the basis of an apparent female predilection for the former – this sex bias is more typical of ET than of CML, although with so few cases of Philadelphia-positive ET it is rather difficult to trust the numbers – and the absence of typical CML-associated features such as splenomegaly, striking leukocytosis, or marked anemia and basophilia [47]. The degree of thrombocytosis in the presence of the Philadelphia chromosome has in some cases been extreme, despite a relatively modest white blood cell count; platelet counts of $5{,}000 \times 10^9/l$ have been reported [56].

But such considerations would seem to reflect the considerable overlap of the clinical and pathological features of the chronic myeloid disorders as well as the heterogeneity of these syndromes rather than marking a special, rare subtype of ET [57]. Extreme thrombocytosis, for example, has been reported in all of the myeloproliferative conditions and is not limited to ET [58]. A low leukocyte alkaline phosphatase score, characteristic of CML, has also been observed in about 10% of cases classifiable as ET [39]. And, of course, leukemic transformation can occur in cases of ET or any myeloproliferative condition, not just CML. In most cases of apparent ET in which the Philadelphia chromosome or a *BCR* rearrangement has been detected, at least some morphological features of CML, such as a mild basophilia and a slight leukocytosis, can be appreciated. In recognition of the initial resemblance of Philadelphia-positive thrombocytosis syndromes to ET but a behavior more like CML, one group has suggested the term 'CML of thrombocythemic onset' to describe such cases [59].

Because ET is a diagnosis of exclusion, the authors of this review consider the presence of the *BCR-ABL* translocation to be diagnostic of CML, even if clinical and pathological features would otherwise be consistent with ET. Many others, including the Polycythemia Vera Study Group, also hold to this position, but there is still some dissent [39, 47, 56]. No data are available yet on the response of such patients to imatinib mesylate (Gleevec; Novartis AG, Basel, Switzerland), a highly specific inhibitor of the aberrant tyrosine kinase encoded by the *BCR-ABL* fusion gene which has dramatic effects in CML [60, 61]. Until larger series are reported and further genetic information is available (e.g. from microarray-based global gene expression profiling, which is likely to prove a very useful tool in disease subclass distinction), there is no convincing reason why *BCR-ABL*-positive cases resembling ET should not be considered to have chronic-phase CML and be treated accordingly.

## Thrombopoietin and c-Mpl Dynamics and Insights from Familial Thrombocythemia

*Thrombopoietin and c-Mpl*

Thrombopoietin (TPO) and its receptor, c-Mpl, appear to be the key regulators of megakaryocytopoiesis in mice and in humans [62]. Since ET has a phenotype of dysregulated megakaryocytopoiesis, interest has focused on the dynamics of the TPO-Mpl ligand-receptor pair in ET in an attempt to achieve some insight into the pathogenesis of the disease and perhaps help with the difficult distinction of ET from reactive thrombocytosis [63].

Serum TPO levels have been demonstrated to be inversely proportional to both the peripheral blood platelet count and the bone marrow megakaryocyte mass in normal persons, and they appear to form a feedback loop, similar to that by which erythropoietin regulates the erythroid mass [64, 65]. The megakaryocyte mass seems to be a key determinant of serum TPO. In aplastic anemia, for instance, in which the megakaryocyte number is low and the platelet count is low, TPO levels are high, but in immune thrombocytopenic purpura, in which the megakaryocyte number is high (but the platelet count is still low), TPO levels are low [66].

When the first report appeared suggesting that TPO levels were high in ET in spite of the characteristic elevated platelet count and megakaryocyte mass, much interest was aroused because dysregulation of the TPO-Mpl signaling cascade appeared to be implicated in the pathogenesis of ET [67, 68]. Several potential explanations of the inappropriately elevated TPO levels in ET were suggested. These included ineffective TPO clearance because of markedly reduced c-Mpl expression in platelets and megakaryocytes (described in more detail below) [46, 64, 69–71], increased bone marrow stromal production of TPO [72, 73], and an alteration in the rate of TPO catabolism [74]. However, subsequent analyses have shown that although there may indeed be a general trend toward higher serum TPO levels in ET when compared with reactive thrombocytosis, there is much overlap between ET and secondary states, and serum TPO levels have not proven to be diagnostically useful [75–78]. In some analyses, there has also been no difference between TPO levels in normal persons and ET patients [69]. Clearly, TPO alterations cannot account for all cases of ET.

Although c-Mpl too has failed to live up to initial hopes, c-Mpl expression has also been studied in ET and may be more promising diagnostically than TPO, at least when used as a complement to morphological assessment. In ET, expression of megakaryocyte and platelet c-Mpl is generally decreased in comparison with that found in normal persons and in patients with reactive thrombocytosis [45, 46, 69, 79]. Not all investigators agree, however; some have reported normal c-Mpl expression in ET [80].

In one large bone marrow immunohistochemical study, megakaryocyte c-Mpl expression in 169 patients with ET was contrasted with that in 19 patients with reactive thrombocytosis; markedly decreased megakaryocyte c-Mpl expression was demonstrated in 68% of the patients with ET but in none of those with reactive thrombocytosis [81]. Although substantial overlap in platelet c-Mpl expression between ET and reactive states suggests that measurement of platelet c-Mpl will not be diagnostically useful [45], megakaryocyte c-Mpl expression, when assessed by immunohistochemistry, may be a helpful complement to the histological distinction between ET and reactive thrombocytosis [46, 82, 83]. Of course, c-Mpl cannot help in distinguishing between ET and other CMPDs, because there is great variability in c-Mpl expression in the myeloid disorders, and c-Mpl expression and activation also appear to be impaired commonly in polycythemia vera [46, 80].

In addition to insights from measurements of expression levels, the fact that the thrombocytosis in ET may be TPO independent but c-Mpl dependent was also suggested by the inhibition of autonomous megakaryocyte growth by soluble c-Mpl receptors but not by antibodies to TPO [84]. Another experiment showed that autonomous megakaryocyte growth in ET diminishes in the presence of antisense oligonucleotides directed at c-Mpl RNA [85]. Nevertheless, the lack of sensitivity and specificity of circulating TPO concentrations and c-Mpl expression patterns suggests that these findings may actually not be pathogenetically relevant and may perhaps be mere epiphenomena [45, 46, 80, 86–89]. Much more work needs to be done in order to clarify these complex relationships.

*Familial Thrombocythemia*

A rare familial syndrome resembling ET exists – more than a dozen affected families have been reported – but the bone marrow karyotype in such patients has generally been normal [90–93]. Recently, several mutations in the 5′ untranslated region of the *TPO* gene have been shown to be responsible for autosomal-dominant familial thrombocythemia in two Japanese families and a Dutch family [93–95]. The three described mutations have included a single base deletion that affects the 5′ untranslated region of the *TPO* mRNA, a G→T point mutation at position 516 in exon 3 (also encoding for a portion of the 5′ untranslated region of *TPO* mRNA), and a nearby G→C point mutation in the splice donor site of intron 3 which results in skipping of the translation of exon 3 [93–95]. These mutations result in increased thrombopoietin synthesis, apparently from a loss of translational inhibition of *TPO* mRNA [96]. Transfection of one of the mutant genes into cell lines resulted in an eightfold increase in TPO protein compared with the wild-type gene [93].

This analysis has provided a clearer understanding of *TPO* dynamics at the genetic level. It appears that effi-

**Table 1.** A compendium of reported chromosomal abnormalities in ET

| Chromosome | Karyotype | Reference |
|---|---|---|
| 1 | 46,XY,dup(1)(q21q32) | 100 |
| | 46,XY,dup(1)(q11q42) | 101 |
| | 46,XX,dup(1)(q11q44) | 102 |
| | 46,XY,t(1;3)(p36;q21) | 103 |
| | 46,XX,+1,der(1;7)(q10;p10)/47,idem,+21 | 104 |
| | 46,XX,+der(1)t(1;7)(p11;p11),–7 | 105 |
| | 46,XX,t(1;17)/47,idem,+8 | 106 |
| 2 | 46,XX,del(2)(q21q23) | 101 |
| | 46,XY,t(2;3)(q23;q25) | 107 |
| | 46,XX,t(2;9)(q?;p?) | 108 |
| 3 | 46,XX,inv(3)(q21q26) | 109 |
| | 46,XY,t(1;3)(p36;q21) | 103 |
| | 46,XY,t(2;3)(q23;q25) | 107 |
| 4 | no reports | |
| 5 | 46,XY,del(5)(q14q22) | 110 |
| | 46,XX,del(5)(q?) | 111 |
| | 46,XX,del(5)(q22q35) | 112 |
| | 46,XX,del(5)(q15q35) | 113 |
| | –46,X,t(X;5)(q13;q33) | 114 |
| 6 | 46,XX,t(6;9)(q21;q34) | 115 |
| 7 | 46,XY,del(7)(q11q22)/45,X,–Y | 116 |
| | 46,XY,del(7)(q21q31) | 117 |
| | 46,XX,+1,der(1;7)(q10;p10)/47,idem,+21 | 104 |
| | 46,XX,+der(1)t(1;7)(p11;p11),7,+21,der(21;21) | 118 |
| | 46,XX,+der(1)t(1;7)(p11;p11),–7 | 105 |
| 8 | 47,XX,+8/47,XX,+9 | 119 |
| | 46,XY,inv(8)(p23q13) | 115 |
| | 47,XX,+8,t(9;22) (2 cases) | 37 |
| | 45,XX,–8 | 106 |
| | 46,XX,t(1;17)/47,idem,+8 | 106 |
| | 47,XX,+8 | 120 |
| 9 | 47,XX,+8/47,XX,+9 | 119 |
| | 47,XX,+9 (2 cases) | 119 |
| | 47,XY,+9 | 119 |
| | 46,XX,del(9) | 101 |
| | 46,XY,inv(9)(p13q21)c,t(13;14)(q32;q32) | 121 |
| | 46,XX,t(2;9)(q?;p?) | 108 |
| | 46,XX,t(6;9)(q21;q34) | 115 |
| | 46,XX,t(9;15;22)(q34;q26;q11-13) | 122 |
| | 46,XX or XY,t(9;22)(q34;q11) | many reports (e.g. 48, 49) |
| 10 | 47,XX,+10 | 123 |
| 11 | 46,XY,del(11)(q21) | 124 |
| | 46,XX,t(11;21)(q25;q11) | 125 |
| | 46,XX,t(11;21)(q25;q21) | 126 |
| 12 | 46,XX,ins(12;5)(q13;q13q31) | 127 |
| 13 | 46,XY,del(13)(q21) | 128 |
| | 46,XY,del(13)(q12q14),del(20)(q11q13),+0–1mar | 129 |
| | 46,XX,del(13)(q13q22) | 37 |
| | 46,XY,inv(9)(p13q21)c,t(13;14)(q32;q32) | 121 |
| 14 | 46,XY,inv(9)(p13q21)c,t(13;14)(q32;q32) | 121 |
| 15 | 46,XX,t(9;15;22)(q34;q26;q11-13) | 122 |
| 16 | 46,XX,del(16)(q?) | 108 |
| 17 | 46,XY,del(17)(q24q25) | 130 |
| | 46,XX,del(17)(q24q25) | 130 |
| | 46,XY,i(17)(q10) | 128 |
| | 42–45,XY,–21/40–45,XY,–17/40–45,X,–Y | 106 |
| | 46,XX,t(1;17)/47,idem,+8 | 106 |
| 18 | no reports | |
| 19 | no reports | |
| 20 | 46,XY,del(20)(q11) | 131 |
| | 46,XX,del(20)(q11) | 132 |
| | 46,XY,del(20)(q11q13) | 130 |
| | 46,XX,ins(12;5),del(20)(q11q13) | 127 |
| | 46,XY,del(13)(q12q14),del(20)(q11q13),+0–1mar | 129 |
| | 45,X,–Y,–20,+mar | 101 |
| | 46,XX,del(20)(q13) | 128 |
| | 46,XY,del(20)(q11q13) | 133 |
| | 46,XY,del(20)(q11) | 115 |
| | 45,XX,–20 | 134 |
| 21 | 46,XX,+der(1)t(1;7),–7,+21,der(21;21)(q10;q10) | 118 |
| | 46,XX,i(21)(q10) | 135 |
| | 47,XX,+21,i(21)(q10)x2 | 135 |
| | 46,XX,+1,der(1;7)(q10;p10)/47,idem,+21 | 104 |
| | 42–45,XY,–21/40–45,XY,–17/40–45,X,–Y | 106 |
| | 46,XX,t(11;21)(q25;q21) | 126 |
| | 46,XX,t(11;21)(q25;q11) | 125 |
| 22 | 46,XX,t(9;15;22)(q34;q26;q11–13) | 122 |
| | 46,XX or XY, t(9;22)(q34;q11) | many reports (e.g. 48, 49) |
| X and Y | 46,XY,del(7)(q11q22)/45,X,–Y | 116 |
| | 45,X,–Y,–20,+mar | 101 |
| | 42–45,XY,–21/40–45,XY,–17/40–45,X,–Y | 106 |
| | 46,X,t(X;5)(q13;q33) | 114 |
| | 46,X,t(X;9;22)(q11;q34;q11) | 52 |
| Hyperdiploid | 177–183,XY?,inc | 136 |
| Unknown | 46,XX,add(2)(q?) | 137 |
| | 46,XY,del(13)(q12q14),del(20)(q11q13),+0–1mar | 129 |
| | 45,X,–Y,–20,+mar | 101 |
| | 48,XX,+2mar | 105 |

Derived from 'Mitelman Database of Chromosome Aberrations in Cancer (2002). Mitelman F, Johansson B, Mertens F (eds), http://cgap.nci.nih.gov/Chromosomes/Mitelman.' Accessed January 21, 2002.

Some patients with chromosomal aberrations involving more than one chromosome are listed more than once in this table (i.e. translocations are listed with each involved partner chromosome).

When breakpoints are not given, that is because they were not provided in the original article.

'Unknown' includes cases with additional genetic material and/or 'marker' chromosomes of uncertain provenance.

The veracity of the ET diagnosis cannot be determined in all cases (see especially the text discussion of the *BCR-ABL* gene in putative ET).

Nomenclature is that of the Standard International System for Human Cytogenetic Nomenclature (ISCN) [ISCN: In Mitelman F (ed): An International System for Human Cytogenetic Nomenclature. Basel, Karger, 1995].

cient translation of *TPO* mRNA may normally be inhibited by repeated elements (seven copies of the AUG codon) in the 5′-untranslated region of the gene; mutations that change the reading frame result in message truncation and release of this inhibition [97].

Other modes of inheritance for familial thrombocytosis which are not TPO dependent appear to exist. *TPO* mutations were not detected in an Arab family who demonstrated familial thrombocythemia with a possibly X-linked inheritance pattern [98]. Of course, familial thrombocythemia represents a very small fraction of patients with primary thrombocytosis, and insights from familial cases will be of narrow interest unless they can be translated into a better understanding of sporadic, acquired ET. To date, mutations in the *TPO* gene have not yet been detected in patients with acquired ET [65]. An initial search for c-Mpl mutations in patients with sporadic ET was also unrevealing [99].

### Cytogenetic Abnormalities Are Uncommon and Variable in ET

Karyotypic abnormalities in ET are uncommon [37]. Fewer than 10% of patients with ET have detectable cytogenetic abnormalities at diagnosis; karyotypic abnormalities generally arise only upon transformation of ET to acute leukemia [37]. No consistent genetic marker for ET exists, and many of the clonal chromosomal lesions that have been observed during the preleukemic phase of ET (e.g. trisomy 8 and deletions of the long arm of chromosome 20) are also observed in other CMPDs. For the majority of these karyotypic abnormalities, the gene or genes responsible for the clinical phenotype in ET and the other CMPDs are not yet known.

Case reports have linked ET with an aberration of nearly every chromosome (table 1). Occasional abnormalities that have been observed in multiple cases, such as duplication of the long arm of chromosome 1 or trisomy 9, are provocative but remain mysterious. Most ET karyotypes have been diploid or only mildly aneuploid, but a case of a near-octaploid karyotype associated with ET was reported in an elderly man and is of uncertain significance [136]. A listing of a large number of cytogenetic lesions associated with ET is provided in table 1.

### The Promise of Microarray Technology: A Global Approach

The recent completion of a draft sequence of the entire human genome heralds a new era in genetic analysis [138, 139]. Until very recently, the expression of only a small number of the approximately 30,000 human genes could be analyzed at one time. But the advent of nucleotide microarray technology (so-called gene chips) permits global analysis of gene expression and function and has the potential to profoundly change how gene function analysis is done [140]. Now, thousands of gene transcripts can be evaluated in a single experiment, and as the complexity of information increases, new patterns begin to emerge [141].

Microarray technology, which has arisen because of the fortunate conjunction of major advances in genome biotechnology with advances in the computer and semiconductor industries, offers great potential for disease class discovery and class distinction. It is likely that current disease classification schema will need to be revised extensively in the coming years on the basis of global profiling results. A number of reviews of microarray techniques and the promise and problems of the new technology have been published [142, 143].

At the time of this writing, no analysis of global gene expression patterns in ET has yet been published. It will be of interest to compare ET mRNA expression profiles with those from normal persons, from patients with other CMPDs, and from patients with apparent secondary thrombocytosis. It may also be instructive to compare polyclonal and clonal ET cases to see whether any differences can be observed. Whether profiling of gene expression specifically in the megakaryocyte-platelet lineage will offer any insights beyond a broader analysis of myeloid cells remains to be determined.

Much has been made of the current problems of microarray technology, such as the difficulty in choosing appropriate comparison groups, the high variability in expression profiles among experiments, the challenge of selecting the appropriate tissue type to study, and the statistical nightmare created by the plethora of data generated [143]. The cost of microarray analysis is also currently prohibitive. Of course, genomic expression profiles also reveal nothing about translation efficiency or the status of post-translational modifications, of which there are myriad types with varying biological relevance. Although global proteomic assays have not yet come as far as global assessments of nucleotide expression, the technology is evolving rapidly [144, 145]. All these problems are certainly real, but the relevant techniques are still in their infancy and will likely be refined in the years to come.

## Summary

Much remains to be learned about the genetics of ET. The recent discovery of 'nonclonal' ET cases challenges the current understanding not only of ET but of other CMPDs as well. In addition, the dynamic TPO-Mpl thrombopoietic regulatory system is just beginning to be understood. Ongoing studies, including microarray analysis of gene expression and protein and global proteomic analysis, will undoubtedly result in more refined diagnostic criteria and perhaps a better, more biologically relevant classification of ET and other CMPDs. This, in turn, should allow more accurate prognostication and tailoring of treatment to individual patients.

## References


1 Murphy S, Peterson P, Iland H, Laszlo J: Experience of the Polycythemia Vera Study Group with essential thrombocythemia: A final report on diagnostic criteria, survival, and leukemic transition by treatment. Semin Hematol 1997;34:29–39.

2 Tefferi A, Murphy S: Current opinion in essential thrombocythemia: Pathogenesis, diagnosis, and management. Blood Rev 2001;15:121–131.

3 Tefferi A: Chronic myeloid disorders: Classification and treatment overview. Semin Hematol 2001;38(suppl 2):1–4.

4 Griesshammer M, Bangerter M, Sauer T, Wennauer R, Bergmann L, Heimpel H: Aetiology and clinical significance of thrombocytosis: Analysis of 732 patients with an elevated platelet count. J Intern Med 1999;245:295–300.

5 Buss DH, Cashell AW, O'Connor ML, Richards F II, Case LD: Occurrence, etiology, and clinical significance of extreme thrombocytosis: A study of 280 cases. Am J Med 1994;96:247–253.

6 Schafer AI: Thrombocytosis and thrombocythemia. Blood Rev 2001;15:159–166.

7 Custodi P, Cerutti A, Balduini CL: Which tests are most useful to distinguish between clonal and reactive thrombocytosis. Am J Med 1996;101:233–235.

8 Tefferi A: Reply – Letter to the Editor: Which tests are most useful to distinguish between clonal and reactive thrombocytosis. Am J Med 1996;101:235.

9 Tefferi A, Hoagland HC: Issues in the diagnosis and management of essential thrombocythemia. Mayo Clin Proc 1994;69:651–655.

10 Nimer SD: Essential thrombocythemia: Another 'heterogeneous disease' better understood? Blood 1999;93:415–416.

11 Murphy S: Diagnostic criteria and prognosis in polycythemia vera and essential thrombocythemia. Semin Hematol 1999;36(suppl 2):9–13.

12 Raskind WH, Steinmann L, Najfeld V: Clonal development of myeloproliferative disorders: Clues to hematopoietic differentiation and multistep pathogenesis of cancer. Leukemia 1998;12:108–116.

13 Dameshek W: Some speculations on the myeloproliferative syndromes (editorial). Blood 1951;6:372–375.

14 Watson JD, Crick FH: Molecular structure of nucleic acids: A structure for deoxyribose nucleic acid. Nature 1953;171:737–738.

15 Tijo JH, Levan A: The chromosome number in man. Hereditas 1956;42:1–6.

16 Fialkow PJ, Faguet GB, Jacobson RJ, Vaidya K, Murphy S: Evidence that essential thrombocythemia is a clonal disorder with origin in a multipotent stem cell. Blood 1981;58:916–919.

17 Fialkow PJ, Jacobson RJ, Papayannopoulou T: Chronic myelocytic leukemia: Clonal origin in a stem cell common to the granulocyte, erythrocyte, platelet and monocyte/macrophage. Am J Med 1977;63:125–130.

18 Racchi O, Mangerini R, Rapezzi D, Rolfo M, Gaetani GF, Ferraris AM: X chromosome inactivation patterns in normal females. Blood Cells Mol Dis 1998;24:439–447.

19 Harrison CN, Gale RE, Machin SJ, Linch DC: A large proportion of patients with a diagnosis of essential thrombocythemia do not have a clonal disorder and may be at lower risk of thrombotic complications. Blood 1999;93:417–424.

20 el-Kassar N, Hetet G, Briere J, Grandchamp B: Clonality analysis of hematopoiesis in essential thrombocythemia: Advantages of studying T lymphocytes and platelets. Blood 1997;89:128–134.

21 Gale RE: Evaluation of clonality in myeloid stem-cell disorders. Semin Hematol 1999;36:361–372.

22 Mitterbauer G, Winkler K, Gisslinger H, Geissler K, Lechner K, Mannhalter C: Clonality analysis using X-chromosome inactivation at the human androgen receptor gene *(HUMARA)*: Evaluation of large cohorts of patients with chronic myeloproliferative diseases, secondary neutrophilia, and reactive thrombocytosis. Am J Clin Pathol 1999;112:93–100.

23 Fey MF, Liechti-Gallati S, von Rohr A, Borisch B, Theilkas L, Schneider V, Oestreicher M, Nagel S, Ziemiecki A, Tobler A: Clonality and X-inactivation patterns in hematopoietic cell populations detected by the highly informative M27 beta DNA probe. Blood 1994;83:931–938.

24 Busque L, Mio R, Mattioli J, Brais E, Blais N, Lalonde Y, Maragh M, Gilliland DG: Nonrandom X-inactivation patterns in normal females: Lyonization ratios vary with age. Blood 1996;88:59–65.

25 Gale RE, Linch DC: Clonality studies in acute myeloid leukemia. Leukemia 1998;12:117–120.

26 Champion KM, Gilbert JG, Asimakopoulos FA, Hinshelwood S, Green AR: Clonal haemopoiesis in normal elderly women: Implications for the myeloproliferative disorders and myelodysplastic syndromes. Br J Haematol 1997;97:920–926.

27 Briere J, el-Kassar N: Clonality markers in polycythaemia and primary thrombocythaemia. Baillières Clin Haematol 1998;11:787–801.

28 Tsukamoto N, Morita K, Maehara T, Okamoto K, Sakai H, Karasawa M, Naruse T, Omine M: Clonality in chronic myeloproliferative disorders defined by X-chromosome linked probes: Demonstration of heterogeneity in lineage involvement. Br J Haematol 1994;86:253–258.

29 Brown CJ, Goss SJ, Lubahn DB, Joseph DR, Wilson EM, French FS, Willard HF: Androgen receptor locus on the human X chromosome: Regional localization to Xq11–12 and description of a DNA polymorphism. Am J Hum Genet 1989;44:264–269.

30 Okamoto T, Okada M, Wada H, Kanamaru A, Kakishita E, Hashimoto T, Furuyama J: Clonal analysis of hematopoietic cells using a novel polymorphic site of the X chromosome. Am J Hematol 1998;58:263–266.

31 el Kassar N, Hetet G, Li Y, Briere J, Grandchamp B: Clonal analysis of haemopoietic cells in essential thrombocythaemia. Br J Haematol 1995;90:131–137.

32 Kopp P, Jaggi R, Tobler A, Borisch B, Oestreicher M, Sabacan L, Jameson JL, Fey MF: Clonal X-inactivation analysis of human tumours using the human androgen receptor gene *(HUMARA)* polymorphism: A non-radioactive and semiquantitative strategy applicable to fresh and archival tissue. Mol Cell Probes 1997;11:217–228.

33 Mutter GL, Boynton KA: PCR bias in amplification of androgen receptor alleles, a trinucleotide repeat marker used in clonality studies. Nucl Acids Res 1995;23:1411–1418.
34 Kubota T, Nonoyama S, Tonoki H, Masuno M, Imaizumi K, Kojima M, Wakui K, Shimadzu M, Fukushima Y: A new assay for the analysis of X-chromosome inactivation based on methylation-specific PCR. Hum Genet 1999; 104:49–55.
35 Uchida T, Ohashi H, Aoki E, Nakahara Y, Hotta T, Murate T, Saito H, Kinoshita T: Clonality analysis by methylation-specific PCR for the human androgen-receptor gene *(HUMARA-MSP)*. Leukemia 2000;14:207–212.
36 Elis A, Amiel A, Manor Y, Tangi I, Fejgin M, Lishner M: The detection of trisomies 8 and 9 in patients with essential thrombocytosis by fluorescence in situ hybridization. Cancer Genet Cytogenet 1996;92:14–17.
37 Sessarego M, Defferrari R, Dejana AM, Rebuttato AM, Fugazza G, Salvidio E, Ajmar F: Cytogenetic analysis in essential thrombocythemia at diagnosis and at transformation: A 12-year study. Cancer Genet Cytogenet 1989; 43:57–65.
38 El-Kassar N, Hetet G, Briere J, Grandchamp B: Clonality analysis of hematopoiesis and thrombopoietin levels in patients with essential thrombocythemia. Leuk Lymphoma 1998;30: 181–188.
39 Murphy S, Iland H, Rosenthal D, Laszlo J: Essential thrombocythemia: An interim report from the Polycythemia Vera Study Group. Semin Hematol 1986;23:177–182.
40 Colombi M, Radaelli F, Zocchi L, Maiolo AT: Thrombotic and hemorrhagic complications in essential thrombocythemia: A retrospective study of 103 patients. Cancer 1991;67:2926–2930.
41 Buss DH, Stuart JJ, Lipscomb GE: The incidence of thrombotic and hemorrhagic disorders in association with extreme thrombocytosis: An analysis of 129 cases. Am J Hematol 1985;20:365–372.
42 Fenaux P, Simon M, Caulier MT, Lai JL, Goudemand J, Bauters F: Clinical course of essential thrombocythemia in 147 cases. Cancer 1990;66:549–556.
43 Besses C, Cervantes F, Pereira A, Florensa L, Sole F, Hernandez-Boluda JC, Woessner S, Sans-Sabrafen J, Rozman C, Montserrat E: Major vascular complications in essential thrombocythemia: A study of the predictive factors in a series of 148 patients. Leukemia 1999;13:150–154.
44 Tefferi A, Fonseca R, Pereira DL, Hoagland HC: A long-term retrospective study of young women with essential thrombocythemia. Mayo Clin Proc 2001;76:22–28.
45 Harrison CN, Gale RE, Pezella F, Mire-Sluis A, MacHin SJ, Linch DC: Platelet c-mpl expression is dysregulated in patients with essential thrombocythaemia but this is not of diagnostic value. Br J Haematol 1999;107:139–147.
46 Yoon SY, Li CY, Tefferi A: Megakaryocyte c-Mpl expression in chronic myeloproliferative disorders and the myelodysplastic syndrome: Immunoperoxidase staining patterns and clinical correlates. Eur J Haematol 2000;65:170–174.
47 Kwong YL, Chiu EK, Liang RH, Chan V, Chan TK: Essential thrombocythemia with *BCR/ABL* rearrangement. Cancer Genet Cytogenet 1996;89:74–76.
48 LeBrun DP, Pinkerton PH, Sheridan BL, Chen-Lai J, Dube ID, Poldre PA: Essential thrombocythemia with the Philadelphia chromosome and *BCR-ABL* gene rearrangement: An entity distinct from chronic myeloid leukemia and Philadelphia chromosome-negative essential thrombocythemia. Cancer Genet Cytogenet 1991;54:21–25.
49 Martiat P, Ifrah N, Rassool F, Morgan G, Giles F, Gow J, Goldman JM: Molecular analysis of Philadelphia positive essential thrombocythemia. Leukemia 1989;3:563–565.
50 Michiels JJ, Prins ME, Hagermeijer A, Brederoo P, van der Meulen J, van Vliet HH, Abels J: Philadelphia chromosome-positive thrombocythemia and megakaryoblast leukemia. Am J Clin Pathol 1987;88:645–652.
51 Verhest A, Monsieur R: Philadelphia chromosome-positive thrombocythemia with leukemic transformation. N Engl J Med 1983;308:1603.
52 Morris CM, Fitzgerald PH, Hollings PE, Archer SA, Rosman I, Beard ME, Heaton DC, Newhook CJ: Essential thrombocythaemia and the Philadelphia chromosome. Br J Haematol 1988;70:13–19.
53 Stoll DB, Peterson P, Exten R, Laszlo J, Pisciotta AV, Ellis JT, White P, Vaidya K, Bozdech M, Murphy S: Clinical presentation and natural history of patients with essential thrombocythemia and the Philadelphia chromosome. Am J Hematol 1988;27:77–83.
54 Tesch H, Zankovich R, Fonatsch C, Diehl V: Analysis of *bcr* rearrangements in primary (essential) thrombocythaemia. Leuk Res 1990;14: 241–246.
55 Richards EM, Bloxham DM, Nacheva E, Marcus RE, Green AR: *BCR* rearrangement in apparent essential thrombocythaemia. Br J Haematol 1993;85:625–626.
56 Cervantes F, Urbano-Ispizua A, Villamor N, Feliu E, Milla F, Lopez-Guillermo A, Ribera J-M, Granada I, Vives-Corrons J-L, Montserrat E, Rozman C: Ph-positive chronic myeloid leukemia mimicking essential thrombocythemia and terminating into megakaryoblastic blast crisis: Report of two cases with molecular studies. Leukemia 1993;7:327–330.
57 Tefferi A: The Philadelphia chromosome negative chronic myeloproliferative disorders: A practical overview. Mayo Clin Proc 1998;73: 1177–1184.
58 Thiele J, Moedder B, Kremer B, Zankovich R, Fischer R: Chronic myeloproliferative diseases with an elevated platelet count (in excess of 1,000,000/μl): A clinicopathological study on 46 patients with special emphasis on primary (essential) thrombocythemia. Hematol Pathol 1987;1:227–237.
59 Cervantes F, Colomer D, Vives-Corrons JL, Rozman C, Montserrat E: Chronic myeloid leukemia of thrombocythemic onset: A CML subtype with distinct hematological and molecular features? Leukemia 1996;10:1241–1243.
60 Druker BJ, Talpaz M, Resta DJ, Peng B, Buchdunger E, Ford JM, Lydon NB, Kantarjian H, Capdeville R, Ohno-Jones S, Sawyers CL: Efficacy and safety of a specific inhibitor of the *BCR-ABL* tyrosine kinase in chronic myeloid leukemia. N Engl J Med 2001;344:1031–1037.
61 Druker BJ, Sawyers CL, Kantarjian H, Resta DJ, Reese SF, Ford JM, Capdeville R, Talpaz M: Activity of a specific inhibitor of the *BCR-ABL* tyrosine kinase in the blast crisis of chronic myeloid leukemia and acute lymphoblastic leukemia with the Philadelphia chromosome. N Engl J Med 2001;344:1038–1042.
62 Gurney AL, de Sauvage FJ: Dissection of c-Mpl and thrombopoietin function: Studies of knockout mice and receptor signal transduction. Stem Cells 1996;14(suppl 1):116–123.
63 Matsumura I, Horikawa Y, Kanakura Y: Functional roles of thrombopoietin-c-mpl system in essential thrombocythemia. Leuk Lymphoma 1999;32:351–358.
64 Fielder PJ, Hass P, Nagel M, Stefanich E, Widmer R, Bennett GL, Keller GA, de Sauvage FJ, Eaton D: Human platelets as a model for the binding and degradation of thrombopoietin. Blood 1997;89:2782–2788.
65 Harrison CN, Gale RE, Wiestner AC, Skoda RC, Linch DC: The activating splice mutation in intron 3 of the thrombopoietin gene is not found in patients with non-familial essential thrombocythaemia. Br J Haematol 1998;102: 1341–1343.
66 Ichikawa N, Ishida F, Shimodaira S, Tahara T, Kato T, Kitano K: Regulation of serum thrombopoietin levels by platelets and megakaryocytes in patients with aplastic anaemia and idiopathic thrombocytopenic purpura. Thromb Haemost 1996;76:156–160.
67 Tahara T, Usuki K, Sato H, Ohashi H, Morita H, Tsumura H, Matsumoto A, Miyazaki H, Urabe A, Kato T: A sensitive sandwich ELISA for measuring thrombopoietin in human serum: Serum thrombopoietin levels in healthy volunteers and in patients with haemopoietic disorders. Br J Haematol 1996;93:783–788.
68 Lugassy G: Essential thrombocythemia – update on pathogenesis and therapy. Cancer 1998;11:57–59.
69 Horikawa Y, Matsumura I, Hashimoto K, Shiraga M, Kosugi S, Tadokoro S, Kato T, Miyazaki H, Tomiyama Y, Kurata Y, Matsuzawa Y, Kanakura Y: Markedly reduced expression of platelet c-mpl receptor in essential thrombocythemia. Blood 1997;90:4031–4038.
70 Stoffel R, Wiestner A, Skoda RC: Thrombopoietin in thrombocytopenic mice: Evidence against regulation at the mRNA level and for a direct regulatory role of platelets. Blood 1996; 87:567–573.

71 Li J, Xia Y, Kuter DJ: Interaction of thrombopoietin with the platelet c-mpl receptor in plasma: Binding, internalization, stability and pharmacokinetics. Br J Haematol 1999;106: 345–356.

72 Sakamaki S, Hirayama Y, Matsunaga T, Kuroda H, Kusakabe T, Akiyama T, Konuma Y, Sasaki K, Tsuji N, Okamoto T, Kobune M, Kogawa K, Kato J, Takimoto R, Koyama R, Niitsu Y: Transforming growth factor-beta1 (TGF-beta1) induces thrombopoietin from bone marrow stromal cells, which stimulates the expression of TGF-beta receptor on megakaryocytes and, in turn, renders them susceptible to suppression by TGF-beta itself with high specificity. Blood 1999;94:1961–1970.

73 Hirayama Y, Sakamaki S, Matsunaga T, Kuga T, Kuroda H, Kusakabe T, Sasaki K, Fujikawa K, Kato J, Kogawa K, Koyama R, Niitsu Y: Concentrations of thrombopoietin in bone marrow in normal subjects and in patients with idiopathic thrombocytopenic purpura, aplastic anemia, and essential thrombocythemia correlate with its mRNA expression of bone marrow stromal cells. Blood 1998;92:46–52.

74 Axelrad AA, Eskinazi D, Correa PN, Amato D: Hypersensitivity of circulating progenitor cells to megakaryocyte growth and development factor (PEG-rHu MGDF) in essential thrombocythemia. Blood 2000;96:3310–3321.

75 Wang JC, Chen C, Novetsky AD, Lichter SM, Ahmed F, Friedberg NM: Blood thrombopoietin levels in clonal thrombocytosis and reactive thrombocytosis. Am J Med 1998;104:451–455.

76 Uppenkamp M, Makarova E, Petrasch S, Brittinger G: Thrombopoietin serum concentration in patients with reactive and myeloproliferative thrombocytosis. Ann Hematol 1998;77: 217–223.

77 Hou M, Carneskog J, Mellqvist UH, Stockelberg D, Hedberg M, Wadenvik H, Kutti J: Impact of endogenous thrombopoietin levels on the differential diagnosis of essential thrombocythaemia and reactive thrombocytosis. Eur J Haematol 1998;61:119–122.

78 Cerutti A, Custodi P, Duranti M, Noris P, Balduini CL: Thrombopoietin levels in patients with primary and reactive thrombocytosis. Br J Haematol 1997;99:281–284.

79 Li J, Xia Y, Kuter DJ: The platelet thrombopoietin receptor number and function are markedly decreased in patients with essential thrombocythaemia. Br J Haematol 2000;111: 943–953.

80 Moliterno AR, Hankins WD, Spivak JL: Impaired expression of the thrombopoietin receptor by platelets from patients with polycythemia vera. N Engl J Med 1998;338:572–580.

81 Mesa RA, Li CY, Yoon SY, Schroeder GS, Tefferi A: Clinical correlates of megakaryocyte C-Mpl expression in essential thrombocythemia: A clinicopathologic study of 169 patients (abstract). Blood 2000;96:742a.

82 Buss DH, O'Connor ML, Woodruff RD, Richards F II, Brockschmidt JK: Bone marrow and peripheral blood findings in patients with extreme thrombocytosis: A report of 63 cases. Arch Pathol Lab Med 1991;115:475–480.

83 Annaloro C, Lambertenghi Deliliers G, Oriani A, Pozzoli E, Lambertenghi Deliliers D, Radaelli F, Faccini P: Prognostic significance of bone marrow biopsy in essential thrombocythemia. Haematologica 1999;84:17–21.

84 Taksin AL, Couedic JP, Dusanter-Fourt I, Masse A, Giraudier S, Katz A, Wendling F, Vainchenker W, Casadevall N, Debili N: Autonomous megakaryocyte growth in essential thrombocythemia and idiopathic myelofibrosis is not related to a c-mpl mutation or to an autocrine stimulation by Mpl-L. Blood 1999;93: 125–139.

85 Li Y, Hetet G, Kiladjian JJ, Gardin C, Grandchamp B, Briere J: Proto-oncogene c-mpl is involved in spontaneous megakaryocytopoiesis in myeloproliferative disorders. Br J Haematol 1996;92:60–66.

86 Wang JC, Chen C, Lou LH, Mora M: Blood thrombopoietin, IL-6 and IL-11 levels in patients with agnogenic myeloid metaplasia. Leukemia 1997;11:1827–1832.

87 Tamura H, Ogata K, Luo S, Nakamura K, Yokose N, Dan K, Tohyama K, Yoshida Y, Hamaguchi H, Sakamaki H, Kuwaki T, Tahara T, Kato T, Nomura T: Plasma thrombopoietin (TPO) levels and expression of TPO receptor on platelets in patients with myelodysplastic syndromes. Br J Haematol 1998;103:778–784.

88 Tefferi A, Yoon SY, Li CY: Immunohistochemical staining for megakaryocyte c-mpl may complement morphologic distinction between polycythemia vera and secondary erythrocytosis. Blood 2000;96:771–772.

89 Le Blanc K, Andersson P, Samuelsson J: Marked heterogeneity in protein levels and functional integrity of the thrombopoietin receptor c-mpl in polycythaemia vera. Br J Haematol 2000;108:80–85.

90 Fickers M, Speck B: Thrombocythaemia: Familial occurrence and transition into blastic crisis. Acta Haematol 1974;51:257–265.

91 Eyster ME, Saletan SL, Rabellino EM, Karanas A, McDonald TP, Locke LA, Luderer JR: Familial essential thrombocythemia. Am J Med 1986;80:497–502.

92 Kikuchi M, Tayama T, Hayakawa H, Takahashi I, Hoshino H, Ohsaka A: Familial thrombocytosis. Br J Haematol 1995;89:900–902.

93 Ghilardi N, Wiestner A, Kikuchi M, Ohsaka A, Skoda RC: Hereditary thrombocythaemia in a Japanese family is caused by a novel point mutation in the thrombopoietin gene. Br J Haematol 1999;107:310–316.

94 Wiestner A, Schlemper RJ, van der Maas AP, Skoda RC: An activating splice donor mutation in the thrombopoietin gene causes hereditary thrombocythaemia. Nat Genet 1998;18:49–52.

95 Kondo T, Okabe M, Sanada M, Kurosawa M, Suzuki S, Kobayashi M, Hosokawa M, Asaka M: Familial essential thrombocythemia associated with one-base deletion in the 5′-untranslated region of the thrombopoietin gene. Blood 1998;92:1091–1096.

96 Ghilardi N, Skoda RC: A single-base deletion in the thrombopoietin (TPO) gene causes familial essential thrombocythemia through a mechanism of more efficient translation of TPO mRNA. Blood 1999;94:1480–1482.

97 Ghilardi N, Wiestner A, Skoda RC: Thrombopoietin production is inhibited by a translational mechanism. Blood 1998;92:4023–4030.

98 Stuhrmann M, Bashawri L, Ahmed MA, Al-Awamy BH, Kuhnau W, Schmidtke J, El-Harith EA: Familial thrombocytosis as a recessive, possibly X-linked trait in an Arab family. Br J Haematol 2001;112:616–620.

99 Kiladjian JJ, Elkassar N, Hetet G, Briere J, Grandchamp B, Gardin C: Study of the thrombopoietin receptor in essential thrombocythemia. Leukemia 1997;11:1821–1826.

100 Knuutila S, Ruutu T, Partanen S, Vuopio P: Chromosome 1q+ in erythroid and granulocyte-monocyte precursors in a patient with essential thrombocythemia. Cancer Genet Cytogenet 1983;9:245–249.

101 Lofvenberg E, Nordenson I, Wahlin A: Cytogenetic abnormalities and leukemic transformation in hydroxyurea-treated patients with Philadelphia chromosome negative chronic myeloproliferative disease. Cancer Genet Cytogenet 1990;49:57–67.

102 Richard C, Conde E, Garijo J, Iriondo A, Bello C, Zubizarreta A: Trisomy 1q in a case of essential thrombocythemia with long survival. Cancer Genet Cytogenet 1987;25:185–186.

103 Secker-Walker LM, Mehta A, Bain B: Abnormalities of 3q21 and 3q26 in myeloid malignancy: A United Kingdom Cancer Cytogenetic Group study. Br J Haematol 1995;91:490–501.

104 Sandberg AA, Morgan R, Hecht BK, Hecht F: Translocation (1;7)(p11;p11): A new myeloproliferative hematologic entity. Cancer Genet Cytogenet 1985;18:199–206.

105 Willem P, Pinto M, Bernstein R: Translocation t(1;7) revisited: Report of three further cases and review. Cancer Genet Cytogenet 1988;36:45–54.

106 Swolin B, Safai-Kutti S, Anghem E, Kutti J: No increased frequency of trisomies 8 and 9 by fluorescence in situ hybridization in untreated patients with essential thrombocythemia. Cancer Genet Cytogenet 2001;126: 56–59.

107 Rios R, Sole F, del Mar Perez M, Gascon F, Garcia F, Gonzalez PM: t(2,3) in a case of blastic transformation of essential thrombocythaemia. Br J Haematol 1996;92:769–770.

108 Kopf I, Swolin B, Weinfeld A: Cytogenetic studies on primary thrombocythaemia. Hereditas 1982;97:217–220.

109 Killick S, Matutes E, Swansbury J, Catovsky D: Case 17: Essential thrombocythaemia with inversion 3 terminating in acute leukaemia. Leuk Lymphoma 1998;30:661–664.

110 Kerkhofs H, Hagemeijer A, Leeksma CH, Abels J, den Ottolander GJ, Somers R, Gerrits WB, Langenhuiyen MM, von dem Borne AE, Van Hemel JO, Geraedts JP: The 5a-chromosome abnormality in haematological disorders: A collaborative study of 34 cases from the Netherlands. Br J Haematol 1982; 52:365–381.
111 Nowell P, Finan J: Chromosome studies in preleukemic states. IV. Myeloproliferative versus cytopenic disorders. Cancer 1978;42: 2254–2261.
112 Reis MD, Sher GD, Lakhani A, Dube ID, Senn JS, Pinkerton PH: Deletion of the long arm of chromosome 5 in essential thrombocythemia. Cancer Genet Cytogenet 1992;61: 93–95.
113 Welborn J: 5q– in essential thrombocythemia. Cancer Genet Cytogenet 1993;67 150.
114 Munro LR, Stevenson DA, Culligan DJ: Translocation (X;5)(q13;q33) in essential thrombocythemia. Cancer Genet Cytogenet 1999;114:78–79.
115 Palka G, Spadano A, Calabrese G, Parruti G, Guanciali Franchi P, Di Sante O, Recchia A, Di Lorenzo R, Torlontano G: Cytogenetic survey of sixty-one patients with preleukemic syndrome including myeloproliferative and myelodysplastic diseases. Ann Genet 1990; 33:214–218.
116 Johansson B, Billstrom R, Broberg K, Fioretos T, Nilsson PG, Ahlgren T, Malm C, Samuelsson BO, Mitelman F: Cytogenetic polyclonality in hematologic malignancies. Genes Chromosomes Cancer 1999;24:222–229.
117 Velloso ER, Michaux L, Ferrant A, Hernandez JM, Meeus P, Dierlamm J, Criel A, Louwagie A, Verhoef G, Boogaerts M, Michaux JL, Bosly A, Mecucci C, Van den Berghe H: Deletions of the long arm of chromosome 7 in myeloid disorders: Loss of band 7q32 implies worst prognosis. Br J Haematol 1996;92:574–581.
118 Horsman DE, Kalousek DK: Acquired Robertsonian translocations in hematologic malignancy: A rare mechanism of clonal evolution. Cancer Genet Cytogenet 1990;45:193–196.
119 Cournoyer D, Noel P, Schmidt MA, Dewald GW: Trisomy 9 in hematologic disorders: Possible association with primary thrombocytosis. Cancer Genet Cytogenet 1987;27:73–78.
120 Uozumi K, Ohno N, Shimotakahara S, Suzuki S, Takeuchi S, Hanada S, Arima T: Trisomy 8 in essential thrombocythemia in leukemic transformation. Cancer Genet Cytogenet 2000;116:84–86.
121 Mitev L, Georgiev G, Petrov A, Monolova Y: Unusual chromosome aberration, t(13;14) (q32;q32.3), in a case of essential thrombocythemia with extreme thrombocytosis. Cancer Genet Cytogenet 1996;91:68–70.
122 Palumbo AP, Boccadoro M, Battaglio S, Corradini P, Giovinazzo B, Avanzi G, Ponzio G: Philadelphia-positive thrombocythemia with a complex translocation involving chromosomes 9, 15, and 22. Cancer Genet Cytogenet 1989;39:77–80.
123 Pedersen B, Andersen CL, Sogaard MM, Norgaard JM, Koch J, Krejci K, Brandsborg M, Clausen N: Trisomy 10 survival: A literature review and presentation of seven new cases. Cancer Genet Cytogenet 1998;103:130–132.
124 Nakamura H, Hayashibara T, Kawachi T, Nagai K, Matsuo T, Sadamori N, Tomonaga M: Chromosome 11 rearrangement at band 11q21 in a patient with essential thrombocythemia. Cancer Genet Cytogenet 1992;58: 105–107.
125 Petit P, Van den Berghe H: A chromosomal abnormality (21q–) in primary thrombocytosis. Hum Genet 1979;50:105–106.
126 Zaccaria A, Baccarani M, Gugliotta L, Guarini A, Betti S, Tura S: 21q-in primary thrombocythemia. Cancer Genet Cytogenet 1980;1: 337–344.
127 Hollings PE: Molecular heterogeneity at the breakpoints of smaller 20q deletions. Genes Chromosomes Cancer 1994;11:21–28.
128 Mertens F, Johansson B, Heim S, Kristoffersson U, Mitelman F: Karyotypic patterns in chronic myeloproliferative disorders: Report on 74 cases and review of the literature. Leukemia 1991;5:214–220.
129 Kurtin PJ, Dewald GW, Shields DJ, Hanson CA: Hematologic disorders associated with deletions of chromosome 20q: A clinicopathologic study of 107 patients. Am J Clin Pathol 1996;106:680–688.
130 Hirst WJ, Czepulkowski B, Mufti GJ: Consistent interstitial chromosomal deletions in myeloid malignancies and their correlation with fragile sites. Cancer Genet Cytogenet 1993; 65:51–57.
131 Aatola M, Armstrong E, Teerenhovi L, Borgstrom GH: Clinical significance of the del(20q) chromosome in hematologic disorders. Cancer Genet Cytogenet 1992;62:75–80.
132 Campbell LJ, Garson OM: The prognostic significance of deletion of the long arm of chromosome 20 in myeloid disorders. Leukemia 1994;8:67–71.
133 Nacheva E, Holloway T, Carter N, Grace C, White N, Green AR: Characterization of 20q deletions in patients with myeloproliferative disorders or myelodysplastic syndromes. Cancer Genet Cytogenet 1995;80:87–94.
134 Wilkens L, Tchinda J, Burkhardt D, Nolte M, Werner M, Georgii A: Analysis of hematologic diseases using conventional karyotyping, fluorescence in situ hybridization (FISH), and comparative genomic hybridization (CGH). Hum Pathol 1998;29:833–839.
135 Huret JL, Schoenwald M, Gabarre J, Vaugier GL, Tanzer J: Two additional cases of isochromosome 21q or translocation 21q21q in hematological malignancies. Acta Haematol 1991;86:111–114.
136 Kwong YL, Chan AY, Wei D, Chan LC: Near-octaploidy in essential thrombocythemia. Cancer Genet Cytogenet 1993;65:74–75.
137 Emilia G, Torelli G, Sacchi S, Donelli A: Chromosomal abnormalities in essential thrombocythemia. Cancer Genet Cytogenet 1985;18:91–93.
138 Venter JC, Adams MD, Myers EW, et al: The sequence of the human genome. Science 2001;291:1304–1351.
139 Lander ES, Linton LM, Birren B, et al: Initial sequencing and analysis of the human genome. Nature 2001;409:860–921.
140 Fields S: The future is function. Nat Genet 1997;15:325–327.
141 Lander ES: The new genomics: Global views of biology. Science 1996;274:536–539.
142 Lockhart DJ, Dong H, Byrne MC, Follettie MT, Gallo MV, Chee MS, Mittmann M, Wang C, Kobayashi M, Horton H, Brown EL: Expression monitoring by hybridization to high-density oligonucleotide arrays. Nat Biotechnol 1996;14:1675–1680.
143 King HC, Sinha AA: Gene expression profile analysis by DNA microarrays: Promise and pitfalls. JAMA 2001;286:2280–2288.
144 Lesley SA: High-throughput proteomics: Protein expression and purification in the postgenomic world. Protein Expr Purif 2001;22: 159–164.
145 Blagoev B, Pandey A: Microarrays go live – new prospects for proteomics. Trends Biochem Sci 2001;26:639–641.

Ayalew Tefferi, MD
Mayo Clinic
200 First Street SW
Rochester, MN 55905 (USA)
E-Mail tefferi.ayalew@mayo.edu

Bain BJ (ed): Chronic Myeloproliferative Disorders.
Basel, Karger, 2003, pp 120–126

# Cytogenetic and Molecular Genetic Abnormalities in Systemic Mastocytosis

Rajeev Gupta[a] Barbara J. Bain[b] Caroline L. Knight[b]

[a]Institute of Cancer Research, Chester Beatty Laboratories, [b]Department of Haematology, Faculty of Medicine, Imperial College of Science, Technology and Medicine, St. Mary's Hospital, London, UK

**Key Words**
Cytogenetics · Molecular genetics · Mastocytosis · Urticaria pigmentosa · c-*KIT*

**Abstract**
Clonal cytogenetic abnormalities similar to those observed in other haematological neoplasms are demonstrable in a proportion of patients with systemic mastocytosis and in a smaller proportion of adults with urticaria pigmentosa without apparent systemic disease. These clonal abnormalities are not likely to represent the primary event in mast cell neoplasms. Although no recurrent cytogenetic abnormalities specific for mastocytosis have yet been recognised, the majority of cases display one of several mutations in c-*KIT*, the gene encoding the receptor for stem cell factor. That most commonly observed is the Asp816Val mutation, which permits proliferation independent of growth factors. c-*KIT* mutations may play a significant role in the biology of mast cell neoplasms, although other mutations may be needed for a malignant phenotype.

This article is based on a review first published in *Acta Haematologica* 2002;107:123–128.

Although the WHO classification of haematological neoplasms [1] has a separate category for mast cell disorders, it is equally valid to classify these conditions with other myeloid neoplasms. Systemic mastocytosis can therefore be regarded as one of the chronic myeloproliferative disorders and acute mast cell leukaemia as a form of acute myeloid leukaemia. It might be expected that mast cell neoplasms would share some cytogenetic and molecular genetic abnormalities with other myeloid neoplasms whilst also showing other genetic alterations more strongly associated with mast cell differentiation. In this review we will assess whether this is, in fact, the case.

Clonal cytogenetic abnormalities have been reported from a significant proportion of patients with systemic mastocytosis (table 1). These have usually been single abnormalities including monosomies, trisomies, deletions and structural rearrangements. Those found include some that have been reported in other myeloid neoplasms such as 5q–, 11q–, 20q– and trisomy 8. The proportion of metaphases showing an abnormality is often low and, not surprisingly, fluorescence in situ hybridisation (FISH) has been found to be more sensitive than conventional cytogenetic analysis [2]. Other cases have shown single metaphases with an abnormality characteristic of a myeloid neoplasm, including 7q–, 11q– and 20q– [3]; these may represent a low level of an abnormal clone. Clonal abnormalities may be present on one occasion but not on another, perhaps indicating the low frequency of the abnormal

**Table 1.** Cytogenetic abnormalities reported in systemic mastocytosis

| Abnormality | Patients | Reference |
|---|---|---|
| Single abnormalities with or without evolution | | |
| 46,XY; trisomy 8 on FISH | 1 | 2 |
| 45X,–Y/46XY; trisomy 9 on FISH | 1 | 2 |
| 46,XX; trisomy 9 on FISH | 1 | 2 |
| 46,XY; trisomy 9 on FISH | 1 | 2 |
| 46XY; trisomy 9 on FISH | 1 | 2 |
| 46,XX; trisomy 9 on FISH | 1 | 2 |
| 47,XY,+14[3]/45,X,–Y[3]/46,XY[19][a] | 1 | 3 |
| 45,XY,–7[4]/46,XY[3] | 1 | 4 |
| 41 ~ 45,XX,–15[4]/42 ~ 45,XX[5]/46,XX[7] | 1 | 3 |
| 45,XY,–22[3]/46,XY[12] | 1 | 3 |
| 46,XY,del(11)(q13q21)[2]/46,XY[9] | 1 | 3, 5 |
| 46,XY,del(13)(q12q22) | 1 | 4 |
| 46,XX,16p–?[8]/47,XX[10] | 1 | 3 |
| 46,XY,del(20)(q12)[3]/46,XY[12] | 1 | 3 |
| 46,XX,t(8;17)(p11;q25)[b] | 1 | 6 |
| 46,XY,?t(11;22)(p15;q12)[5]/46,XX[10] | 1 | 3 |
| 43 ~ 45,XX,1qh+,9qh+[7]/46,XY[10] | 1 | 3 |
| 47,Y,(X;8)(q2?6;q21.3),+der(8)t(X;8)(q2?6;q21.3) | 1 | 7 |
| Complex[c] | | |
| 48,XY,+8,+11 | 1 | 4 |
| 51,XY,+6,+8,+10,+15,+19[15]/46,XY[5] | 1 | 4 |
| Two apparently independent clones | | |
| 46,XY,–16,+mar[6]/46,XY,+3,–6[2]/46,XY[20] | 1 | 3, 5 |
| 45,XY,–19[5]/46,XX,del(5)(q21q31)[2]/46,XX[18] | 1 | 3 |

[a] The –Y is likely to be an age-related change rather than a clonal abnormality so this karyotype has been regarded as showing a single clonal abnormality.
[b] At the time of evolution to Ph-negative chronic myeloid leukaemia.
[c] Two or more unrelated abnormalities.

clone rather than clonal evolution since the observation of an abnormal karyotype may be followed by normal cytogenetic analysis. Cytogenetic changes are more often observed in cases of systemic mastocytosis with associated myeloproliferative features in other lineages or with more rapidly progressive disease. This may indicate a sampling bias since, in indolent disease, mast cells usually comprise a low percentage of bone marrow cells and, in addition to this, are under-represented in the aspirate because of their paratrabecular and perivascular position and associated fibrosis. However it is also the case that clonal cytogenetic abnormalities often indicate disease evolution and are therefore found in more aggressive disease.

Several patients have been described with two apparently independent clones [3]. It is likely that that neither clonal chromosomal abnormality represents the initial mutational event, but rather that there are two daughter clones resulting from further evolution within an original neoplastic clone that was abnormal at a molecular but not at a cytogenetic level.

Cytogenetic abnormalities have been detected not only in systemic mastocytosis but also in patients who appear to have uncomplicated urticaria pigmentosa; 2 of 6 patients studied by Swolin et al. [3, 5] were found to have clonal abnormalities, 9p+ and +22/–20, respectively. This suggests that, in at least some adult patients with urticaria pigmentosa, the condition is neoplastic in nature. In another patient with an initial diagnosis of urticaria pigmentosa and a normal karyotype, repeat examination 4 years later showed mast cell infiltration of the bone marrow leading to reclassification as systemic mastocytosis. On this second occasion, a clonal abnormality, monosomy 22, was present in 3 of 15 metaphases. This sequence of events could represent either clonal evolution or sampling error with a very low burden of neoplastic mast cells not having been detected on the first occasion.

**Table 2.** Cytogenetic abnormalities reported in aggressive systemic mast cell disease or mast cell leukaemia

| Abnormality | Patients | Reference |
|---|---|---|
| t(8;21)(q22;q22),+4 | 1 | 8 |
| 46,XY,dir ins(10;16)(q2?2;q13q22)[a] | 1 | 9 |
| Complex including near-haploid, near-diploid, polyploid, 17p+, dicentric chromosomes, rings and double minutes | 1 | 10, 11 |
| Complex including near-haploid with structural as well as numerical abnormalities and a minor hyperdiploid clone | 1 | 12 |

[a] Detected in a cell line.

The recurrent cytogenetic abnormalities in systemic mastocytosis have been those commonly observed in acute myeloid leukaemia and the myelodysplastic syndromes, and give no clue as to why mast cell differentiation has occurred. No abnormalities have been identified that have led to the demonstration of an underlying genetic mechanism of the disease. This information has come directly from molecular genetic studies.

Cytogenetic abnormalities have been described in only a small number of patients with mast cell leukaemia or other very aggressive mast cell disease (table 2). In addition, in 1 further case a cytogenetically abnormal cell line was established from the patient's cells [4]. It is of interest that, in contrast to the findings in more indolent mast cell disease, 2 of these patients had a highly complex karyotype with near-haploid clones [10–12]. As for systemic mastocytosis, the abnormalities observed have not generally given any indication of the basis for the prominent mast cell differentiation. An exception is the report of mast cell differentiation occurring as an unusual features of FAB M2 category of acute myeloid leukaemia associated with t(8;21)(q22;q22) [8]. In this patient, clonal evolution had occurred with development of trisomy 4. This chromosome bears the locus for the c-*KIT* gene and this patient was demonstrated to have a mutation of c-*KIT* (see below). Interestingly, a second patient has been reported with coexisting systemic mastocytosis and acute myeloid leukaemia associated with t(8;21) [13]; in this case, the relationship of the two disorders was not apparent.

In view of the lack of recurrent cytogenetic abnormalities in mast cell neoplasms, the search for molecular markers of these diseases has relied on studies of mutations in genes expressed in normal mast cells, and analysis of known markers of mast cell tumours. Attention in this type of endeavour is generally focussed on signalling proteins or transcription factors. There are few unique markers for normal cells of this lineage which are not merely structural proteins or enzymes involved in mast cell function. A recent study of mast cells in bone marrow trephine sections from patients with a variety of mast cell disorders has shown invariant expression of the receptor tyrosine kinase CD117 (c-KIT) and mast cell tryptase [14]. Characteristic aberrant immunophenotypic features include the expression of the anti-apoptotic protein bcl-x(L), the tyrosine phosphatase CD45, the sialoprotein CD68R, the T-cell marker CD2 and the interleukin (IL)-2 receptor, CD25 [15].

The c-*KIT*-encoded transmembrane tyrosine kinase receptor for stem cell factor (SCF) is required for normal haematopoiesis, melanogenesis and gametogenesis and is a marker for mast cells [16]. Normal c-KIT activity is tightly regulated by SCF; ligand binding results in receptor dimerization and autophosphorylation. In turn this leads to increased c-KIT kinase activity, which triggers a cascade of events leading to increased mast cell proliferation and reduced apoptosis [17]. A domain called the activation loop, the conformation of which changes on ligand binding, regulates the kinase activity of the c-KIT protein.

The activation loop mutation Asp816Val in c-*KIT* is a recurrent abnormality recognised for some time in adult mastocytosis (table 3) [31]. Levels of circulating c-KIT protein have been shown to correlate positively with tumour burden and disease activity in mast cell neoplasms [32]. The link of c-*KIT* mutations with childhood mastocytosis, which represents a broader spectrum of disorders, some of which may be reactive, is weaker. The mutation can be detected almost universally in DNA extracted from lesional skin biopsies, but when studies are performed with peripheral blood or bone marrow DNA, the frequency of detection is much lower. This is consis-

**Table 3.** c-*KIT* mutations in mast cell neoplasms

| Mutation | Site of defect | Tissue examined | Haematological diagnosis | Positive cases | Additional features in +ve cases | References |
|---|---|---|---|---|---|---|
| c-*KIT* Asp816Val | A-loop | lesional skin biopsy | adult urticaria pigmentosa | 17/17 | does not predict clinical severity or outcome | 18, 19 |
| | | | childhood urticaria pigmentosa | 2/26 | associated with severe systemic and/or cutaneous disease | 18–20 |
| | | peripheral blood | adult mastocytosis | 19/73 | seen in patients with and without other haematological manifestations | 21–23 |
| | | | childhood urticaria pigmentosa | 0/2 | | |
| | | bone marrow | adult mastocytosis and MDS or AML | 2/7 | | 24 |
| | | bone marrow mast cells | adult mastocytosis | 1/1 | mutation found on other BMNCs | 25 |
| | | | adult mastocytosis and AML | 2/2 | mutations not found in AML blasts | 26 |
| | | | adult mastocytosis and CMML | 1/1 | mutation also found in microdissected CMML infiltrates | 27 |
| | | | adult mastocytosis | 4/4 | mutation found in peripheral blood monocytes in 1/4 | 28 |
| c-*KIT* Asp816Tyr | A-loop | lesional skin biopsy | childhood systemic mastocytosis | 2/14 | severe systemic disease | 18 |
| | | bone marrow | adult AML with either inv(16) or t(8;21) | 1/15 | heavy mast cell involvement | 29 |
| c-*KIT* Asp816His | A-loop | bone marrow | adult mastocytosis and MDS or AML | 4/7 | | 24 |
| c-*KIT* Asp816Phe | A-loop | lesional skin biopsy | childhood urticaria pigmentosa | 1/14 | extensive cutaneous disease without systemic involvement | 18 |
| c-*KIT* Asp820Gly | phospho-transferase domain | bone marrow | adult mastocytosis | 1/1 | aggressive systemic disease; not seen in peripheral blood cells | 30 |
| c-*KIT* Val560Gly | JM domain | lesional skin biopsy | adult mastocytosis | 2/4 | both patients also had Asp816Val | 19 |
| c-*KIT* Glu839Lys | kinase domain | lesional skin biopsy | childhood urticaria pigmentosa | 3/7 | indolent disease | 18 |

MDS = Myelodysplastic syndrome; AML = acute myeloid leukaemia; BMNC = bone marrow mononuclear cells; CMML = chronic myelomonocytic leukaemia.

tent with the mutation being restricted to mast cells, which constitute only a minority of bone marrow mononuclear cells. In some patients with systemic mastocytosis, the Asp816Val mutation is also detectable in non-mast cell lineages, e.g. peripheral blood B cells and bone marrow cells bearing the lineage commitment markers CD14 (monocytes) and CD19 (B cells) [28, 33]. These cells do not express c-KIT protein, and this may be why they are spared any phenotypic consequences of the mutation. This situation parallels findings in chronic granulocytic leukaemia when analysis for the Philadelphia chromosome is carried out in non-myeloid cells, the clonal abnormality being detectable in a proportion of B cells [34].

Other mis-sense mutations involving Asp816 and mutations affecting the kinase and juxtamembrane (JM) domains of c-*KIT* have been reported (table 3), but these have generally been small series or single cases reported by individual groups. Mutant c-KIT molecules are found in a variety of haematopoietic and other neoplasms. In functional terms they are thought to have transforming potential because they have receptor-independent kinase (and hence signalling) activity. Biochemically, this may be demonstrated by constitutive receptor dimerization and/or phosphorylation.

The functional properties of the c-KIT Asp816 mutant protein are the most extensively studied. Many workers have investigated the analogous mutations at Asp814 in the mouse gene. However, Longley et al. [18] were able to demonstrate ligand-independent autophosphorylation of human Asp816Val, Asp816Tyr and Asp816Phe in vitro. Mouse Asp814Val mutant c-KIT proteins are not constitutively dimerised in the absence of ligand [35], though some evidence is emerging for a novel form of receptor association of Asp814Val via an as yet ill-defined cytoplasmic interaction specific to this particular mutant [36]. C-KIT Asp814Val appears to target a different spectrum of substrate proteins to wild-type c-KIT and cells expressing this mutant rapidly turn over the tyrosine protein phosphatase, SHP-1 [37]. The human Asp816Val c-KIT appears to specifically target the PI-3 kinase-signalling cascade, unlike the wild-type protein [38]. Asp816Val c-KIT and its mouse homologue have both been shown to cause factor-independent growth in haematopoietic cells [35, 39]. In addition, transgenic mice bearing the mouse Asp814Val *c-KIT* gene have also been shown to develop acute leukaemia [35].

Much less information is currently available regarding other mutations. The mouse homologue of the JM domain mutant Val560Gly protein, Val559Gly, is known to be constitutively phosphorylated and dimerised [35]. In addition retroviral transfection experiments have shown that it can cause factor-independent growth and that it can induce acute leukaemias when such transfected cells are transplanted into syngeneic recipients. The human Glu839Lys kinase domain mutant is unusual in that it is not constitutively phosphorylated even in the presence of an Asp816Val protein. The functional significance of this is unclear [18].

The tyrosine kinase inhibitor imatinib mesylate (STI571) has been shown to diminish ligand-induced phosphorylation of wild-type c-KIT, and also inhibit the SCF-dependent proliferation of an acute myeloid leukaemia cell line expressing wild-type c-KIT [40]. Studies of the human mast cell leukaemia line HMC1 (which has one wild-type, and one double mutant *c-KIT* allele containing Asp816Val and Val560Gly substitutions) have shown that STI 571 inhibits the constitutive autophosphorylation of c-KIT and also inhibits ligand independent cellular proliferation in this system [40]. Subsequent studies of simian fibroblast (COS) cells transfected with mutated human *c-KIT* complementary DNAs encoding single amino acid substitutions have suggested that STI571 and the tyrosine kinase inhibitor SU9529 inhibit the ligand-independent autophosphorylation of the JM mutant protein Val560Gly, but not that of the Asp816Val protein [41]. The same authors have studied subclones of the HMC-1 line carrying the double mutant or a Val560Gly *c-KIT* allele, respectively, and have suggested that the autophosphorylation of Val560Gly c-KIT but not the double mutant protein is sensitive to these agents. They have also provided evidence that the subclones expressing Val560Gly *c-KIT* but not those expressing the double mutant protein undergo apoptosis in response to tyrosine kinase inhibitors. There are at present no reports of the effect of tyrosine kinase inhibitors on a mast cell line expressing a single Asp816 mutation *c-KIT* or on a cell line that has been transformed by a *c-KIT* allele encoding a single Asp816 mutation. The potential efficacy of these agents in mastocytosis associated with Asp816 mutations is unknown.

In the case of Asp816Val *c-KIT* at least, it would appear that the same mutation can give rise to mast cell neoplasms with vastly differing phenotypes. Clearly, mutations or genetic polymorphisms in other genes must have a role to play in determining disease severity. One such gene is the IL-4 receptor. IL-4 is known to induce apoptosis in mast cells [42]. The Gln576Arg polymorphism in the IL-4 receptor occurs in 20% of the Caucasian population and is associated with atopy [43]. Although not associated with an increased incidence of mastocytosis, this polymorphism is commoner in patients with mast cell neoplasms and solely cutaneous disease (69%) when compared to those with systemic disease (9%) [44]. There are as yet no data that analyse the effect of this polymorphism in patients who have been proven to have the *c-KIT* Asp816Val mutation.

## Conclusion

Mast cell neoplasms comprise a heterogeneous group of clonal haematopoietic disorders ranging from indolent cutaneous disease to highly aggressive mast cell leukaemia. There are no recognised recurrent cytogenetic abnormalities associated with these conditions, and it appears likely that, at least in some cases, the clonal cytogenetic abnormality demonstrated is not the primary event. The receptor tyrosine kinase, c-KIT, is constitutively expressed in normal mast cells and appears to be an invariant marker of neoplastic mast cells. Point mutations involving the Asp816 codon are frequently observed in mast cell neoplasms with differing clinical and biological phenotypes. In view of this heterogeneity, it appears likely that this mutation is one event in the development of mast cell neoplasia, with subsequent events being important in those cases with a more malignant disease phenotype.

## References


1 Jaffe ES, Harris NL, Stein H, Vardiman J: The World Health Organization Classification of Tumours: Pathology and Genetics of Tumours of Haematopoietic and Lymphoid Tissues. Lyon, IARC Press, 2001.

2 Lishner M, Confino-Cohen R, Mekori YA, Feigin M, Manor Y, Goldberg A, Ravid M, Amiel M: Trisomies 9 and 8 detected by fluorescence in situ hybridization in patients with systemic mastocytosis. J Allergy Clin Immunol 1996;98:199–204.

3 Swolin B, Rodjer S, Roupe G: Cytogenetic studies in patients with mastocytosis. Cancer Genet Cytogenet 2000;120:131–135.

4 Travis WD, Li C-Y, Yam T, Bergstralh EJ, Swee RG: Significance of systemic mast cell disease with associated haematologic diagnosis. Cancer 1988;62:965–972.

5 Swolin B, Rodjer S, Roupe G: Cytogenetic studies and in vitro colony growth in patients with mastocytosis. Blood 1987;70:1928–1932.

6 Lewis JP, Welborn JL, Meyers FJ, Levy NB, Roschak T: Mast cell disease followed by leukemia with clonal evolution. Leukemia Res 1987;11:769–773.

7 Zhang Y, Schlegelberger B, Weber-Matthiesen K, Grote W, Bartels H: Translocation (X;8)(q2?6;q21.3) in a case of systemic mastocytosis. Cancer Genet Cytogenet 1994;78:236–241.

8 Beghini A, Cairoli R, Morra E, Larizza L: In vivo differentiation of mast cells from AML M2 blasts carrying a novel activating ligand-independent c-KIT mutation. Br J Haematol 1998;102:13.

9 Travis WD, Li C-Y, Hoagland HC, Travis LB, Banks PM: Mast cell leukemia: Report of a case and review of the literature. Mayo Clinic Proc 1986;61:957–966.

10 Bauchinger M, Mezger J: A case of malignant mastocytosis with near-haploid, near-diploid, and polyploid cells in the bone marrow. Cancer Genet Cytogenet 1990;48:13–21.

11 Mezger J, Permanetter W, Gerhartz H, Bartl R, Bauchinger M, Schmetzer H, Sauer H: Philadelphia chromosome-negative acute hemopoietic malignancy: Ultrastructural, cytochemical and immunocytochemical evidence of mast cell and basophil differentiation. Leukemia Res 1990;14:169–175.

12 Shekter-Levin S, Ball E. Swerdlow SH, Li WV, Kapadia SB, Sherer ME, Wald N, Gollin SM: A near-haploid marrow in systemic mast cell disease: Is it characteristic of the disease or an incidental finding? Cancer Genet Cytogenet 1998;103:124–129.

13 Wong KF, Chan JK, Chan JC, Kwong YL, Ma SK, Chow TC: Concurrent acute myeloid leukemia and systemic mastocytosis. Am J Hematol 1991;38:243–244.

14 Jordan JH, Walchshofer S, Jurecka W, Mosberger I, Sperr WR, Wolff K, Chott A, Buhring HJ, Lechner K, Horny HP, Valent P: Immunohistochemical properties of bone marrow mast cells in systemic mastocytosis: Evidence for expression of CD2, CD117/Kit, and bcl-x(L). Hum Pathol 2001;32:545–552.

15 Escribano L, Orfao A, Villarrubia J, Martin F, Madruga JI, Cuevas M, Velasco JL, Rios A, San Miguel JF: Sequential immunophenotypic analysis of mast cells in a case of systemic mast cell disease evolving to a mast cell leukemia. Cytometry 1997;30:98–102.

16 Nocka K, Majumder S, Chabot B, Ray P, Cervone M, Bernstein A, Besmer P: Expression of c-kit gene products in known cellular targets of W mutations in normal and W mutant mice – evidence for an impaired c-kit kinase in mutant mice. Genes Dev 1989;3:816–826.

17 Mekori YA, Oh CK, Metcalfe DD: IL-3-dependent murine mast cells undergo apoptosis on removal of IL-3. Prevention of apoptosis by c-kit ligand. J Immunol 1993;151:3775–3784.

18 Longley BJ Jr, Metcalfe DD, Tharp M, Wang X, Tyrrell L, Lu SZ, Heitjan D, Ma Y: Activating and dominant inactivating c-KIT catalytic domain mutations in distinct clinical forms of human mastocytosis. Proc Natl Acad Sci USA 1999;96:1609–1614.

19 Buttner C, Henz BM, Welker P, Sepp NT, Grabbe J: Identification of activating c-kit mutations in adult-, but not in childhood-onset indolent mastocytosis: A possible explanation for divergent clinical behavior. J Invest Dermatol 1998;111:1227–1231.

20 Nagata H, Okada T, Worobec AS, Semere T, Metcalfe DD: c-kit mutation in a population of patients with mastocytosis. Int Arch Allergy Immunol 1997;113:184–186.

21 Nagata H, Worobec AS, Oh CK, Chowdhury BA, Tannenbaum S, Suzuki Y, Metcalfe DD: Identification of a point mutation in the catalytic domain of the protooncogene c-kit in peripheral blood mononuclear cells of patients who have mastocytosis with an associated hematologic disorder. Proc Nat Acad Sci USA 1995;92:10560–10564.

22 Worobec AS, Semere T, Nagata H, Metcalfe DD: Clinical correlates of the presence of the Asp816Val c-kit mutation in the peripheral blood mononuclear cells of patients with mastocytosis. Cancer 1998;83:2120–2129.

23 Sato-Matsumura KC, Matsumura T, Koizumi H, Sato H, Nagashima K, Ohkawara A: Analysis of c-kit exon 11 and exon 17 of urticaria pigmentosa that occurred in monozygotic twin sisters. Br J Dermatol 1999;140:1130–1132.

24 Pullarkat VA, Pullarkat ST, Calverley DC, Brynes RK: Mast cell disease associated with acute myeloid leukemia: Detection of a new c-kit mutation Asp816His. Am J Hematol 2000;65:307–309.

25 Afonja O, Amorosi E, Ashman L, Takeshita K: Multilineage involvement and erythropoietin-'independent' erythroid progenitor cells in a patient with systemic mastocytosis. Ann Hematol 1998;77:183–186.

26 Sperr WR, Walchshofer S, Horny HP, Fodinger M, Simonitsch I, Fritsche-Polanz R, Schwarzinger I, Tschachler E, Sillaber C, Hagen W, Geissler K, Chott A, Lechner K, Valent P: Systemic mastocytosis associated with acute myeloid leukaemia: Report of two cases and detection of the c-kit mutation Asp-816 to Val. Br J Haematol 1998;103:740–749.

27 Sotlar K, Marafioti T, Griesser H, Theil J, Aepinus C, Jaussi R, Stein H, Valent P, Horny HP: Detection of c-kit mutation Asp 816 to Val in microdissected bone marrow infiltrates in a case of systemic mastocytosis associated with chronic myelomonocytic leukaemia. Mol Pathol 2000;53:188–193.

28 Jordan J-H, Fritsche-Polanz R, Sperr WR, Mitterbauer G, Fodinger M, Schernthaner G-H, Bankl HC, Gebhart W, Chott A, Lechner K, Valent P: A case of 'smouldering' mastocytosis with high mast cell burden, monoclonal myeloid cells and c-KIT mutation Asp-816-Val. Leukemia Res 2001;25:627–634.

29 Beghini A, Peterlongo P, Ripamonti CB, Larizza L, Cairoli R, Morra E, Mecucci C: C-kit mutations in core binding factor leukemias. Blood 2000;95:726–727.

30 Pignon JM, Giraudier S, Duquesnoy P, Jouault H, Imbert M, Vainchenker W, Vernant JP, Tulliez M: A new c-kit mutation in a case of aggressive mast cell disease. Br J Haematol 1997;96:374–376.

31 Bain BJ: Systemic mastocytosis and other mast cell neoplasms. Br J Haematol 1999;106:9–17.

32 Akin C, Schwartz LB, Kitoh T, Obayashi H, Worobec AS, Scott LM, Metcalfe DD: Soluble stem cell factor receptor (CD117) and IL-2 receptor alpha chain (CD25) levels in the plasma of patients with mastocytosis: Relationships to disease severity and bone marrow pathology. Blood 2000;96:1267–1273.

33 Akin C, Kirshenbaum AS, Semere T, Worobec AS, Scott LM, Metcalfe DD: Analysis of the surface expression of c-kit and occurrence of the c-kit Asp816Val activating mutation in T cells, B cells and myelomonocytic cells in patients with mastocytosis. Exp Hematol 2000; 28:140–147.
34 Takahashi N, Miura I, Saitoh K, Miura AB: Lineage involvement of stem cells bearing the Philadelphia chromosome in chronic myeloid leukemia in the chronic phase as shown by a combination of fluorescence-activated cell sorting and fluorescence in situ hybridization. Blood 1998;92:4758–4763.
35 Kitayama H, Tsujimura T, Matsumura I, Oritani K, Ikeda H, Ishikawa J, Okabe M, Suzuki M, Yamamura K, Matsuzawa Y, Kitamura Y, Kanakura Y: Neoplastic transformation of normal hematopoietic cells by constitutively activating mutations of c-kit receptor tyrosine kinase. Blood 1996;88:995–1004.
36 Tsujimura T, Hashimoto K, Kitayama H, Ikeda H, Sugahara H, Matsumura I, Kaisho T, Terada N, Kitamura Y, Kanakura YZ: Activating mutation in the catalytic domain of c-kit elicits hematopoietic transformation by receptor self-association not at the ligand-induced dimerization site. Blood 1999;93:1319–1329.
37 Piao X, Paulson R, van der Geer P, Pawson T, Bernstein A: Oncogenic mutation in the Kit receptor tyrosine kinase alters substrate specificity and induces degradation of the protein tyrosine phosphatase SHP-1. Proc Natl Acad Sci USA 1996;93:14665–14669.
38 Chian R, Young S, Danilkovitch-Miagkova A, Ronnstrand L, Leonard E, Ferrao P, Ashman L, Linnekin D: PI3 kinase mediates transformation of hematopoietic cells by the V816 c-kit mutant. Exp Haematol 2000;28:1491.
39 Ashman LK, Ferrao P, Cole SR, Cambareri AC: Effects of mutant c-kit in early myeloid cells. Leuk Lymphoma 2000;37:233–243.
40 Heinrish MC, Griffith DJ, Druker BJ, Wait CL, Ott KA, Zigler AJ: Inhibition of c-kit receptor tyrosine kinase activity by STI571, a selective tyrosine kinase inhibitor. Blood 2000; 96:925–932.
41 Ma Y, Zeng S, Metcalfe DD, Akin C, Dimitrijevic S, Butterfield JH, McMahon G, Longley BJ: The c-KIT mutation causing human mastocytosis is resistant to STI571 and other KIT inhibitors: Kinases with enzymatic site mutations show different inhibitor sensitivity profiles than wild-type kinases and those with regulatory-type mutations. Blood 2002;99:1741–1744.
42 Nilsson G, Miettinen U, Ishizaka T, Ashman LK, Irani AM, Schwartz LB: Interleukin-4 inhibits the expression of Kit and tryptase during stem cell factor-dependent development of human mast cells from fetal liver cells. Blood 1994;84:1519–1527.
43 Khurana Hershey GK, Friedrich MF, Esswein LA, Thomas ML, Chatila TA: The association of atopy with a gain-of-function mutation in the alpha subunit of the interleukin-4 receptor. N Engl J Med 1997;337:1720–1725.
44 Daley T, Metcalfe DD, Akin C: Association of the Q576R polymorphism in the interleukin-4 receptor alpha chain with indolent mastocytosis limited to the skin. Blood 2001;98:880–882.



Dr. Rajeev Gupta
Section of Gene Function and Regulation
Institute of Cancer Research
Chester Beatty Laboratories
237 Fulham Road, London SW3 6JB (UK)
E-Mail rajeevg@icr.ac.uk

# Author Index

Bain, B.J. 1, 120
Barnes, D.J. 9
Chase, A. 32
Cross, N.C.P. 52, 69
Fioretos, T. 44
Goldman, J.M. 32
Gupta, R. 120
Hall, G.W. 80
Johansson, B. 44
Kaeda, J. 32
Knight, C.L. 120
Macdonald, D. 62
Melo, J.V. 9
Mitelman, F. 44
Pahl, H.L. 90
Reilly, J.T. 102
Reiter, A. 62
Steer, E.J. 69
Steensma, D.P. 109
Tefferi, A. 109

# Subject Index


*ABL*, *see BCR-ABL* fusion
Allogeneic stem cell transplantation
  cytogenetic evolution effects on Philadelphia chromosome-positive CML 53
  8p11 myeloproliferative syndrome management 64
  minimal residual disease monitoring in chronic myeloid leukemia
    cytogenetic analysis 34
    reverse transcriptase-polymerase chain reaction 39, 40
Apoptosis, Bcr-Abl inhibition 23

*BCR-ABL* fusion
  *ABL*
    breakpoints 11
    protein structure and function 2, 16, 17
    tyrosine kinase activity deregulation 17, 18
  *BCR*
    breakpoints 11–13
    protein structure and function 14–16
  cytogenetic evolution of Philadelphia chromosome-positive chronic myeloid leukemia
    banding era studies 48, 49, 51
    overview 44, 45
    pre-banding era studies 45, 47
    prognostic impact of secondary genetic changes 55, 56
    treatment effects 53, 54
  essential thrombocythemia 111, 112
  functional effects 2
  fusion protein
    apoptosis inhibition 23
    cell adhesion effects 20, 21
    mitogenic signaling pathway activation 21–23
    proteasomal degradation effects 24
    structures 18–20
    therapeutic targeting 24–26
  genotype-phenotype correlations 18–20, 54, 55
  leukemia types 2
  mechanism of translocation 10, 11
  minimal residual disease, *see* Minimal residual disease, chronic myeloid leukemia monitoring
  molecular genetic evolution of Philadelphia chromosome-positive chronic myeloid leukemia 51–53
  Philadelphia chromosome discovery 1, 9, 45
  research prospects 26
  transcript detection in healthy individuals 13, 14
  unusual breakpoints 13
*BCR-FGFR1* fusion
  chronic myeloid leukemia 5
  8p11 myeloproliferative syndrome 65, 66
*BCR-JAK2* fusion, chronic myeloid leukemia 4
Bone marrow transplantation, *see* Allogeneic stem cell transplantation

*CDKN2A*, chronic myeloid leukemia mutations 52
*CEP110-FGFR1* fusion
  chronic myeloid leukemia 4
  8p11 myeloproliferative syndrome 65, 66
CGP57418B, *see* STI571
Childhood myeloproliferative/myelodysplastic disorders, *see* Chronic idiopathic myelofibrosis, Chronic myeloid leukemia, 8p11 myeloproliferative syndrome, Essential thrombocythemia, Juvenile myelomonocytic leukemia, Polycythemia vera, Transient myeloproliferative disorder
Chromosomal translocation, *see also* specific translocations
  chronic myeloid leukemia 1–6, 9–26
  platelet-derived growth factor-β fusions, *see PDGFRB*
Chronic idiopathic myelofibrosis, childhood disease 80, 81
Chronic myeloid leukemia
  atypical leukemia in children 81
  childhood disease 80, 81
  chromosomal translocations, *see also* specific translocations
    cytogenetics at diagnosis 33
    homologous acute leukemia translocations 6
    Philadelphia chromosome, *see BCR-ABL* fusion
    variant translocations 47
  minimal residual disease, *see* Minimal residual disease, chronic myeloid leukemia monitoring
c-KIT, mastocytosis mutations 122–125
c-Mpl, *see* Thrombopoietin receptor
Cytogenetics
  childhood myeloproliferative/myelodysplastic disorders 80–86
  chronic myeloid leukemia 1, 9, 33, 34
  essential thrombocythemia 114, 115
  evolution of Philadelphia chromosome-positive leukemia 44–56
  idiopathic myelofibrosis 102–106
  mastocytosis 120–125

minimal residual disease monitoring in chronic myeloid leukemia
metaphase cytogenetics 33
treatment responses
allogeneic stem cell transplantation 34
imatinib mesylate 35
interferon-α 34, 35
polycythemia vera
chromosome 9 loss of heterozygosity 95, 96
chromosome 20q deletions 95

DNA microarray, essential thrombocythemia analysis 115, 116
DNA-PKcs, Bcr-Abl downregulation 24

(EMS) 8p11 myeloproliferative syndrome
childhood disease 81, 82
chromosomal translocations
*BCR-FGFR1* fusion 65, 66
*CEP110-FGFR1* fusion 65, 66
*FOP-FGFR1* fusion 65
overview 63
*ZNF198-FGFR1* fusion 64, 65
clinical features 63
lymphadenopathy and non-Hodgkin lymphoma 63, 64
natural history 64
therapeutic targets 66
treatment 64
Endogenous erythroid colonies, *see* Polycythemia vera
Eosinophilia, *see* 8p11 myeloproliferative syndrome, *ETV6-PDGFRB* fusion
Essential thrombocythemia
*BCR-ABL* fusion 111, 112
childhood disease 80, 81
chromosomal abnormalities 114, 115
clinical features 109, 110
clonality determination 110, 111
diagnosis 110
DNA microarray analysis of gene expression 115, 115
familial thrombocythemia 113, 115
thrombopoietin and receptor dynamics 112, 113
*ETV6-ABL* fusion, chronic myeloid leukemia 3
*ETV6-EVI1* fusion, chronic myeloid leukemia 3
*ETV6-JAK2* fusion, chronic myeloid leukemia 3, 4
*ETV6-PDGFRB* fusion
chronic myeloid leukemia 2, 69
clinical features
eosinophilia 71
initial diagnoses 71
myeloid disorders 71
progression and survival 73
splenomegaly and monocytosis 71, 72
male bias 73
therapeutic targeting 75, 76
treatment of disease 72, 73

*FGFR1*
activation consequences 66
fusions, *see BCR-FGFR1, CEP110-FGFR1, FOP-FGFR1, ZNF198*
5q31–35 translocations, *see also ETV6-PDGFRB* fusion, *PDGFRB*
clinical features 73, 74
skin involvement 74
target genes 70, 71
Fluorescence in situ hybridization, minimal residual disease monitoring in chronic myeloid leukemia
hypermetaphase 34
interphase cells 33, 34
*FOP-FGFR1* fusion
chronic myeloid leukemia 4
8p11 myeloproliferative syndrome 65

Glucose-6-phosphate dehydrogenase, isoforms in polycythemia vera 91–93

*HIP1-PDGFRB* fusion, chronic myeloid leukemia 2

Idiopathic myelofibrosis
clinical features 102
cytogenetics
deletions
13q 103, 104
20q 104
overview 103
prognosis 105
trisomies
1q 104
8 104
hematopoietic clonality 102, 103
incidence 102
molecular genetics studies 105, 106
research prospects 106
Imatinib mesylate, *see* STI571
Interferon-α
cytogenetic evolution effects on Philadelphia chromosome-positive chronic myeloid leukemia 53, 54
minimal residual disease monitoring in chronic myeloid leukemia
cytogenetic analysis 34, 35
reverse transcriptase-polymerase chain reaction 40

Janus kinases, Bcr-Abl activation 22
Juvenile myelomonocytic leukemia
clinical presentation 82
clonality 82, 83
cytogenetics 83
diagnostic criteria 82
molecular genetics 83–85
pathobiology 83–85
research prospects 86

Mastocytosis
*c-KIT* mutations 122–125
classification 120
cytogenetics 120–125
STI571 therapy 126
Minimal residual disease, chronic myeloid leukemia monitoring
cytogenetic analysis of treatment response
allogeneic stem cell transplantation 34
imatinib mesylate 35
interferon-α 34, 35
fluorescence in situ hybridization
hypermetaphase 34
interphase cells 33, 34
metaphase cytogenetics 33
rationale 32
reverse transcriptase-polymerase chain reaction
cure definition 38, 39
molecular remission definition 40
overview 35, 36
prospects 41
quantitative technique 36–38
relapse definition 40
sensitivity 36
treatment response
allogeneic stem cell transplantation 39, 40
imatinib mesylate 41
interferon-α 40
Southern blot 35
Western blot 35
Mitogen-activated protein kinase, Bcr-Abl activation 21, 22
Molecular genetics
childhood myeloproliferative/myelodysplastic disorders 80–86
chronic myeloid leukemia 1, 9
evolution of Philadelphia chromosome-positive leukemia 51–53
idiopathic myelofibrosis 105, 106
mastocytosis 120–125
Myc, Bcr-Abl activation 22, 23, 51
Myeloproliferative disorder, *see also* specific diseases
definition 62
molecular etiology and classification 98, 99
types 62

*NUP98-HOXA9* fusion, chronic myeloid leukemia 5, 53

Oncogenes, chronic myeloid leukemia 1, 9

*PDGFRB, see also ETV6-PDGFRB* fusion
alternative target genes at 5q31–35 70, 71, 73
constitutive activity of fusion proteins 75
myeloproliferative disease translocations 69, 70
protein
function 75
signaling 75
structure 74, 75
therapeutic targeting 75, 76
Philadelphia chromosome, *see BCR-ABL* fusion
Phosphatidylinositol-3 kinase, Bcr-Abl activation 22
Polycythemia vera
childhood disease 80, 81
cytogenetic markers
chromosome 9 loss of heterozygosity 95, 96
chromosome 20q deletions 95
endogenous erythroid colonies
marker utilization 93, 94
origins 91
potential for formation 91–93
endogenous megakaryocytic colonies as markers 94
glucose-6-phosphate dehydrogenase isoforms 91–93
growth factor independence and hypersensitivity 94, 95
history of study 90
PRV-1
clinical classification of myeloproliferative disorders 98
function 97, 98
molecular etiology of myeloproliferative disorders 98, 99
transcript quantification in diagnosis 97
stem cell studies 90, 91
thrombopoietin receptor reduction and diagnostics 96, 97
Polymerase chain reaction, *see also* Real time polymerase chain reaction, Reverse transcriptase-polymerase chain reaction
competitive 36
quantitative, for minimal residual disease detection 36–38
PRV-1
clinical classification of myeloproliferative disorders 98
function 97, 98
molecular etiology of myeloproliferative disorders 98, 99
polycythemia vera, transcript quantification in diagnosis 97

*Rabaptin5-PDGFRB* fusion, chronic myeloid leukemia 3
Ras
Bcr-Abl activation 21, 22
chronic myeloid leukemia mutations 52
*RB1*, chronic myeloid leukemia mutations 52
Real time polymerase chain reaction, quantitative polymerase chain reaction for chronic myeloid leukemia minimal residual disease detection 36–38
Reverse transcriptase-polymerase chain reaction
minimal residual disease monitoring in chronic myeloid leukemia
cure definition 38, 39
molecular remission definition 40
overview 35, 36
prospects 41
quantitative technique 36–38
relapse definition 40
sensitivity 36

treatment response
allogeneic stem cell transplantation 39, 40
imatinib mesylate 41
interferon-α 40
PRV-1, transcript quantification in polycythemia vera diagnosis 97

Southern blot, minimal residual disease monitoring in chronic myeloid leukemia 35
STAT5
activation in leukemias 4
Bcr-Abl activation 22
STI571
chronic myeloid leukemia management 24–26
combination therapy 76
cytogenetic evolution effects on Philadelphia chromosome-positive chronic myeloid leukemia 54
dermatofibrosarcoma protuberans management 76
hypereosinophilic syndrome management 76
mastocytosis management 124
minimal residual disease monitoring in chronic myeloid leukemia
cytogenetic analysis 35
reverse transcriptase-polymerase chain reaction 41
Systemic mastocytosis, *see* Mastocytosis

Thrombopoietin, essential thrombocythemia dynamics and mutations 112, 113, 115
Thrombopoietin receptor
essential thrombocythemia dynamics 112, 113
polycythemia vera reduction of levels and diagnostics 96, 97
*TP53*, chronic myeloid leukemia mutations 52
Transient myeloproliferative disorder
clinical presentation 85
molecular genetics 85, 86
pathobiology 85, 86
research prospects 86

Urticaria pigmentosa 120

Western blot, minimal residual disease monitoring in chronic myeloid leukemia 35

*ZNF198-FGFR1* fusion
chronic myeloid leukemia 4
8p11 myeloproliferative syndrome 64, 65